Acute Brain Impairment

Scientific Discoveries and Translational Research

Drug Discovery Series

Editor-in-Chief:
David Thurston, *King's College, UK*

Series Editors:
David Fox, *Vulpine Science and Learning, UK*
Ana Martinez, *Centro de Investigaciones Biologicas-CSIC, Spain*
David Rotella, *Montclair State University, USA*
Sarah Skerratt, *Vertex Pharmaceuticals, UK*

Editorial Advisor:
Hong Shen, *Roche Innovation Center Shanghai, China*

Titles in the Series:
1: Metabolism, Pharmacokinetics and Toxicity of Functional Groups
2: Emerging Drugs and Targets for Alzheimer's Disease; Volume 1
3: Emerging Drugs and Targets for Alzheimer's Disease; Volume 2
4: Accounts in Drug Discovery
5: New Frontiers in Chemical Biology
6: Animal Models for Neurodegenerative Disease
7: Neurodegeneration
8: G Protein-Coupled Receptors
9: Pharmaceutical Process Development
10: Extracellular and Intracellular Signaling
11: New Synthetic Technologies in Medicinal Chemistry
12: New Horizons in Predictive Toxicology
13: Drug Design Strategies: Quantitative Approaches
14: Neglected Diseases and Drug Discovery
15: Biomedical Imaging
16: Pharmaceutical Salts and Cocrystals
17: Polyamine Drug Discovery
18: Proteinases as Drug Targets
19: Kinase Drug Discovery
20: Drug Design Strategies: Computational Techniques and Applications
21: Designing Multi-Target Drugs
22: Nanostructured Biomaterials for Overcoming Biological Barriers
23: Physico-Chemical and Computational Approaches to Drug Discovery
24: Biomarkers for Traumatic Brain Injury
25: Drug Discovery from Natural Products
26: Anti-Inflammatory Drug Discovery
27: New Therapeutic Strategies for Type 2 Diabetes: Small Molecules
28: Drug Discovery for Psychiatric Disorders
29: Organic Chemistry of Drug Degradation
30: Computational Approaches to Nuclear Receptors
31: Traditional Chinese Medicine
32: Successful Strategies for the Discovery of Antiviral Drugs

33: Comprehensive Biomarker Discovery and Validation for Clinical Application
34: Emerging Drugs and Targets for Parkinson's Disease
35: Pain Therapeutics; Current and Future Treatment Paradigms
36: Biotherapeutics: Recent Developments using Chemical and Molecular Biology
37: Inhibitors of Molecular Chaperones as Therapeutic Agents
38: Orphan Drugs and Rare Diseases
39: Ion Channel Drug Discovery
40: Macrocycles in Drug Discovery
41: Human-based Systems for Translational Research
42: Venoms to Drugs: Venom as a Source for the Development of Human Therapeutics
43: Carbohydrates in Drug Design and Discovery
44: Drug Discovery for Schizophrenia
45: Cardiovascular and Metabolic Disease: Scientific Discoveries and New Therapies
46: Green Chemistry Strategies for Drug Discovery
47: Fragment-Based Drug Discovery
48: Epigenetics for Drug Discovery
49: New Horizons in Predictive Drug Metabolism and Pharmacokinetics
50: Privileged Scaffolds in Medicinal Chemistry: Design, Synthesis, Evaluation
51: Nanomedicines: Design, Delivery and Detection
52: Synthetic Methods in Drug Discovery: Volume 1
53: Synthetic Methods in Drug Discovery: Volume 2
54: Drug Transporters: Role and Importance in ADME and Drug Development
55: Drug Transporters: Recent Advances and Emerging Technologies
56: Allosterism in Drug Discovery
57: Anti-aging Drugs: From Basic Research to Clinical Practice
58: Antibiotic Drug Discovery: New Targets and Molecular Entities
59: Peptide-based Drug Discovery: Challenges and New Therapeutics
60: Drug Discovery for Leishmaniasis
61: Biophysical Techniques in Drug Discovery
62: Acute Brain Impairment: Scientific Discoveries and Translational Research

How to obtain future titles on publication:
A standing order plan is available for this series. A standing order will bring delivery of each new volume immediately on publication.

For further information please contact:
Book Sales Department, Royal Society of Chemistry, Thomas Graham House, Science Park, Milton Road, Cambridge, CB4 0WF, UK
Telephone: +44 (0)1223 420066, Fax: +44 (0)1223 420247,
Email: booksales@rsc.org
Visit our website at www.rsc.org/books

Acute Brain Impairment
Scientific Discoveries and Translational Research

Edited by

Philip V. Peplow
University of Otago, New Zealand
Email: phil.peplow@otago.ac.nz

Svetlana A. Dambinova
DeKalb Medical Center, Decatur, GA, USA
Email: dambinova@aol.com

Thomas A. Gennarelli
George Washington University, USA
Email: tgennarelli@att.net

and

Bridget Martinez
University of California, Merced, USA
Email: bmartinez26@ucmerced.edu

ROYAL SOCIETY OF CHEMISTRY

THE QUEEN'S AWARDS FOR ENTERPRISE: INTERNATIONAL TRADE 2013

Drug Discovery Series No. 62

Print ISBN: 978-1-78262-950-4
PDF ISBN: 978-1-78801-253-9
EPUB ISBN: 978-1-78801-337-6
ISSN: 2041-3203

A catalogue record for this book is available from the British Library

The Royal Society of Chemistry is a charity, registered in England and Wales, Number 207890, and a company incorporated in England by Royal Charter (Registered No. RC000524), registered office: Burlington House, Piccadilly, London W1J 0BA, UK, Telephone: +44 (0) 207 4378 6556.

For further information see our web site at www.rsc.org

Printed in the United Kingdom by CPI Group (UK) Ltd, Croydon, CR0 4YY, UK

Preface

Acute brain injury is associated with considerable short-term and long-term morbidity and mortality. The increasing effort devoted to understanding the cellular pathways and key molecules involved in acute brain injury, together with identifying markers that can improve diagnostic and prognostic accuracy, has provided the impetus for the construction of this book.

It is the goal of this book to provide a forum for clinical experts to present recent data on biomarkers and advanced neuroimaging modalities for acute brain injury. It also brings together scientific experts to present the state of the art in their fields. It is my hope that the topics covered herein will provide new information on the cellular and biochemical mechanisms involved in acute brain injury and that this knowledge will lead to a more effective approach to recognition and clinical management.

I wish to express my deep appreciation to each of the chapter authors for the time and effort spent on writing informative reviews on their respective areas of clinical and research interest. Also, I wish to express my gratitude to Professor Svetlana Dambinova for helping me to plan and organize the book chapters, and to my co-editors Professor Svetlana Dambinova, Professor Thomas A. Gennarelli, and Dr Bridget Martinez. I also wish to thank Rowan Frame and Dr Hannah Aitchison of the Royal Society of Chemistry for their help, support, and patience in the course of putting together this book.

Philip V. Peplow

Drug Discovery Series No. 62
Acute Brain Impairment: Scientific Discoveries and Translational Research
Edited by Philip V. Peplow, Svetlana A. Dambinova, Thomas A. Gennarelli and Bridget Martinez

Published by the Royal Society of Chemistry, www.rsc.org

Dedication

This book is dedicated to Dr Bridget Martinez, whose friendship is a joy and inspiration to me. In loving memory of Bridget's father, Elmer Martinez, whose memory and spirit have guided her through life and an incredible journey in science, research and medicine. It is also dedicated to Bridget's mother, Eda Luz Noguera, who has been a source of love and motivation for her.

Philip Peplow

To my teacher, Acad. Natalia P. Bechtereva, an outstanding neurologist and neurophysiologist who unlocked to me the wonders of research in executive brain functions.

Svetlana Dambinova

My gratitude to my colleagues, friends and family who have supported me in this effort.

Thomas Gennarelli

Drug Discovery Series No. 62
Acute Brain Impairment: Scientific Discoveries and Translational Research
Edited by Philip V. Peplow, Svetlana A. Dambinova, Thomas A. Gennarelli and Bridget Martinez

Published by the Royal Society of Chemistry, www.rsc.org

Contents

Drug Discovery Series No. 62
Acute Brain Impairment: Scientific Discoveries and Translational Research
Edited by Philip V. Peplow, Svetlana A. Dambinova, Thomas A. Gennarelli and Bridget Martinez

Published by the Royal Society of Chemistry, www.rsc.org

CHAPTER 1

Transient Ischemic Attack, Traumatic Brain Injury, and Ischemic Stroke: Risk Factors and Treatments

PHILIP V. PEPLOW* AND JACQUELINE J. T. LIAW

Department of Anatomy, University of Otago, Dunedin, New Zealand
*Email: phil.peplow@otago.ac.nz

1.1 Introduction

The overall burden of stroke in the number of people affected is increasing, especially in the younger age groups and in lower-to-middle-income countries.[1–3] Despite some improvements in stroke prevention and management in high-income countries, the growth and aging of the global population is leading to a rise in the number of young and old patients with stroke. Closely related to this is the increasing number of people, including young children, who are overweight or obese.[4] Obesity is a risk factor for stroke on account of it leading to atherosclerosis, which promotes thrombosis and obstruction to the flow of blood.[5] Moreover, there is an increasing prevalence of hypertension, which also leads to atherosclerosis, and this in turn can lead to blockage of small blood vessels in the brain, causing stroke.[6–8]

In people experiencing a transient ischemic attack (TIA), the incidence of subsequent stroke is as high as 11% over the next 7 days and 24–29% over the following 5 years.[9] Unlike a stroke, the symptoms of a TIA can resolve

Drug Discovery Series No. 62
Acute Brain Impairment: Scientific Discoveries and Translational Research
Edited by Philip V. Peplow, Svetlana A. Dambinova, Thomas A. Gennarelli and Bridget Martinez

Published by the Royal Society of Chemistry, www.rsc.org

within a few minutes or within 24 hours. However, brain injury may still occur in a TIA lasting only a few minutes.[10] A strong association has been shown between traumatic brain injury (TBI) and later development of ischemic stroke, which remained significant even after several potential confounders, such as vascular risk factors and comorbidities, were taken into consideration. The association was similar in magnitude to that of hypertension, which is the leading stroke risk factor.[8,11]

1.2 Definition of TIA, TBI, and Stroke

Early definitions of stroke and TIA focused on the duration of symptoms and signs. More recently, using clinical observation and modern brain imaging, it has been shown that the duration and reversibility of brain ischemia are variable. While brain tissue deprived of needed nutrients undergoes irreversible damage (infarction) in most individuals, in some patients it can survive for a considerable period of time, which may be several hours or even, rarely, days. In 2009, an expert committee of the American Heart Association/American Stroke Association (AHA/ASA) published a scientific statement defining TIA as a transient episode of neurological dysfunction caused by focal brain, spinal cord, or retinal ischemia without acute infarction.[12] The word 'transient' indicates a lack of permanence. However, modern brain imaging has shown that many patients in whom symptoms and signs of brain ischemia are clinically transient have evidence of brain infarction. Similarly, ischemia may produce symptoms and signs that are prolonged (and would qualify as strokes in older definitions) and yet no permanent brain infarction has occurred. In 2013, the AHA/ASA published a scientific statement defining ischemic stroke as an episode of neurological dysfunction caused by focal cerebral, spinal, or retinal infarction. Stroke caused by intracerebral hemorrhage was defined as rapidly developing clinical signs of neurological dysfunction attributable to a focal collection of blood within the brain parenchyma or ventricular system that is not caused by trauma. Likewise, stroke caused by subarachnoid hemorrhage was defined as rapidly developing clinical signs of neurological dysfunction and/or headache because of bleeding into the subarachnoid space between the arachnoid membrane and the pia mater of the brain or spinal cord, which is not caused by trauma.[13] TBI has been defined as a non-degenerative, non-congenital insult to the brain from an external mechanical force, possibly leading to permanent or temporary impairment of cognitive, physical, and psychosocial functions, with an associated diminished or altered state of consciousness.[14] It usually results from a violent blow or jolt to the head or body. The resulting brain damage can be focal or diffuse. A focal TBI is usually caused by sudden contact; diffuse injury is more likely to be caused by an acceleration/deceleration trauma. The severity of TBI is determined by the nature, speed, and location of the impact, and by complications such as hypotension, intracranial hemorrhage, or increased intracranial pressure. These complications may cause secondary injury hours or even days after the trauma.[15]

1.3 Incidence of TIA, TBI, and Stroke

Precise estimates of the incidence and prevalence of TIAs are difficult to determine because of varying criteria used in epidemiological studies to identify TIA. Failure of recognition by both the public and health professionals of the transitory focal neurological symptoms associated with TIAs may also lead to significant underestimates. Given these limitations, the incidence of TIA in the United States has been estimated to be 200 000–500 000 per year, with a population prevalence of 2.3%, which equated to 5 million individuals in 1999.[12,16] TIA incidence markedly increases with age and varies by race–ethnicity. The prevalence rate varies depending on the age distribution of the study population. The Cardiovascular Health Study estimated a prevalence of TIA in men of 2.7% for 65–69 years of age and 3.6% for 75–79 years of age. For women, TIA prevalence was 1.6% for 65–69 years of age and 4.1% for 75–79 years of age.[17] In the Atherosclerosis Risk in Communities Study, the overall prevalence of TIAs among adults 45–64 years of age was 0.4%.[18] Among patients who present with stroke, the prevalence of prior TIA may range from 7% to 40% depending on how TIA is defined, which stroke subtypes are evaluated, and whether the study is a population-based series or a hospital-based series. Variability in the use of brain imaging and the type of diagnostic imaging used can also affect estimates of the incidence and prevalence of TIAs. For example, a revision of the TIA definition to include the absence of changes on magnetic resonance imaging could lead to a decrease in the incidence of TIAs by 33% and a resultant 7% increase in the number of cases labeled as stroke.[19]

Approximately 795 000 strokes occur each year in the United States.[20] In total, 10–20% of patients have a stroke within 90 days following a TIA, and in up to 50% of these patients the stroke occurs within 24–48 hours.[21] In 2000, there were 167 661 deaths in the United States attributable to stroke, which was nearly 24% of stroke patients.[22] The AHA/ASA reported that among non-Hispanic blacks, the relative risk of stroke is four-times higher than among whites at 35–54 years of age and three-times higher at 55–64 years of age. Among Hispanics and American Indians/Alaska natives, the relative risk is about 1.3-times higher than in whites at 35–64 and 45–54 years of age, respectively.[22,23] Ischemic stroke accounts for 87% of all strokes and can be divided into two main types: thrombotic and embolic. Thrombotic disease accounts for about 60% of acute ischemic strokes.[24] It is estimated that 14–30% of ischemic strokes are cardioembolic in origin.[25,26] The incidence of cardioembolic strokes increases with age.[25]

TBI is the greatest cause of mortality and disability in young adults in modern Western societies. In the United States, 1.6 million people sustain a TBI each year, approximately 50 000 people die from a TBI, and 125 000 people are disabled 1 year after injury. However, the exact facts and figures on the incidence, prevalence, and long-term consequences of TBI are uncertain, and it is likely the incidence figures have to be multiplied by 5 or even 10 in order to include every unregistered patient.[15] TBI is strongly associated with several neurological disorders 6 months or more after injury.[27] Seizures are associated with most types of TBI. About 25% of patients with brain contusions or hematomas and about 50% of patients with penetrating head injuries will develop seizures within the first 24 hours of the injury.[15]

1.4 Risk Factors for TIA and Stroke

1.4.1 Age

The most important of the non-modifiable risk factors associated with stroke is age. Stroke is most prevalent among the elderly and the majority of strokes occur in those aged >65 years.[26] Age-related risk also increases for TIA, except for those in the oldest category (≥85 years), where it may decrease slightly.[28]

1.4.2 Gender

Men are at greater risk of stroke, with the incidence rate being 1.25-times higher than in women.[29] Some differences between the genders have also been noted for stroke subtype. Men have about a fourfold greater age-adjusted incidence rate of ischemic stroke due to large vessel atherosclerosis than women, which may account for their higher rate of undergoing carotid endarterectomy.[30]

1.4.3 Race

Significant differences in stroke frequency have been found for race. Black Americans have a higher incidence of stroke than whites. The age- and gender-adjusted ischemic stroke incidence for blacks is 246 per 100 000, compared with 147 per 100 000 for whites.[30] Racial differences have also been found for the distribution of stroke subtype. Stroke due to large vessel atherosclerosis occurs more often among whites (27 per 100 000) than blacks (17 per 100 000). Cardioembolic and small vessel infarcts were the two most important identifiable causes of ischemic stroke among black Americans, and most small vessel strokes in blacks can be attributed to hypertension and diabetes.[31]

1.4.4 Hypertension

Hypertension is the most important modifiable stroke risk factor, with an age-adjusted relative risk of approximately 3 and an attributable risk that may be as high as about 50%, depending on age.[32] The risk for stroke increases proportionately with increasing blood pressure, with systolic blood pressure ≥160 mm Hg or diastolic blood pressure ≥95 mm Hg constituting a relative risk of approximately 4. Even small improvements in the control of hypertension can reduce stroke risk significantly.[32] In the United States in 2011–2012, 29% of adults aged ≥18 years had hypertension, and this was similar among men and women. This figure varied considerably depending on ethnicity and race, with the age-adjusted prevalence being 42.1% for non-Hispanic blacks, 28.0% for non-Hispanic whites, 26.0% for Hispanics, and 24.7% for non-Hispanic Asians. Hypertension increases with age, and according to estimates in the United States in 2011–2012, in adults aged ≥60 years the prevalence was 65.0%.[33] The percentage of those with hypertension controlled to a blood pressure of <140 mm Hg systolic and <90 mm Hg diastolic was <50%.[34,35]

1.4.5 Diabetes Mellitus

Diabetes mellitus is a potentially modifiable risk factor with a relative risk of ischemic stroke of 1.8–3.0.[36] In the United States in 2012, 29.1 million inhabitants had diabetes, of which 21.0 million were diagnosed and 8.1 million were undiagnosed. The rates of diagnosed diabetes varies in different subgroups of the population and was 15.9% for American Indians/Alaskan Natives, 13.2% for non-Hispanic blacks, 12.8% for Hispanics, 9.0% for Asian Americans, and 7.6% for non-Hispanic whites.[37] Diabetes is associated with the development of atherosclerosis, hypertension, obesity, abnormal blood lipid levels, and ultimately stroke.

1.4.6 Smoking

Active cigarette smoking has been attributed to causing approximately 18% of ischemic strokes, and the risk increases twofold among heavy smokers compared with light smokers. Former smoking has an attributable risk of 6%, with the level of risk varying according to the time since quitting, and major risk reduction occurs within 2–4 years of smoking cessation.[38] Cigarette smoke may enhance platelet aggregation, increase coagulability, blood viscosity, and fibrinogen levels, and raise blood pressure.[39]

1.4.7 Previous Stroke

Previous stroke is a major risk factor. The cumulative risk for recurrent stroke is 3–10% in the first 30 days and 5–14% in the first year. The 5-year estimated recurrence rate is 25–29% or higher.

1.4.8 TIA

TIA is a significant independent risk factor for ischemic stroke and has an average annual risk of approximately 4%.[40] The greatest risk is in the first year following a TIA. In addition to TIA, the presence of other risk factors, including older age (>60 years), diabetes mellitus, duration of a TIA event >10 minutes, and signs or symptoms of weakness and speech impairment are associated with increased risk of subsequent stroke. Half of all strokes occur within 23 days of the TIA, and the 90-day stroke risk is 11%.

1.4.9 Atrial Fibrillation

The risk of stroke in patients with non-valvular atrial fibrillation is approximately 3–5% per year, and approximately two-thirds of strokes are cardioembolic.[38] Factors that increase risk in patients with atrial fibrillation include older age, prior TIA or stroke, hypertension, impaired left ventricular function, and diabetes mellitus.[38] Atrial fibrillation is implicated in approximately 24% of strokes in patients aged 80–89 years.[41]

1.4.10 Atherosclerosis of Carotid Arteries

Atherosclerosis of carotid arteries is implicated in 20–30% of ischemic strokes.[42] Both TIA and ischemic stroke are more frequent in patients with severe carotid artery stenosis (>75%), with an annual risk of 3% in those with 60–90% stenosis.[38]

1.4.11 Coronary Artery Disease

In the Framingham study, ischemic stroke occurred in 8% of men and 11% of women within 6 years of an acute myocardial infarction. Coronary artery bypass procedures and open heart surgery carry a risk of stroke, and perioperative stroke occurs in 1–7% of patients undergoing cardiac surgery.[38]

1.4.12 Obesity

Obesity and in particular abdominal obesity are associated with greater stroke risk, with the risk rising as body weight increases. The relative risk in overweight women ranges from 1.8 for a body mass index (BMI) of 27–29 $kg\,m^{-2}$ to a risk of 2.4 for a BMI of ≥32 $kg\,m^{-2}$.[38]

1.4.13 Diet

Increased salt intake is associated with hypertension, and a decrease in salt intake may lower both blood pressure and the associated stroke risk.[39] A higher cereal fiber intake has been associated with a lower risk of total stroke and ischemic stroke.[43]

1.4.14 Physical Activity

Physical activity reduces the risk of cardiovascular disease. Light to moderate physical activities such as walking, jogging, and swimming provide a stroke reduction benefit.[38] Physical exercise may also have a beneficial effect on other risk factors, such as body weight and blood pressure.

1.4.15 Blood Lipids

Greater high-density lipoprotein cholesterol levels reduce the occurrence of both TIA and ischemic stroke. Elevated total serum cholesterol increases the incidence of thromboembolic stroke, while very low cholesterol levels are associated with an increase in hemorrhagic stroke risk.[44]

1.4.16 Hormone Replacement Therapy

In the Women's Health Initiative Study, a 41% increase in stroke rate occurred among women using hormone replacement therapy, beginning in the second year of therapy and persisting beyond the fifth year.[45]

1.5 Treatment of TIA and Ischemic Stroke

1.5.1 Thrombolysis

Two thrombolytic strategies have been efficacious in improving outcome after ischemic stroke: intravenous-administered tissue plasminogen activator (tPA)[46] and intra-arterial prourokinase.[47] Currently, tPA is the only FDA-approved treatment for ischemic stroke. Thrombolysis is necessary to restore blood flow to the affected area of the brain and prevent greater cell death. However, following reperfusion, there is frequently some cell death due to breakdown of the blood–brain barrier, excitotoxicity, infiltration of leukocytes, and production of free radicals. Neuroprotective agents reduce the loss of neuronal cells by being delivered to the ischemic penumbra, and are less likely to suffer irreversible injury at early time points than neurons in the infarct core. Intravenous tPA thrombolysis is recommended for all eligible patients. However, the criteria for eligibility for intravenous tPA are numerous and strict, which accounts for $<10\%$ of patients being eligible, and treatment should be started within 4.5 hours of symptom onset.[48] The use of intra-arterial tPA is recommended for patients who are no longer eligible for intravenous administration of tPA due to the time window restraints, but who are still within the 6-hour cut-off time for intra-arterial treatment. Also, patients who are excluded from intravenous tPA due to contraindications such as recent surgery may be eligible for intra-arterial treatment in the case of occlusion of middle cerebral artery (MCA) or another proximal cerebral artery.[49] Treatment of patients with intra-arterial prourokinase within 6 hours of the onset of acute ischemic stroke caused by MCA occlusion significantly improved clinical outcomes at 90 days, despite an increased frequency of early symptomatic intracranial hemorrhage.[47]

1.5.2 Antiplatelet Therapy

The role of aspirin in the prophylaxis of ischemic cerebrovascular events and stroke has been well documented. Numerous studies and reviews have shown a highly significant risk reduction of 13% in the incidence of acute ischemic stroke when daily low-dose aspirin is taken, without a greatly significant increase in the incidence of hemorrhagic complications including stroke.[50] In 2012, a revised set of guidelines was published by the American College of Chest Physicians in which starting aspirin at doses of between 160 and 325 mg daily within 48 hours of the onset of symptoms of stroke in adults is recommended.[51] The general consensus is that an initial 325 mg dose of aspirin should be given to most patients suffering from a stroke or TIA within 24 hours of the onset of stroke or as early as possible, but not before 24 hours have elapsed since thrombolytic therapy, except when contraindicated by evidence of intracranial hemorrhage, bleeding diathesis, recent surgery, and sensitivity to aspirin, among others. After the initial higher dose, subsequent daily low-dose aspirin might be more adequate than the higher dose, as there is no evidence suggesting that the higher dose provides better protection from further strokes and there is an associated greater risk of intracranial bleeding with chronic use of high-dose aspirin therapy compared to low-dose therapy.[49]

1.5.3 Anticoagulant Therapy

The use of anticoagulants in the first stages of acute ischemic stroke has met with little success. The International Stroke Trial and the National Institute of Neurological Disorders and Stroke recommend against the use of anticoagulants such as heparin within 24 hours of treatment with tPA.[51] This is due to the marked increase in symptomatic intracranial hemorrhage observed in the trials testing anticoagulants for acute ischemic stroke. Currently, no anticoagulant is recommended in the treatment of the acute stages of acute ischemic stroke. Patients with cardioembolic stroke need to receive oral prophylactic anticoagulation, particularly when associated with atrial fibrillation. Patients with mild stroke or TIA may be started on warfarin or newer agents such as dabigatran, titrating the dose to an international normalized ratio of between 2.0 and 3.0 after 48 hours if there is no contraindication. Patients with moderate to severe strokes should not receive anticoagulants after 2–4 weeks have passed.[52]

1.5.4 Statin Therapy

Due to a lack of large randomized clinical trials, current guidelines do not provide specific recommendations on statin initiation in acute ischemic stroke. A recent meta-analysis indicates that pre-stroke statin is associated with milder initial stroke severity, good functional outcome, and lower mortality. In addition, in-hospital statin use is associated with good functional outcome and lower mortality. In patients treated with thrombolysis, statin use is associated with good functional outcome, despite an increased risk of symptomatic hemorrhagic transformation (HT).[53] While these findings support the use of statin in acute ischemic stroke, the findings were mostly from observational studies at risk of bias.[53]

1.5.5 Neuroprotective Therapy

Despite widespread interest and the large number of published studies, no neuroprotective agents have passed clinical trials with the same observable effect seen in animal models.

1.5.6 Thrombectomy

Endovascular thrombectomy for large vessel ischemic stroke substantially reduces disability.

1.6 Hemorrhagic Transformation

Intravenous tPA is the most effective treatment of acute ischemic stroke. However, the most significant concern about the use of tPA is an increased risk of hemorrhage. Many studies have shown thrombolysis is independently associated with HT.[54] HT is a spectrum of ischemia-related brain hemorrhage and is a frequent spontaneous complication of ischemic stroke, especially after thrombolytic therapy.[55] HT can be divided into hemorrhagic infarction (HI) and parenchymal hematoma (PH).[56] HI is a heterogeneous hyperdensity occupying a portion of an ischemic infarct zone on computed tomography (CT) images, whereas PH refers to a more homogeneous, dense hematoma with mass effect. The incidence of spontaneous HT ranges from 13% to 43% in CT studies, whereas the incidence of symptomatic HT is from 0.6% to 7% in the placebo group and from 6.4% to 20% in the thrombolysed group.[57,58] The incidence depends upon many factors such as age, blood glucose level, thrombolytic agent used, route of administration, and time window for the initiation of therapy.[59,60] Symptomatic HT implies a clear causal relationship between clinical deterioration and HT. The rate of occurrence of HI is higher than that of PH; in a large group of consecutive patients with acute ischemic stroke, the incidence of HI was about 9%, whereas that of PH was about 3%.[54] The risk of HT increases remarkably when massive cerebral infarction is present.[61,62] Massive cerebral infarction is often accompanied by substantial brain edema, which results in compression of the peripheral vasculature. The increased vascular permeability because of prolonged ischemia and hypoxia caused by vascular compression greatly increases the likelihood of HT after release of the edema. HT often occurs in the gray matter, especially in the cerebral cortex, because of its abundant collateral vessels. Gray matter infarction is often due to a large artery occlusion and can lead to massive edema, causing ischemic injury by compressing the surrounding blood vessels. Most instances of white matter infarction are of lacunar type and are caused by the terminal vascular occlusion.

Atrial fibrillation and cerebral embolism are associated with an increased risk of HT.[63–65] Hyperglycemia has a major role in post-ischemic HT. In a transient MCA occlusion rat model, acute hyperglycemia consistently resulted in HT.[66] Clinical trials also showed a close association between HT and high blood glucose.[67,68] Hyperglycemia during acute ischemic stroke predisposes to PH, which in turn determines a non-favorable outcome at 3 months.[67] Other clinical studies indicate that lower low-density lipoprotein cholesterol and lower total cholesterol are associated with all the types of HT and symptomatic HT, respectively.[69–71] High-density lipoprotein cholesterol, cholesterol, and triglycerides are not linked to the HT risk.[69]

1.7 Treatment of TBI

Mild TBI usually requires no treatment other than rest and analgesics to treat a headache. However, a person with a mild TBI needs to be monitored closely at home for any persisting, worsening, or new symptoms.

Immediate emergency care for moderate to severe traumatic brain injuries, as classified on the Glasgow Coma Scale,[72] focuses on ensuring the person has an adequate oxygen and blood supply, maintaining blood pressure, and preventing any further injury to the head or neck. Additional treatments in the emergency room or intensive care unit of a hospital are directed at minimizing secondary damage due to inflammation, bleeding, or reduced oxygen supply to the brain. Medications to limit secondary damage to the brain immediately after an injury may include:

Diuretics: diuretics given intravenously to people with TBI help to reduce pressure inside the brain.

Anti-seizure drugs: an anti-seizure drug may be given during the first week to avoid any additional brain damage that might be caused by a seizure. Additional anti-seizure drug treatments are only used if seizures occur.

Coma-inducing drugs: sometimes drugs are used to induce temporary comas because a comatose brain needs less oxygen to function. This is especially beneficial if blood vessels compressed by increased pressure in the brain are unable to deliver the normal amount of nutrients and oxygen to brain cells.

Emergency surgery may be required to minimize additional damage to brain tissues. This could involve:

Controlling bleeding: if there is bleeding in the skull cavity, this may be surgically removed or drained. Bleeding vessels or tissue may need to be repaired. Bleeding outside or within the brain can result in a hematoma that puts pressure on the brain and damages brain tissue.

Repairing skull fractures: in an open head injury, skull fractures may need to be repaired and damaged tissue removed.

Creating a window in the skull: surgery may be used to relieve pressure inside the skull by draining accumulated cerebral spinal fluid or creating a window in the skull to provide more space for swollen tissues.[73]

The overall goal of all surgical treatment is to prevent secondary injury by helping to maintain blood flow and oxygen to the brain and minimizing swelling and pressure.

Most people who have had a significant brain injury will require rehabilitation. The aim is to improve their ability to perform daily activities. Therapy usually begins in the hospital and continues at an inpatient rehabilitation unit, a residential treatment facility, or through outpatient services. The type and duration of rehabilitation varies by individual, depending on the severity of the brain injury and which part of the brain is injured.[73,74]

1.8 Conclusion

The incidence of stroke is increasing in both high-income and lower-to-middle-income countries. Relevant to this is the increasing number of people, including young children, who are overweight or obese. There is also an increasing prevalence of hypertension, which is a major risk factor for stroke. A strong association has been shown between TIA or TBI and subsequent stroke. The risk for stroke is related, among other factors, to age, gender, and race. Existing comorbidities, such as hypertension, diabetes mellitus, atrial fibrillation, coronary artery disease, and obesity, significantly increase stroke risk. Treatment using intravenous tPA or intra-arterial tPA thrombolysis has proven to be efficacious in improving outcome after ischemic stroke. Other treatments that have been shown to provide benefit include aspirin and statins. HT is a frequent spontaneous complication of ischemic stroke, especially after thrombolytic therapy. The risk of HT increases considerably when massive cerebral infarction is present. Atrial fibrillation and cerebral embolism are associated with an increased risk of HT. Hyperglycemia has a major role in post-ischemic HT. Lower levels of low-density lipoprotein cholesterol and lower total cholesterol are associated with all the types of HT and symptomatic HT, respectively.

References

1. V. L. Feigin, M. H. Forouzanfar, R. Krishnamurthi, D. A. Bennett and A. E. Moran, *et al.*, *Lancet*, 2014, **383**, 245.
2. M. Giroud, A. Jacquin and Y. Bejot, *Lancet*, 2014, **383**, 195.
3. R. V. Krishnamurthi, V. L. Feigin, M. H. Forouzanfar, G. A. Mensah, M. Connor, D. A. Bennett and A. E. Moran, *et al.*, *Lancet Global Health*, 2013, **1**, e259.
4. Obesity and overweight. WHO Fact sheet No 311. January 2015. http://www.who.int/mediacentre/factsheets/fs311/en/ (accessed 4 May 2017).
5. A. B. Mitchell, J. W. Cole, P. F. McArdle, Y. C. Cheng, K. A. Ryan, M. J. Sparks, B. D. Mitchell and S. J. Kittner, *Stroke*, 2015, **46**, DOI: 10.1161/STROKEAHA.115.008940.

6. World health statistics, 2012 report: increase of hypertension and diabetes. http://www.euro.who.int/en/health-topics/noncommunicable-diseases/diabetes/news/news/2012/5/world-health-statistics-2012-report-increase-of-hypertension-and-diabetes (accessed 9 May 2017).
7. New data highlight increases in hypertension, diabetes incidence. http://www.who.int/mediacentre/news/releases/2012/world_health_statistics_20120516/en/ (accessed 9 May 2017).
8. B. B. Johansson, *Clin. Exp. Pharmacol. Physiol.*, 1999, **26**, 563.
9. A. Nanda, Transient ischemic attack, 2015. http://emedicine.medscape.com/article/1910519-overview (accessed 9 May 2017).
10. Transient ischemic attack. https://en.wikipedia.org/wiki/Transient_ischemic_attack (accessed 9 May 2017).
11. J. F. Burke, J. L. Stule, L. E. Skolarus, E. D. Sears, D. B. Zahuranec and L. B. Morgenstern, *Neurology*, 2013, **81**, 33.
12. J. D. Easton, J. L. Saver, G. W. Albers, M. J. Alberts, S. Chaturvedi, E. Feldmann and T. S. Hatsukami, *et al.*, *Stroke*, 2009, **409**, 2276.
13. R. L. Sacco, S. E. Kasner, J. P. Broderick, L. R. Caplan, J. J. Connors, A. Culebras and M. S. Elkind, *et al.*, *Stroke*, 2013, **44**, 2064.
14. S. T. Dawodu, Traumatic brain injury (TBI) - definition and pathophysiology.http://emedicine.medscape.com/article/326510-overview (accessed 9 May 2017).
15. G. M. Ribbers, in Brain Injury: Long Term Outcome After Traumatic Brain Injury, ed. J. H. Stone and M. Blouin, International Encyclopedia of Rehabilitation, 2010. http://cirrie.buffalo.edu/encyclopedia/en/article/338/ (accessed 4 May 2017).
16. S. C. Johnston, P. B. Fayad, P. B. Gorelich, D. F. Hanley, P. Shwayder, D. van Husen and T. Weiskopf, *Neurology*, 2003, **60**, 1429.
17. T. R. Price, B. Psaty, D. O'Leary, G. Burke and J. Gardin, *Ann. Epidemiol.*, 1993, **3**, 504.
18. J. F. Toole, L. E. Chambless, G. Heiss, H. A. Tyroler and C. C. Paton, *Ann. Epidemiol.*, 1993, **3**, 500.
19. B. Ovbiagele, C. S. Kidwell and J. L. Saver, *Stroke*, 2003, **34**, 919.
20. AHS/ASA Statistical Fact Sheet, 2016, Update http://www.strokeassociation.org/STROKEORG/AboutStroke/Impact-of-Stroke-Stroke-statistics_UCM_310728_Article.jsp#.WQptsoWlNsM (accessed 4 May 2017).
21. G. W. Albers, L. R. Caplan, J. D. Easton, P. B. Fayad, J. P. Mohr, J. L. Saver and D. G. Sherman, *N. Eng. J. Med.*, 2002, **347**, 1713.
22. P. B. Gorelick, *Clin. Cardiol.*, 2004, **27**(Suppl II), II–4.
23. S. Cruz-Flores, A. Rabinstein, J. Biller, M. S. Elkind, P. Griffith, P. B. Gorelick and G. Howard, *et al.*, *Stroke*, 2011, **42**, 2091.
24. The New Jersey Comprehensive Stroke Center at University Hospital. Types of Stroke. http://www.uhnj.org/stroke/types.htm (accessed 4 May 2017).
25. A. Arboix and J. Alio, *Curr. Cardiol. Rev.*, 2010, **6**, 150.
26. R. L. Sacco, in *Current Review of Cerebrovascular Disease*, ed. M. Fisher and J. Bogousslavsky, Current Medicine, Philadelphia, PA, 1993, pp. 3–14.

27. J. J. Bazarian, I. Cernak, L. Noble-Haeusslein, D. S. Potolicchio and N. Temkin, *J. Head Trauma Rehabil.*, 2009, **24**, 439.
28. R. D. Brown, G. W. Petty, M. O'Fallon, D. O. Wiebers and J. P. Whisnant, *Stroke*, 1998, **29**, 2109.
29. G. Howard and V. J. Howard, in *The Prevention of Stroke*, ed. P. B. Gorelich and M. Alter, Parthenon Publishing, New York, NY, 2002, pp. 1–10.
30. G. W. Petty, R. D. Brown, J. P. Whisnant, J. D. Sicks, W. M. O'Fallon and D. O. Wiebers, *Stroke*, 1999, **30**, 2513.
31. D. Woo, J. Gebel, R. Miller, R. Kothari, T. Brott, J. Khoury and S. Salisbury, *et al.*, *Stroke*, 1999, **30**, 2517.
32. R. L. Sacco, *Neurology*, 1997, **49**(5 Suppl. 4), S39.
33. CDC National Center for Health Statistics. Hypertension among adults in the United States: National Health and Nutrition Examination Survey, 2011–2012. https://www.cdc.gov/nchs/products/databriefs/db133.htm (accessed 9 May 2017).
34. D. L. Cohen and R. R. Townsend, *J. Clin. Hypertens.*, 2013, **15**, 143.
35. The seventh report of the Joint National Committee on Prevention, Detection, Evaluation, and Treatment of High Blood Pressure. http://www.ncbi.nlm.nih.gov/books/NBK9626/ (accessed 4 May 2017).
36. C. M. Helgason and P. A. Wolf, *Circulation*, 1997, **96**, 701.
37. American Diabetes Association. Statistics about diabetes. http://www.diabetes.org/diabetes-basics/statistics/ (accessed 9 May 2017).
38. L. B. Goldstein, R. Adams, K. Becker, C. D. Furberg, P. B. Gorelick, G. Hademenos and M. Hill, *et al.*, *Circulation*, 2001, **103**, 163.
39. P. B. Gorelich, R. L. Sacco, D. Smith, M. Alberts, L. Mustone-Alexander, D. Radar and J. L. Ross, *et al.*, *JAMA*, 1999, **281**, 1112.
40. R. L. Sacco, E. J. Benjamin, J. P. Broderick, M. Dyken, J. D. Easton, W. M. Feinberrg and L. B. Goldstein, *et al.*, *Stroke*, 1997, **28**, 1507.
41. P. A. Wolf, R. D. Abbott and W. B. Kannel, *Stroke*, 1991, **22**, 983.
42. Z. G. Nadareishvili, P. M. Rothwell, V. Beletsky, A. Pagniello and J. W. Norris, *Arch. Neurol.*, 2002, **59**, 1162.
43. D. Mozaffarian, S. K. Kumanyika, R. N. Lemaitre, J. L. Olson, G. L. Burke and D. S. Siscovick, *JAMA*, 2003, **289**, 1659.
44. H. Iso, D. R. Jacobs, D. Wentworth, J. D. Neaton and J. D. Cohen, *N. Eng. J. Med.*, 1989, **320**, 904.
45. A. Pradhan, J. E. Manson, J. E. Rossouw, D. S. Siscovick, C. P. Mouton, N. Rifai, R. B. Wallace, R. D. Jackson, M. B. Pettinger and P. M. Ridker, *JAMA*, 2002, **288**, 980.
46. K. M. Chapman, A. R. Woolfenden, D. Graeb, D. C. Johnston, J. Beckman, M. Schulzer and P. A. Teal, *Stroke*, 2000, **31**, 2920.
47. A. Furlan, R. Higashida, L. Wechsler, M. Gent, H. Rowley, C. Kase and M. Pessin, *et al.*, *JAMA*, 1999, **282**, 2003.
48. G. J. del Zoppo, J. L. Saver, E. C. Jauch and H. P. Adams, *Stroke*, 2009, **40**, 2945.
49. H. Mestre, Y. Cohen-Minian, D. Zajarias-Fainsoid and A. Ibarra, in *Neurodegenerative Diseases*, ed. Uday Kishore, Intech, 2013, p. 581.

50. J. G. Tijssen, *Neurology*, 1998, **51**(3 Suppl. 3), S15.
51. M. G. Lansberg, M. J. O'Donnell, P. Khatri, E. S. Lang, M. N. Nguyen-Huynh, N. E. Schwartz and F. A. Sonnenberg, *et al.*, *Chest*, 2012, **141**(2 Suppl.), e601S.
52. C. S. Miller, S. M. Grandi, A. Shimony, K. B. Filion and M. J. Eisenberg, *Am. J. Cardiol.*, 2012, **110**, 453.
53. K. S. Hong and J. S. Lee, *J Stroke*, 2015, **17**, 282.
54. M. Paciaroni, G. Agnelli, F. Corea, W. Ageno, A. Alberti, A. Lanarti and V. Caso, *et al.*, *Stroke*, 2008, **39**, 2249.
55. E. S. Sussman and E. S. Connolly, *Front. Neurol.*, 2013, **4**, 69.
56. P. Trouillas and R. von Kummer, *Stroke*, 2006, **37**, 556.
57. A. Jaillard, C. Cornu, A. Durieux, T. Moulin, F. Boutitie, K. R. Lees and M. Hommel, *Stroke*, 1999, **30**, 1326.
58. O. Y. Bang, J. L. Saver, S. J. Kim, G. M. Kim, C. S. Chung, B. Ovbiagele and K. H. Lee, *et al.*, *Stroke*, 2011, **42**, 2235.
59. C. S. Kidwell, J. L. Saver, J. Carneado, J. Sayre, S. Starkman, G. Duckwiler and Y. P. Gobin, *et al.*, *Stroke*, 2002, **33**, 717.
60. The NINDS t-PA Stroke Study Group, Intracerebral hemorrhage after intravenous t-PA therapy for ischemic stroke, *Stroke*, 1997, **28**, 2109.
61. B. G. Wang, N. Yang, M. Lin and B. Lu, *Cell. Biochem. Biophys.*, 2014, **70**, 135.
62. L. Kerenyi, L. Kardos, J. Szasz, S. Szatmari, D. Bereczki, K. Hegedus and L. Csiba, *Eur. J. Neurol.*, 2006, **13**, 1251.
63. S. Tan, D. Wang, M. Liu, S. Zhang, B. Wu and B. Liu, *J. Neurol.*, 2014, **261**, 905.
64. R. G. Nogueira, R. Gupta, T. G. Jovin, E. I. Levy, D. S. Liebeskind, O. O. Zaidat and A. Rai, *et al.*, *J. Neurointervent. Surg.*, 2015, 7, 16.
65. J. H. Lee, K. Y. Park, J. H. Shin, J. K. Cha, H. Y. Kim, J. H. Kwon and H. G. Oh, *et al.*, *Eur. Neurol.*, 2010, **64**, 193.
66. Y. Xing, X. Jiang, Y. Yang and G. Xi, *Acta Neurochir. Suppl.*, 2011, **111**, 49.
67. M. Paciaroni, G. Agnelli, V. Caso, F. Corea, W. Ageno, A. Alberti and A. Lanari *et al.*, *Cerebrovasc. Dis.*, 2009, **28**, 119.
68. H. Kunte, M. A. Busch, K. Trostdorf, B. Vollnberg, L. Harms, R. I. Mehta and R. J. Castellani, *et al.*, *Ann. Neurol.*, 2012, **72**, 799.
69. K. Nardi, D. Leys, P. Eusebi, C. Cordonnier, S. Gautier, H. Henon and R. Bordet, *Cerebrovac. Dis. Extra*, 2011, **1**, 130.
70. O. Y. Bang, J. L. Saver, D. S. Liebeskind, S. Starkman, P. Villablanca, N. Salamon and B. Buck *et al.*, *Neurology*, 2007, **68**, 737.
71. M. D'Amelio, V. Terruso, G. Famoso, P. Ragonese, P. Aridon and G. Savettieri, *Cerebrovasc. Dis.*, 2011, **32**, 234.
72. Glasgow Coma Scale – Adult. http://emedicine.medscape.com/article/2172603-overview (accessed 9 May 2017).
73. Traumatic brain injury. Treatment and drugs. http://www.mayoclinic.org/diseases-conditions/traumatic-brain-injury/basics/treatment/con-20029302 (accessed 9 May 2017).
74. Treatments for TBI. http://www.traumaticbraininjury.com/treatments-for-tbi/ (accessed 9 May 2017).

CHAPTER 2

Inflammatory Changes in Cerebral Ischemic Injury: Cellular and Molecular Involvement

HIROAKI OOBOSHI

Department of Internal Medicine, Fukuoka Dental College Medical and Dental Hospital, Fukuoka 814-0193, Japan
Email: ooboshi@college.fdcnet.ac.jp

2.1 Introduction

Post-ischemic inflammation is an essential step in ischemic brain injury.[1,2] The progression of the infarct lesion is directly linked to the prognosis of ischemic stroke patients. The cerebrovascular units, which consist of glial cells, endothelial cells, pericytes, and neurons,[3–5] are damaged by brain ischemia, leading to blood–brain barrier (BBB) breakdown. In combination with the dysfunction of the BBB, various blood-borne immune cells infiltrate into the ischemic brain from disrupted vessels. These infiltrating immune cells and the injured brain cells produce inflammatory mediators that exaggerate brain edema or directly promote the death of brain cells in the penumbra, resulting in the secondary progression of the infarct lesion.

Effectors of innate immunity, especially macrophages and T cells, appear to play a pivotal role in post-ischemic inflammation.[6,7] It is known that macrophages and neutrophils are key players in the various processes of

Drug Discovery Series No. 62
Acute Brain Impairment: Scientific Discoveries and Translational Research
Edited by Philip V. Peplow, Svetlana A. Dambinova, Thomas A. Gennarelli and Bridget Martinez

Published by the Royal Society of Chemistry, www.rsc.org

post-ischemic inflammation, but the mechanism responsible for their activation is still unknown. In addition, T cells have also been reported to participate in the delayed cerebral inflammation. This chapter focuses on the potential mechanisms underlying the post-ischemic inflammation induced by infiltrating immune cells and explores potential approaches for developing neuroprotective therapies based on regulating post-ischemic inflammation.

2.2 Damage-associated Molecular Patterns in Ischemic Brain Tissue

Brain ischemia induces the production of reactive oxygen species (ROS) from injured brain cells, and ROS activate platelets and endothelial cells, leading to microvascular occlusion. Oxidative stress and the inflammatory cascade alter the permeability of the BBB and exacerbate leukocyte extravasation. Recently, ROS and the regulators of their production have been implicated in the activation of infiltrating leukocytes,[8] and several molecules have been regarded as activators of infiltrating immune cells in the ischemic brain.[2] Certain endogenous molecules must be released from the necrotic brain to activate infiltrating immune cells. Such endogenous molecules are called damage-associated molecular patterns (DAMPs) or alarmins and are considered to be danger signals or alarm molecules that alert immune cells to the presence of brain tissue injury (Figure 2.1).

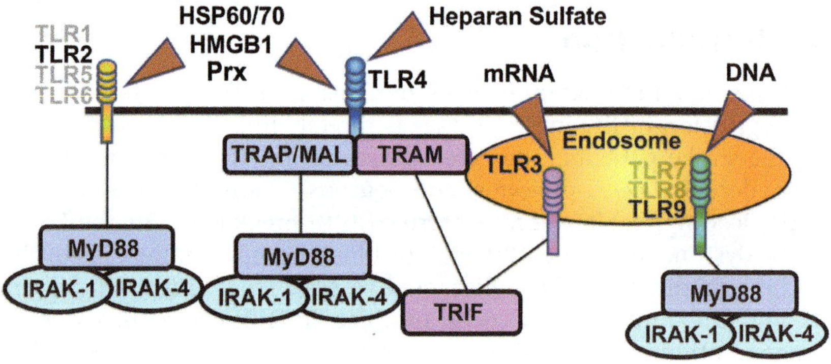

Figure 2.1 Brain ischemia causes the release of damage-associated molecular patterns (DAMPs) from brain cells. In those DAMPs, HMGB1, heat shock protein (HSP), peroxiredoxin (Prx), and heparan sulfate bind to Toll-like receptor 2 (TLR2) and TLR4, mRNA complexes to TLR3, and DNA complexes to TLR9. While TLR3 activates TRIF as the downstream adaptor, TRL2, TLR4, and TLR9 activate MyD88, forming a complex with interleukin-1 receptor-associated kinase (IRAK).
Adapted from *Neuroscience*, **158**, B. J. Marsh, R. L. Williams-Karnesky and M. P. Stenzel-Poore, Toll-like receptor signaling in endogenous neuroprotection and stroke, 14, Copyright (2009), with permission from Elsevier.

So far, heat shock proteins (HSPs), β-amyloid, hyaluronan, heparin sulfate, DNA or RNA immune complexes, oxidized low-density lipoproteins, and several other molecules have been considered as possible DAMPs in the ischemic brain.[9–13] Interestingly, many reports have also demonstrated that HSPs exert neuroprotective roles in post-ischemic injury.[14–16]

Despite numerous studies, it is not yet clear which molecule is the most important for triggering inflammation and activating leukocytes in the ischemic brain. Among these DAMPs, high mobility group box 1 (HMGB1) is one of the most investigated DAMPs in ischemic brain injury.[17–19] Under ischemic conditions, HMGB1 locates into the cytosol from the nucleus and is then released into the extracellular compartment. Such an extracellular release of HMGB1 is observed within 2–4 hours after ischemia–reperfusion. HMGB1 increases the vascular permeability and promotes BBB breakdown.[20] The neutralization of extracellular HMGB1 is reported to improve the BBB permeability and reduce the infarct volume in mice and rats.[18,20,21] Therefore, HMGB1 may be one of the most important DAMPs in the very early phase of ischemic brain injury. In contrast, it is also demonstrated that the extracellular release of HMGB1 is diminished within 12 hours after ischemia–reperfusion.[22] Because the infiltration of immune cells and the production of inflammatory mediators become prominent thereafter, the direct activation of infiltrating immune cells may not be achieved by HMGB1. Thus, it may not be an ideal therapeutic target for a neuroprotective agent in the delayed phase of brain ischemia.

As potential DAMPs in brain ischemia, peroxiredoxin (Prx) family proteins have been identified as strong inducers of inflammatory cytokines in infiltrating macrophages (Figure 2.1).[21] The Prx family proteins are known as ubiquitous antioxidant enzymes and are more abundant in the brain than in other tissues. Prx proteins contain one or two cysteine residues that allow them to scavenge ROS in cooperation with thioredoxin, another important redox modulator.[23,24] In various brain tissue injuries, the increased expression of Prx proteins within brain cells has been observed and considered to diminish ROS.[25] In fact, such an intracellular expression of Prx proteins has been reported to play a neuroprotective role.[26] These results indicate that Prx family members improve the survival of injured cells by strongly inducing inside the cell. However, once the cells are on the verge of apoptosis, Prx proteins are released into the extracellular compartment and the Prx proteins function as DAMPs. Therefore, Prx proteins can exert two opposing functions: one inside and one outside the brain cells.

Compared with HMGB1, it may take more time for Prx proteins to function in the ischemic brain. In ischemic brain tissue, the Prx expression is evident on day 1, but disappears by day 4 after ischemia–reperfusion in mice.[21] The highly expressed Prx is released into the extracellular compartment around injured brain cells and co-localizes with the membranes of infiltrating macrophages. An increased amount of Prx proteins in the extracellular fluid of the ischemic brain has been also reported in rats and stroke patients.[27,28] Thus, the extracellularly released Prx from injured brain cells appears to activate infiltrating macrophages and promotes post-ischemic inflammation. Neutralization of extracellular Prx proteins decreases the expression of inflammatory cytokines by infiltrating inflammatory cells, thereby attenuating ischemic damage in mice.

2.3 Toll-like Receptors as DAMP Receptors

Toll-like receptors (TLRs) are essential receptors involved in the innate immune response to general pathogens such as bacteria and viruses (Figure 2.1). Recently, TLR2 and TLR4 have been reported to contribute to non-infectious immune-mediated injury, including ischemic brain injury.[29,30] Indeed, HMGB1, Prx proteins, and other DAMPs can stimulate TLR2 and TLR4. The activation of macrophages and T cells through the TLR pathway induces strong inflammatory responses. Post-ischemic inflammation and subsequent ischemic damage depend on TLR2 and TLR4, but not TLR9,[31] because TLR2 or TLR4 deficiency significantly attenuates ischemic brain damage and suppresses inflammatory cytokine expression in infiltrating immune cells 1 day after ischemia–reperfusion in mice.[21] The clinical relevance of TLR2 and TLR4 in stroke patients has also been demonstrated.[32] Despite these findings about the importance of TLRs, the mechanisms responsible for activating them in infiltrating immune cells, such as macrophages and T cells, are not yet fully understood. CD36, receptor for advanced glycation end-products (RAGE), scavenger receptor A (SRA), macrophage-inducible Ca^{2+}-dependent lectin (Mincle), and other pattern recognition receptors are also possible receptors for DAMPs.[33]

In addition, the fact that TLRs are expressed on both leukocytes and brain cells makes it difficult to fully understand the complex of inflammatory mechanisms and TLR activation in the ischemic brain. It remains to be clarified whether the effect of TLRs on brain cells is neurotoxic or neuroprotective. Brain cells, including astrocytes, oligodendrocytes, endothelial cells, and pericytes, constitute the neurovascular unit and maintain the neuronal microenvironment by increasing their antioxidant activities to scavenge ROS and support the essential metabolic needs of neurons.[34,35] On the other hand, these brain cells, which are activated by ischemic injury, also contribute to post-ischemic inflammation by producing inflammatory mediators. Tumor necrosis factor-α (TNF-α), interleukin-1β (IL-1β), nitric oxide, and matrix metalloproteinases (MMPs) produced from brain cells regulate the cerebrovascular permeability and exaggerate brain edema.[36,37] Microglia, which are resident macrophages in the brain, are also activated through the TLR signaling pathway, and act as inflammatory mediators by producing both neurotoxic cytokines and neurotrophic factors for tissue repair.[38]

Reports using bone marrow chimeric knockout (KO) mice have revealed that microglial TLR2 and TLR4 play minor roles in ischemic injury.[21,39] Interestingly, mice lacking MyD88, the adaptor protein required for almost all TLR signaling cascades (other than TLR3), showed either no improvement or exaggeration of ischemic brain injury.[40] These results indicate that the function of TLRs is dependent of the cell types in the brain and the timing of their activation. Recently, the administration of an anti-TLR blocking antibody has emerged as a neuroprotective therapy for ischemic myocardial or renal injury.[41,42] To explore the therapeutic potential of this strategy, more detailed knowledge about the functions of TLRs in the ischemic brain is needed.

2.4 Inflammatory Cytokines and Mediators

Activated infiltrating immune cells and injured brain cells produce various inflammatory cytokines and mediators. IL-1β is expressed in the ischemic brain within 30 minutes after ischemia–reperfusion.[43] Recently, the previously unknown mechanism of IL-1β production and caspase-1 activation mediated by the inflammasome has attracted particular attention. IL-1β is produced in an inactive form, pro-IL-1β, whose mRNA expression is regulated by TLRs or other pattern recognition receptor signaling pathways (signal 1). Pro-IL-1β is cleaved to become an active 17 kDa form by caspase-1 inside multiprotein complexes called inflammasomes, which are activated by hypoxia, ATP, or endogenous molecules from damaged or dying cells (signal 2).[44,45] The inflammasome is present in brain cells (neurons, astrocytes, and microglia) and macrophages.[46,47] Several types of inflammasome have been discovered, and these are denoted NLRP1, NLRP3, AIM, and so on, but the specific type that is most important in ischemic brain injury remains unknown. Recently, Bruton's tyrosine kinase has been reported to play important roles in the activation of NLRP3 and infarct growth.[48] IL-1β is considered to be a neurotoxic mediator, given that the loss of IL-1β function is reported to reduce infarct size.[49] IL-1β directly induces neuronal cell death and enhances the expression of chemokines in microglia and astrocytes.[50] Inhibiting IL-1β or its induction may represent a possible therapeutic approach.

TNF-α is another essential cytokine involved in ischemic brain injury. TNF-α is expressed in ischemic brain tissue within 1 hour after ischemia–reperfusion.[51] TNF-α exerts neurotoxic effects by promoting neuronal cell death and the expression of major histocompatibility complex (MHC) class II molecules and intercellular adhesion molecule-1 (ICAM-1) in astrocytes, resulting in leukocyte infiltration and BBB breakdown. The neuroprotective effect of TNF-α gene deficiency or the administration of an anti-TNF-α neutralizing antibody has been demonstrated. On the other hand, TNFR KO mice, which lack both p75 and p50 genes, exhibit enlargement of the infarct volume, indicating that TNF-α can be considered to function as both a neurotoxic and a neuroprotective mediator.[52] TNF-α promotes post-ischemic inflammation, but also participates in a negative feedback loop to suppress inflammatory signal cascades, and it controls the duration of post-ischemic inflammation by regulating these two functions. It appears that which of the opposing functions of TNF-α—toxic or protective—is active depends on the type of brain cells. Further elucidation of the various functions of TNF-α is needed for the application of anti-TNF-α neutralizing therapy for ischemic stroke.

IL-6 is an important cytokine in various types of inflammation, but ischemic brain damage is not attenuated by IL-6 deficiency or administration of an anti-IL-6R antagonistic antibody.[53] However, it has recently been reported that IL-6 produced from brain cells contributes to angiogenesis and neuronal survival through STAT3 activation.[54,55] Consistent with these observations, the inhibition of the JAK/STAT pathway or the enhancement of SOCS3 (a negative regulator of the JAK/STAT pathway) has been reported to promote neuronal cell death.[56] Thus, it is possible that IL-6 contributes to the process of tissue repair after ischemic brain injury.

Chemokines are also important enhancers of post-ischemic inflammation. RANTES, MCP-1, and IL-8 have been reported to promote leukocyte infiltration into the ischemic brain and to increase infarct growth.[57–59] The overexpression of dominant negative MCP-1 has been demonstrated to attenuate macrophage infiltration and to reduce the infarct size.[60] Although the chemokines for T cell infiltration remain unknown, CCL12, CCL20, and their receptor, CCR6, have been reported to be essential for the induction of IL-17-producing helper T cells in experimental autoimmune encephalomyelitis (EAE).[61,62] Nevertheless, the mechanism of T cell infiltration in the ischemic brain tissue remains to be elucidated.

ICAM-1 is essential for chemotaxis and the infiltration of leukocytes. Increased expression of ICAM-1 in cerebrovascular endothelial cells is observed in the ischemic brain, and ICAM-1 deficiency or neutralization of ICAM-1 by an antibody leads to the improvement of ischemic brain damage.[63,64] MMPs are also important mediators that promote post-ischemic inflammation by enhancing the breakdown of the BBB. MMP-9-deficient mice exhibit a smaller infarct size compared with control mice,[65] and the neurotoxic function of MMP-9 has been established.[66]

Sphingosine-1-phosphate (S1P) is a bioactive phospholipid at sites of tissue injury. S1P is mainly released from platelets and mediates its effect *via* the activation of cell-surface S1P receptors, which are ubiquitously expressed in neurons, astrocytes, and microglial cells.[67] S1P directly acts on brain cells and induces astrocyte proliferation and migration, oligodendrocyte differentiation and survival, neurite outgrowth, and neurogenesis. S1P receptors are also present on the surface of T cells; therefore, S1P is considered to be essential for the infiltration of T cells. FTY720 (fingolimod), a functional S1P receptor antagonist, has recently been examined in intravenous administration and was found to decrease the number of infiltrating T cells in the ischemic brain and to reduce infarct size.[68,69] Therefore, S1P is considered to be an important inflammatory mediator in ischemic brain injury. Because a clinical trial of FTY720 for multiple sclerosis showed favorable results,[70] FTY720 may be a safe and promising new drug for the treatment of ischemic stroke.

2.5 Infiltrating Immune Cells in the Ischemic Brain

Macrophages are the main inflammatory effectors among the various infiltrating immune cells from blood. The infiltration of macrophages becomes evident from 12 to 24 hours after ischemia–reperfusion and reaches a peak on day 3.[71] Infiltrating macrophages produce various inflammatory cytokines, such as IL-1β, TNF-α, and IL-23, and they promote post-ischemic inflammation. Mice deficient in CD11b, a specific marker of macrophages, have been reported to exhibit attenuated ischemic damage.[72]

T cells are also important effectors in the delayed phase of brain ischemia.[73] The number of infiltrating T cells in the ischemic brain increases over 24 hours after ischemia–reperfusion and reaches a peak in the delayed phase (around day 3).[74,75] T cells appear to be localized to the infarct boundary zones, typically close to blood vessels. T cells consist of ~30–40% CD4+ helper T cells, ~20–30% γδ T cells, and ~20–30% CD8+ cytotoxic T cells. Although the function of T cells in the ischemic brain has not been fully clarified, on the whole, T cells are considered to exert neurotoxic effects based on the following observations: first, recombination activating gene (RAG)-deficient mice and severe combined immunodeficient (SCID) mice, both of which lack T and B cells, show a significant reduction of infarct volume compared to wild-type mice.[73,76] Second, the depletion of CD4+ helper T cells or CD8+ cytotoxic T cells, but not of B cells, led to a significant reduction of the infarct volume. Thus, specific T cells are promising targets for the development of a neuroprotective treatment with a prolonged therapeutic time window.

B cells, natural killer (NK) cells, and natural killer T (NKT) cells are also observed in ischemic brain tissue. However, their function in the ischemic brain remains to be clarified. One report showed that regulatory B cells protect the brain from ischemic damage.[77] Another report revealed a protective effect of hepatic invariant NKT cells against stroke-associated infections.[78] It is important to elucidate the roles of these immune cells in ischemic brain injury.

2.6 T Cells in Ischemic Brain Injury

Various cytokines produced from T cells are strong regulators of post-ischemic inflammation. These cytokines seem to function as innate immune mediators, given that the needs of antigen-specific T cell receptor (TCR) stimulation have not been demonstrated for the production of these cytokines. Interferon-γ (IFN-γ) and IL-4 are well-known classical cytokines released from T cells. IFN-γ is mainly produced from CD4+ helper T cells (Th1) and may be neurotoxic, as it directly acts on neurons and induces neuronal cell death.[79] However, the neuroprotective effect of IFN-γ deficiency is controversial.[73,79] IL-12 is produced from myeloid cells such as macrophages, dendritic cells, and neutrophils, and is important for the differentiation of IFN-γ-producing helper T cells (Th1). In the ischemic brain, IL-12 is produced from infiltrating macrophages, but its function has not been fully elucidated. IL-4 may have a neuroprotective function by promoting tissue repair, given that IL-4 deficiency exacerbates ischemic brain damage and neurological deficits in mice.[80] Although previously unknown functions of IL-4 in tissue repair have been receiving increasing attention,[81] the involvement of IL-4 in tissue repair after brain ischemia remains to be clarified.

IL-10 is an immunosuppressive cytokine and is thought to have a neuroprotective effect against ischemic brain injury.[82] Although the IL-10-producing T cell population is small in the ischemic brain, a neuroprotective therapy may be developed by enhancing IL-10 production or promoting differentiation into IL-10-producing T cells. Regulatory T cells are reported to be important IL-10-producing cells in ischemic brain injury and may exert a neuroprotective effect by suppressing the neurotoxic functions of TNF-α and IFN-γ.[83] Recently, the transcription factor E4BP4 and the expression of lymphocyte activation gene 3 (LAG-3) have been discovered to be essential for the regulation of IL-10 production from T cells.[84,85] Because TGF-β, an important cytokine for IL-10 production from T cells, is highly expressed in the brain tissue, infiltrating T cells may easily differentiate into IL-10-producing T cells in the brain. The elucidation of the detailed mechanism(s) responsible for IL-10 production in the ischemic brain has the potential to lead to the development of a neuroprotective therapy.

IL-17 is an emerging therapeutic target for various organ injuries. IL-23 has been reported to be essential for IL-17 induction from CD4+ helper T cells, $\gamma\delta$ T cells, and NKT cells in EAE.[86] In ischemic brain injury, IL-23 is produced by the infiltrating macrophage on day 1 after ischemia–reperfusion,[68] and IL-23 induces IL-17 production from $\gamma\delta$ T cells in the delayed phase (Figure 2.2). Either IL-23 or IL-17 deficiency prevents infarct growth in the delayed phase (days 1–4) in mice. IL-23 exerts its primary effects earlier in the pathological course, given that IL-23 deficiency or the neutralization of IL-23 attenuate ischemic damage on day 1.[68,87] The IL-17 receptor is ubiquitously expressed in brain cells and modifies various inflammatory responses in the central nervous system. IL-17 has been reported to promote the expression of inflammatory cytokines and chemokines from macrophages.[88] IL-17 also modulates the epithelial barrier function by promoting the expression of MMPs and ICAM-1.[89,90] Although it remains unknown whether IL-17 directly affects neurons, it appears to be a pivotal inflammatory cytokine in the delayed phase of brain ischemia. The main source of IL-17 in the ischemic brain is $\gamma\delta$ T cells.[68] This can be reasonably explained by the fact that the IL-17 production from $\gamma\delta$ T cells requires only IL-1β and IL-23 stimulation without specific TCR stimulation, while IL-6 and TGF-β stimulation are indispensable for the IL-17-producing CD4+ helper T cell differentiation.[91] Therefore, compared with helper T cells, IL-23-induced IL-17-producing $\gamma\delta$ T cells are rapid inflammatory effectors involved in the delayed phase (days 1–4) of brain ischemia. As clinical trials of IL-23 or IL-17 neutralizing antibodies are now ongoing for the treatment of various inflammatory diseases, targeting IL-23- or IL-17-producing $\gamma\delta$ T cells may be a promising therapeutic strategy for ischemic stroke (Figure 2.3).

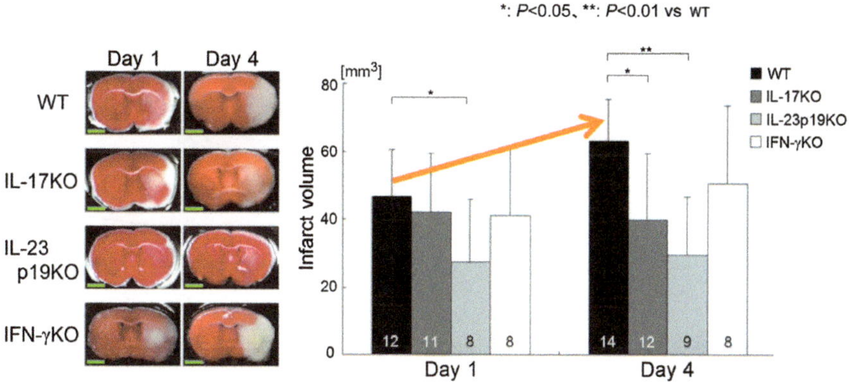

Figure 2.2 Brain infarction evolves from day 1 to day 4 in wild-type (WT) mice. Knockout (KO) of interferon-γ (IFN-γ) has the least effect on infarct size at day 1 and day 4. While IL-23p19KO mice reduce the infarct volume on days 1 and 4, IL-17KO mice show a significant reduction of infarct volume on day 4, but not on day 1, suggesting that IL-23 functions at a very early phase of injury and IL-17 plays a role in a later phase of damage.
Adapted by permission from Macmillan Publishers Ltd: *Nature Medicine* (ref. 68), copyright (2009).

Because post-ischemic inflammation occurs in the acute phase, T cells are considered to act mainly as effectors of an antigen-independent innate inflammatory response. However, it is possible that acquired immunity by T cells also plays a role in ischemic brain injury, even though it has not been clarified whether a specific antigen in the ischemic brain is involved in the activation of T cells. Some reports suggest the importance of antigen recognition by T cells in ischemic brain injury. Treatment with a TCR ligand, which consisted of major histocompatibility complex class II molecules bound to myelin peptides, was protective against ischemic brain injury.[92] In myelin basic protein-tolerized animals, a reduction of the infarct size has been demonstrated.[93] Therefore, the possibility exists that some T cell subsets specifically tolerized to neuronal proteins could be neuroprotective.[94] This idea is supported by the recent finding that regulatory T cells are protective against ischemic brain injury.[83]

2.7 Resolution of Post-ischemic Inflammation

Endogenous mechanisms for the resolution of inflammation associated with infection or atherosclerosis have been proposed.[95,96] Since the resolution of inflammation is often associated with tissue repair and regeneration,[97–99] elucidation of the molecular mechanisms underlying the resolution of cerebral post-ischemic inflammation would lead to the establishment of novel therapies for ischemic stroke.

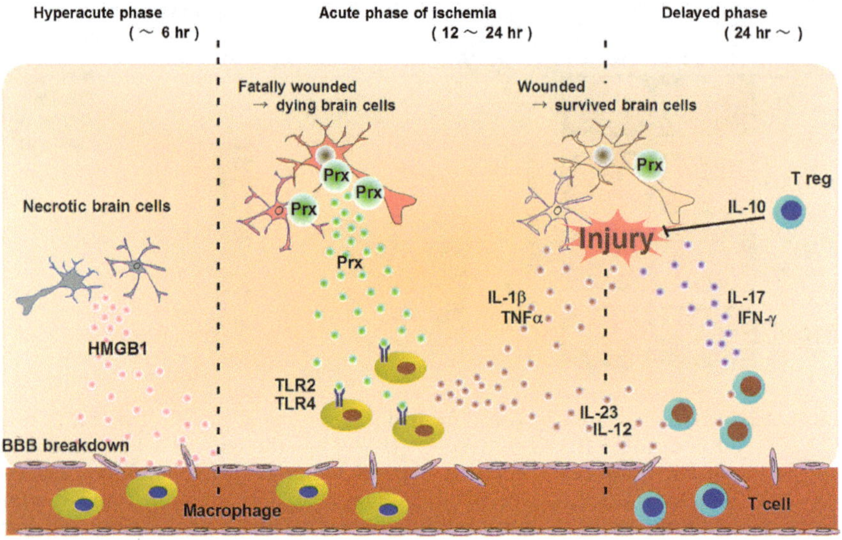

Figure 2.3 Post-ischemic inflammation is triggered by damage-associated molecular patterns (DAMPs). During the hyperacute phase of brain ischemia (within 6 hours after stroke onset), high mobility group box 1 (HMGB1) is released from the damaged brain cells and induces blood–brain barrier (BBB) breakdown. Following this, macrophages begin to infiltrate into the ischemic brain tissue *via* disrupted vessels during the acute phase of ischemia (12–24 hours after stroke onset). In this phase, the expression of peroxiredoxin (Prx) is induced in the wounded brain cells by ischemic stress to catalyze the destruction of reactive oxygen species in order to improve their survival. However, when the brain cells are about to die, Prx is released into the extracellular compartment. The extracellular Prx activates infiltrating macrophages through the Toll-like receptor (TLR) 2 and TLR4 signaling pathways, leading to the production of various inflammatory cytokines. Interleukin-1β (IL-1β) and tumor necrosis factor-α (TNF-α) directly act on neuronal cells in the penumbral region. The IL-23 and IL-12 produced from infiltrating macrophages affect infiltrating T cells during the delayed phase (more than 24 hours after stroke onset) and induce IL-17 and interferon-γ (IFN-γ) production. These inflammatory cytokines promote post-ischemic inflammation and neuronal damage. On the other hand, regulatory T cells (Tregs) produce IL-10 and exert an anti-inflammatory effect.

Acute inflammatory responses triggered by DAMPs exaggerate tissue injury and eventually lead to the further release of DAMPs by promoting cell death.[100] Thus, the clearance of DAMPs from injured tissue is required for the resolution of inflammation. The well-characterized DAMPs (Prx proteins and HMGB1) are internalized more efficiently by macrophages compared with neutrophils or microglia in the ischemic brains of mice.[101] The class A scavenger receptor *Msr1* is the key molecule in the internalization of those DAMPs, and the transcription factor

Mafb is critical for the upregulation of MSR1 expression in infiltrating macrophages. A retinoic acid receptor (RAR) agonist, Am80, has reinforced MSR1 expression in cerebral myeloid cells through *Mafb* induction, and its therapeutic effects last until 24 hours after stroke onset, suggesting that DAMP clearance may be a promising strategy to improve ischemic brain injury.

2.8 Conclusion

Recent research is gradually shedding light on the mechanisms of cerebral post-ischemic inflammation mediated by immune cells. Further understanding of the immune processes would lead to the development of novel treatments with a prolonged therapeutic time window.

References

1. H. K. Eltzschig and T. Eckle, Ischemia and reperfusion—from mechanism to translation, *Nat. Med.*, 2011, **17**, 1391–1401.
2. C. Iadecola and J. Anrather, The immunology of stroke: from mechanisms to translation, *Nat. Med.*, 2011, **17**, 796–808.
3. G. J. del Zoppo, Stroke and neurovascular protection, *N. Engl. J. Med.*, 2006, **354**, 553–555.
4. C. Iadecola, Neurovascular regulation in the normal brain and in Alzheimer's disease, *Nat. Rev. Neurosci.*, 2004, **5**, 347–360.
5. M. Kamouchi, T. Ago and T. Kitazono, Brain pericytes: emerging concepts and functional roles in brain homeostasis, *Cell. Mol. Neurobiol.*, 2011, **31**, 175–193.
6. F. C. Barone and G. Z. Feuerstein, Inflammatory mediators and stroke: new opportunities for novel therapeutics, *J. Cereb. Blood Flow Metab.*, 1999, **19**, 819–834.
7. R. Macrez, C. Ali, O. Toutirais, B. Le Mauff, G. Defer, U. Dirnagl and D. Vivien, Stroke and the immune system: from pathophysiology to new therapeutic strategies, *Lancet Neurol.*, 2011, **10**, 471–480.
8. G. Y. Chen and G. Nuñez, Sterile inflammation: sensing and reacting to damage, *Nat. Rev. Immunol.*, 2010, **10**, 826–837.
9. H. Yanai, T. Ban, Z. Wang, M. K. Choi, T. Kawamura, H. Negishi, M. Nakasato, Y. Lu, S. Hangai, R. Koshiba, D. Savitsky, L. Ronfani, S. Akira, M. E. Bianchi, K. Honda, T. Tamura, T. Kodama and T. Taniguchi, HMGB proteins function as universal sentinels for nucleic-acid-mediated innate immune responses, *Nature*, 2009, **462**, 99–103.
10. B. J. Marsh, R. L. Williams-Karnesky and M. P. Stenzel-Poore, Toll-like receptor signaling in endogenous neuroprotection and stroke, *Neuroscience*, 2009, **158**, 1007–1020.

11. C. R. Stewart, L. M. Stuart, K. Wilkinson, J. M. van Gils, J. Deng, A. Halle, K. J. Rayner, L. Boyer, R. Zhong, W. A. Frazier, A. Lacy-Hulbert, J. El Khoury, D. T. Golenbock and K. J. Moore, CD36 ligands promote sterile inflammation through assembly of a Toll-like receptor 4 and 6 heterodimer, *Nat. Immunol.*, 2010, **11**, 155–161.
12. Q. Zhang, M. Raoof, Y. Chen, Y. Sumi, T. Sursal, W. Junger, K. Brohi, K. Itagaki and C. J. Hauser, Circulating mitochondrial DAMPs cause inflammatory responses to injury, *Nature*, 2010, **464**, 104–107.
13. S. Rivest, Regulation of innate immune responses in the brain, *Nat. Rev. Immunol.*, 2009, **9**, 429–439.
14. R. A. Stetler, Y. Gao, L. Zhang, Z. Weng, F. Zhang, X. Hu, S. Wang, P. Vosler, G. Cao, D. Sun, S. H. Graham and J. Chen, Phosphorylation of HSP27 by protein kinase D is essential for mediating neuroprotection against ischemic neuronal injury, *J. Neurosci.*, 2012, **32**, 2667–2682.
15. K. Zheng, J. Y. Kim, H. Ma, J. E. Lee and M. A. Yenari, Anti-inflammatory effects of the 70 kDa heat shock protein in experimental stroke, *J. Cereb. Blood Flow Metab.*, 2008, **28**, 53–63.
16. L. Whitesell, R. Bagatell and R. Falsey, The stress response: implications for the clinical development of hsp90 inhibitors, *Curr. Cancer Drug Targets*, 2003, **3**, 349–358.
17. J. B. Kim, J. Sig Choi, Y. M. Yu, K. Nam, C. S. Piao, S. W. Kim, M. H. Lee, P. L. Han, J. S. Park and J. K. Lee, HMGB1, a novel cytokine-like mediator linking acute neuronal death and delayed neuroinflammation in the postischemic brain, *J. Neurosci.*, 2006, **26**, 6413–6421.
18. K. Liu, S. Mori, H. K. Takahashi, Y. Tomono, H. Wake, T. Kanke, Y. Sato, N. Hiraga, N. Adachi, T. Yoshino and M. Nishibori, Anti-high mobility group box 1 monoclonal antibody ameliorates brain infarction induced by transient ischemia in rats, *FASEB J.*, 2007, **21**, 3904–3916.
19. K. Hayakawa, J. Qiu and E. H. Lo, Biphasic actions of HMGB1 signaling in inflammation and recovery after stroke, *Ann. N. Y. Acad. Sci.*, 2010, **1207**, 50–57.
20. J. Zhang, H. K. Takahashi, K. Liu, H. Wake, R. Liu, T. Maruo, I. Date, T. Yoshino, A. Ohtsuka, S. Mori and M. Nishibori, Anti-high mobility group box-1 monoclonal antibody protects the blood-brain barrier from ischemia-induced disruption in rats, *Stroke*, 2011, **42**, 1420–1428.
21. T. Shichita, E. Hasegawa, A. Kimura, R. Morita, R. Sakaguchi, I. Takada, T. Sekiya, H. Ooboshi, T. Kitazono, T. Yanagawa, T. Ishii, H. Takahashi, S. Mori, M. Nishibori, K. Kuroda, S. Akira, K. Miyake and A. Yoshimura, Peroxiredoxin family proteins are key initiators of post-ischemic inflammation in the brain, *Nat. Med.*, 2012, **18**, 911–918.
22. J. Qiu, M. Nishimura, Y. Wang, J. R. Sims, S. Qiu, S. I. Savitz, S. Salomone and M. A. Moskowitz, Early release of HMGB-1 from neurons after the onset of brain ischemia, *J. Cereb. Blood Flow Metab.*, 2008, **28**, 927–938.

23. T. Ago, T. Liu, P. Zhai, W. Chen, H. Li, J. D. Molkentin, S. F. Vatner and J. Sadoshima, A redox-dependent pathway for regulating class II HDACs and cardiac hypertrophy, *Cell*, 2008, **133**, 978–993.
24. Z. A. Wood, E. Schröder, J. Robin Harris and L. B. Poole, Structure, mechanism and regulation of peroxiredoxins, *Trends Biochem. Sci.*, 2003, **28**, 32–40.
25. A. Patenaude, M. R. Murthy and M. E. Mirault, Emerging roles of thioredoxin cycle enzymes in the central nervous system, *Cell. Mol. Life Sci.*, 2005, **62**, 1063–1080.
26. J. Rashidian, M. W. Rousseaux, K. Venderova, D. Qu, S. M. Callaghan, M. Phillips, R. J. Bland, M. J. During, Z. Mao, R. S. Slack and D. S. Park, Essential role of cytoplasmic cdk5 and Prx2 in multiple ischemic injury models, in vivo, *J. Neurosci.*, 2009, **29**, 12497–12505.
27. J. L. Chou, C. H. Wu, C. Y. Tsai, A. Y. Chang and S. H. Chan, Proteomic investigation of a neural substrate intimately related to brain death, *Proteomics*, 2011, **11**, 239–248.
28. L. Dayon, N. Turck, T. Garcí-Berrocoso, N. Walter, P. R. Burkhard, A. Vilalta, J. Sahuquillo, J. Montaner and J. C. Sanchez, Brain extracellular fluid protein changes in acute stroke patients, *J. Proteome Res.*, 2011, **10**, 1043–1051.
29. C. J. Chen, H. Kono, D. Golenbock, G. Reed, S. Akira and K. L. Rock, Identification of a key pathway required for the sterile inflammatory response triggered by dying cells, *Nat. Med.*, 2007, **13**, 851–856.
30. S. C. Tang, T. V. Arumugam, X. Xu, A. Cheng, M. R. Mughal, D. G. Jo, J. D. Lathia, D. A. Siler, S. Chigurupati, X. Ouyang, T. Magnus, S. Camandola and M. P. Mattson, Pivotal role for neuronal Toll-like receptors in ischemic brain injury and functional deficits, *Proc. Natl. Acad. Sci. U. S. A.*, 2007, **104**, 13798–13803.
31. K. Hyakkoku, J. Hamanaka, K. Tsuruma, M. Shimazawa, H. Tanaka, S. Uematsu, S. Akira, N. Inagaki, H. Nagai and H. Hara, Toll-like receptor 4 (TLR4), but not TLR3 or TLR9, knock-out mice have neuroprotective effects against focal cerebral ischemia, *Neuroscience*, 2010, **171**, 258–267.
32. D. Brea, M. Blanco, P. Ramos-Cabrer, O. Moldes, S. Arias, M. Pérez-Mato, R. Leira, T. Sobrino and J. Castillo, Toll-like receptors 2 and 4 in ischemic stroke: outcome and therapeutic values, *J. Cereb. Blood Flow Metab.*, 2011, **31**, 1424–1431.
33. S. Akashi-Takamura and K. Miyake, TLR accessory molecules, *Curr. Opin. Immunol.*, 2008, **20**, 420–425.
34. K. Arimura, T. Ago, M. Kamouchi, K. Nakamura, K. Ishitsuka, J. Kuroda, H. Sugimori, H. Ooboshi, T. Sasaki and T. Kitazono, PDGF receptor β signaling in pericytes following ischemic brain injury, *Curr. Neurovasc. Res.*, 2012, **9**, 1–9.
35. P. A. Fraser, The role of radical generation in increasing cerebrovascular permeability, *Free Radical Biol. Med.*, 2011, **51**, 967–977.

36. T. Takano, N. Oberheim, M. L. Cotrina and M. Nedergaard, Astrocytes and ischemic injury, *Stroke*, 2009, **40**, S8–S12.
37. A. Morancho, A. Rosell, L. Garcia-Bonilla and J. Montaner, Metalloproteinase and stroke infarct size: role for anti-inflammatory treatment?, *Ann. N. Y. Acad. Sci.*, 2010, **1207**, 123–133.
38. A. Y. Lai and K. G. Todd, Microglia in cerebral ischemia: molecular actions and interactions, *Can. J. Physiol. Pharmacol.*, 2006, **84**, 49–59.
39. Q. W. Yang, F. L. Lu, Y. Zhou, L. Wang, Q. Zhong, S. Lin, J. Xiang, J. C. Li, C. Q. Fang and J. Z. Wang, HMBG1 mediates ischemia-reperfusion injury by TRIF-adaptor independent Toll-like receptor 4 signaling, *J. Cereb. Blood Flow Metab.*, 2011, **31**, 593–605.
40. B. M. Famakin, Y. Mou, C. A. Ruetzler, J. Bembry, D. Maric and J. M. Hallenbeck, Disruption of downstream MyD88 or TRIF Toll-like receptor signaling does not protect against cerebral ischemia, *Brain Res.*, 2011, **1388**, 148–156.
41. C. A. Farrar, B. Keogh, W. McCormack, A. O'Shaughnessy, A. Parker, M. Reilly and S. H. Sacks, Inhibition of TLR2 promotes graft function in a murine model of renal transplant ischemia-reperfusion injury, *FASEB J.*, 2012, **26**, 799–807.
42. F. Arslan, M. B. Smeets, L. A. O'Neill, B. Keogh, P. McGuirk, L. Timmers, C. Tersteeg, I. E. Hoefer, P. A. Doevendans, G. Pasterkamp and D. P. de Kleijn, Myocardial ischemia/reperfusion injury is mediated by leukocytic toll-like receptor-2 and reduced by systemic administration of a novel anti-toll-like receptor-2 antibody, *Circulation*, 2010, **121**, 80–90.
43. S. E. Lakhan, A. Kirchgessner and M. Hofer, Inflammatory mechanisms in ischemic stroke: therapeutic approaches, *J. Transl. Med.*, 2009, **7**, 97.
44. F. Martinon, K. Burns and J. Tschopp, The inflammasome: a molecular platform triggering activation of inflammatory caspases and processing of proIL-1beta, *Mol. Cell*, 2002, **10**, 417–426.
45. V. A. Rathinum, S. K. Vanaja and K. A. Fitzgerald, Regulation of inflammasome signaling, *Nat. Immunol.*, 2012, **13**, 333–342.
46. D. P. Abulafia, J. P. de Rivero Vaccari, J. D. Lozano, G. Lotocki, R. W. Keane and W. D. Dietrich, Inhibition of the inflammasome complex reduces the inflammatory response after thromboembolic stroke in mice, *J. Cereb. Blood Flow Metab.*, 2009, **29**, 534–544.
47. S. Chakraborty, D. K. Kaushik, M. Gupta and A. Basu, Inflammasome signaling at the heart of central nervous system pathology, *J. Neurosci. Res.*, 2010, **88**, 1615–1631.
48. M. Ito, T. Shichita, M. Okada, R. Komine, Y. Noguchi, A. Yoshimura and R. Morita, Bruton's tyrosine kinase is essential for NLRP3 inflammasome activation and contributes to ischaemic brain injury, *Nat. Commun.*, 2015, **6**, 7360, DOI: 10.1038.
49. H. Boutin, R. A. LeFeuvre, R. Horai, M. Asano, Y. Iwakura and N. J. Rothwell, Role of IL-1alpha and IL-1beta in ischemic brain damage, *J. Neurosci.*, 2001, **21**, 5528–5534.

50. S. M. Allan, P. J. Tyrrell and N. J. Rothwell, Interleukin-1 and neuronal injury, *Nat. Rev. Immunol.*, 2005, **5**, 629–640.
51. T. Liu, R. K. Clark, P. C. McDonnell, P. R. Young, R. F. White, F. C. Barone and G. Z. Feuerstein, Tumor necrosis factor-alpha expression in ischemic neurons, *Stroke*, 1994, **25**, 1481–1488.
52. J. M. Hallenbeck, The many faces of tumor necrosis factor in stroke, *Nat. Med.*, 2002, **8**, 1363–1368.
53. T. Yamashita, K. Sawamoto, S. Suzuki, N. Suzuki, K. Adachi, T. Kawase, M. Mihara, Y. Ohsugi, K. Abe and H. Okano, Blockade of interleukin-6 signaling aggravates ischemic cerebral damage in mice: possible involvement of Stat3 activation in the protection of neurons, *J. Neurochem.*, 2005, **94**, 459–468.
54. K. Gertz, G. Kronenberg, R. E. Kälin, T. Baldinger, C. Werner, M. Balkaya, G. D. Eom, J. Hellmann-Regen, J. Kröber, K. R. Miller, U. Lindauer, U. Laufs, U. Dirnagl, F. L. Heppner and M. Endres, Essential role of interleukin-6 in post-stroke angiogenesis, *Brain*, 2012, **135**, 1964–1980.
55. J. E. Jung, G. S. Kim and P. H. Chan, Neuroprotection by interleukin-6 is mediated by signal transducer and activator of transcription 3 and antioxidative signaling in ischemic stroke, *Stroke*, 2011, **42**, 3574–3579.
56. A. Yadav, A. Kalita, S. Dhillon and K. Banerjee, JAK/STAT3 pathway is involved in survival of neurons in response to insulin like growth factor and negatively regulated by suppressor of cytokine signal-3, *J. Biol. Chem.*, 2005, **280**, 31830–31840.
57. J. K. Strecker, J. Minnerup, B. Gess, E. B. Ringelstein, W. R. Schäbitz and M. Schilling, Monocyte chemoattractant protein-1-deficiency impairs the expression of IL-6, IL-1b and G-CSF after transient focal ischemia in mice, *PLoS One*, 2011, **6**, e25863.
58. S. Terao, G. Yilmaz, K. Y. Stokes, J. Russell, M. Ishikawa, T. Kawase and D. N. Granger, Blood cell-derived RANTES mediates cerebral microvascular dysfunction, inflammation, and tissue injury after focal ischemia-reperfusion, *Stroke*, 2008, **39**, 2560–2570.
59. Y. Terao, H. Ohta, A. Oda, Y. Nakagaito, Y. Kiyota and Y. Shintani, Macrophage inflammatory protein-3alpha plays a key role in the inflammatory cascade in rat focal cerebral ischemia, *Neurosci. Res.*, 2009, **64**, 75–82.
60. Y. Kumai, H. Ooboshi, J. Takada, M. Kamouchi, T. Kitazono, K. Egashira, S. Ibayashi and M. Iida, Anti-monocyte chemoattractant protein-1 gene therapy protects against focal brain ischemia in hypertensive rats, *J. Cereb. Blood Flow Metab.*, 2004, **24**, 1359–1368.
61. A. Reboldi, C. Coisne, D. Baumjohann, F. Benvenuto, D. Bottinelli, S. Lira, A. Uccelli, A. Lanzavecchia, B. Engelhardt and F. Sallusto, C-C chemokine receptor 6-regulated entry of TH-17 cells into the CNS through the choroid plexus is required for the initiation of EAE, *Nat. Immunol.*, 2009, **10**, 514–523.

62. B. Martin, K. Hirota, D. J. Cua, B. Stockinger and M. Veldhoen, Interleukin-17-producing gammadelta T cells selectively expand in response to pathogen products and environmental signals, *Immunity*, 2009, **31**, 321–330.
63. E. S. Connolly Jr., C. J. Winfree, T. A. Springer, Y. Naka, H. Liao, S. D. Yan, D. M. Stern, R. A. Solomon, J. C. Gutierrez-Ramos and D. J. Pinsky, Cerebral protection in homozygous null ICAM-1 mice after middle cerebral artery occlusion. Role of neutrophil adhesion in the pathogenesis of stroke, *J. Clin. Invest.*, 1996, **97**, 209–216.
64. A. Liesz, W. Zhou, É. Mracskó, S. Karcher, H. Bauer, S. Schwarting, L. Sun, D. Bruder, S. Stegemann, A. Cerwenka, C. Sommer, A. H. Dalpke and R. Veltkamp, Inhibition of lymphocyte trafficking shields the brain against deleterious neuroinflammation after stroke, *Brain*, 2011, **134**, 704–720.
65. M. Asahi, K. Asahi, J. C. Jung, G. J. del Zoppo, M. E. Fini and E. H. Lo, Role for matrix metalloproteinase 9 after cerebral ischemia: effects of gene knockout and enzyme inhibition with BB-94, *J. Cereb. Blood Flow Metab.*, 2000, **20**, 1681–1689.
66. B. Q. Zhao, S. Wang, H. Y. Kim, H. Storrie, B. R. Rosen, D. J. Mooney, X. Wang and E. H. Lo, Role of matrix metalloproteinases in delayed cortical responses after stroke, *Nat. Med.*, 2006, **12**, 441–445.
67. K. K. Dev, F. Mullershausen, H. Mattes, R. R. Kuhn, G. Bilbe, D. Hoyer and A. Mir, Brain sphingosine-1-phosphate receptors: implication for FTY720 in the treatment of multiple sclerosis, *Pharmacol. Ther.*, 2008, **117**, 77–93.
68. T. Shichita, Y. Sugiyama, H. Ooboshi, H. Sugimori, R. Nakagawa, I. Takada, T. Iwaki, Y. Okada, M. Iida, D. J. Cua, Y. Iwakura and A. Yoshimura, Pivotal role of cerebral interleukin-17-producing gammadeltaT cells in the delayed phase of ischemic brain injury, *Nat. Med.*, 2009, **15**, 946–950.
69. Y. Hasegawa, H. Suzuki, T. Sozen, W. Rolland and J. H. Zhang, Activation of sphingosine 1-phosphate receptor-1 by FTY720 is neuroprotective after ischemic stroke in rats, *Stroke*, 2010, **41**, 368–374.
70. J. A. Cohen, F. Barkhof, G. Comi, H. P. Hartung, B. O. Khatri, X. Montalban, J. Pelletier, R. Capra, P. Gallo, G. Izquierdo, K. Tiel-Wilck, A. de Vera, J. Jin, T. Stites, S. Wu, S. Aradhye and L. Kappos, TRANSFORMS Study Group, Oral fingolimod or intramuscular interferon for relapsing multiple sclerosis, *N. Engl. J. Med.*, 2010, **362**, 402–415.
71. B. H. Clausen, K. L. Lambertsen, A. A. Babcock, T. H. Holm, F. Dagnaes-Hansen and B. Finsen, Interleukin-1beta and tumor necrosis factor-alpha are expressed by different subsets of microglia and macrophages after ischemic stroke in mice, *J. Neuroinflammation*, 2008, **5**, 46, DOI: 10.1186.
72. S. G. Soriano, A. Coxon, Y. F. Wang, M. P. Frosch, S. A. Lipton, P. R. Hickey and T. N. Mayadas, Mice deficient in Mac-1 (CD11b/CD18)

are less susceptible to cerebral ischemia/reperfusion injury, *Stroke*, 1999, **30**, 134–139.
73. G. Yilmaz, T. V. Arumugam, K. Y. Stokes and D. N. Granger, Role of T lymphocytes and interferon-gamma in ischemic stroke, *Circulation*, 2006, **113**, 2105–2112.
74. M. Schroeter, S. Jander, O. W. Witte and G. Stoll, Local immune responses in the rat middle cerebral artery occlusion, *J. Neuroimmunol.*, 1994, **55**, 195–203.
75. S. Jander, M. Karemer, M. Schroeter, O. W. Witte and G. Stoll, Lymphocytic infiltration and expression of intercellular adhesion molecule-1 in photochemically induced ischemia of the rat cortex, *J. Cereb. Blood Flow Metab.*, 1995, **15**, 42–51.
76. P. D. Hurn, S. Subramanian, S. M. Parker, M. E. Afentoulis, L. J. Kaler, A. A. Vandenbark and H. Offner, T- and B-cell-deficient mice with experimental stroke have reduced lesion size and inflammation, *J. Cereb. Blood Flow Metab.*, 2007, **27**, 1798–1805.
77. X. Ren, K. Akiyoshi, S. Dziennis, A. A. Vandenbark, P. S. Herson, P. D. Hurn and H. Offner, Regulatory B cells limit CNS inflammation and neurologic deficits in murine experimental stroke, *J. Neurosci.*, 2011, **31**, 8556–8563.
78. C. H. Wong, C. N. Jenne, C. Leger and P. Kubes, Functional innervation of hepatic iNKT cells is immunosuppressive following stroke, *Science*, 2011, **334**, 101–105.
79. K. L. Lambertsen, R. Gregersen, M. Meldgaard, B. H. Clausen, E. K. Heibøl, R. Ladeby, J. Knudsen, A. Frandsen, T. Owens and B. Finsen, A role for interferon-gamma in focal cerebral ischemia in mice, *J. Neuropathol. Exp. Neurol.*, 2004, **63**, 942–955.
80. X. Xiong, G. E. Barreto, L. Xu, Y. B. Ouyang, X. Xie and R. G. Giffard, Increased brain injury and worsened neurological outcome in interleukin-4 knockout mice after transient focal cerebral ischemia, *Stroke*, 2011, **42**, 2026–2032.
81. F. Chen, Z. Liu, W. Wu, C. Rozo, S. Bowdridge, A. Millman, N. Van Rooijen, J. F. Urban Jr., T. A. Wynn and W. C. Gause, An essential role for TH2-type responses in limiting acute tissue damage during experimental helminth infection, *Nat. Med.*, 2012, **18**, 260–266.
82. H. Ooboshi, S. Ibayashi, T. Shichita, Y. Kumai, J. Takada, T. Ago, S. Arakawa, H. Sugimori, M. Kamouchi, T. Kitazono and M. Iida, Postischemic gene transfer of interleukin-10 protects against both focal and global brain ischemia, *Circulation*, 2006, **111**, 913–919.
83. A. Liesz, E. Suri-Payer, C. Veltkamp, H. Doerr, C. Sommer, S. Rivest, T. Giese and R. Veltkamp, Regulatory T cells are key cerebroprotective immunomodulators in acute experimental stroke, *Nat. Med.*, 2009, **15**, 192–199.
84. Y. Motomura, H. Kitamura, A. Hijikata, Y. Matsunaga, K. Matsumoto, H. Inoue, K. Atarashi, S. Hori, H. Watarai, J. Zhu, M. Taniguchi and

M. Kubo, The transcription factor E4BP4 regulates the production of IL-10 and IL-13 in CD4 T cells, *Nat. Immunol.*, 2011, **12**, 450–459.
85. T. Okamura, K. Fujio, M. Shibuya, S. Sumitomo, H. Shoda, S. Sakaguchi and K. Yamamoto, CD4 + CD25-LAG3 + regulatory T cells controlled by the transcription factor Egr-2, *Proc. Natl. Acad. Sci. U. S. A.*, 2009, **106**, 13974–13979.
86. D. J. Cua, J. Sherlock, Y. Chen, C. A. Murphy, B. Joyce, B. Seymour, L. Lucian, W. To, S. Kwan, T. Churakova, S. Zurawski, M. Wiekowski, S. A. Lira, D. Gorman, R. A. Kastelein and J. D. Sedgwick, Interleukin-23 rather than interleukin-12 is the critical cytokine for autoimmune inflammation of the brain, *Nature*, 2003, **421**, 744–748.
87. F. Konoeda, T. Shichita, H. Yoshida, Y. Sugiyama, G. Muto, E. Hasegawa, R. Morita, N. Suzuki and A. Yoshimura, Therapeutic effect of IL-12/23 and their signaling pathway blockade on brain ischemia model, *Biochem. Biophys. Res. Commun.*, 2010, **402**, 500–506.
88. F. Fossiez, O. Djossou, P. Chomarat, L. Flores-Romo, S. Ait-Yahia, C. Maat, J. J. Pin, P. Garrone, E. Garcia, S. Saeland, D. Blanchard, C. Gaillard, B. Das Mahapatra, E. Rouvier, P. Golstein, J. Banchereau and S. Lebecque, T cell interleukin-17 induces stromal cells to produce proinflammatory and hematopoietic cytokines, *J. Exp. Med.*, 1996, **183**, 2593–2603.
89. H. Kebir, K. Kreymborg, I. Ifergan, A. Dodelet-Devillers, R. Cayrol, M. Bernard, F. Giuliani, N. Arbour, B. Becher and A. Prat, Human TH17 lymphocytes promote blood-brain barrier disruption and central nervous system inflammation, *Nat. Med.*, 2007, **13**, 1173–1175.
90. I. Ifergan, H. Kébir, M. Bernard, K. Wosik, A. Dodelet-Devillers, R. Cayrol, N. Arbour and A. Prat, The blood-brain barrier induces differentiation of migrating monocytes into Th17-polarizing dendritic cells, *Brain*, 2008, **131**, 785–799.
91. C. E. Sutton, S. J. Lalor, C. M. Sweeney, C. F. Brereton, E. C. Lavelle and K. H. Mills, Interleukin-1 and IL-23 induce innate IL-17 production from gammadelta T cells, amplifying Th17 responses and autoimmunity, *Immunity*, 2009, **31**, 331–341.
92. S. Subramanian, B. Zhang, Y. Kosaka, G. G. Burrows, M. R. Grafe, A. A. Vandenbark, P. D. Hurn and H. Offner, Recombinant T cell receptor ligand treats experimental stroke, *Stroke*, 2009, **40**, 2539–2545.
93. K. Becker, D. Kindrick, R. McCarron, J. Hallenbeck and R. Winn, Adoptive transfer of myelin basic protein-tolerized splenocytes to naive animals reduces infarct size: a role for lymphocytes in ischemic brain injury?, *Stroke*, 2003, **34**, 1809–1815.
94. K. J. Becker, Sensitization and tolerization to brain antigens in stroke, *Neuroscience*, 2009, **158**, 1090–1097.
95. S. Zimmer, A. Grebe, S. S. Bakke, N. Bode, B. Halvorsen, T. Ulas, M. Skjelland, D. De Nardo, L. I. Labzin, A. Kerksiek, C. Hempel, M. T. Heneka, V. Hawxhurst, M. L. Fitzgerald, J. Trebicka, I. Björkhem, J. Gustafsson, M. Westerterp, A. R. Tall, S. D. Wright, T. Espevik,

J. L. Schultze, G. Nickenig, D. Lütjohann and E. Latz, Cyclodextrin promotes atherosclerosis regression *via* macrophage reprogramming, *Sci. Transl. Med.*, 2016, **8**, 333ra50.

96. J. Dalli, N. Chiang and C. Serhan, Elucidation of novel 13-series resolvins that increase with atorvastatin and clear infections, *Nat. Med.*, 2015, **21**, 1071–1075.
97. C. D. Buckley, D. W. Gilroy, C. N. Serhan, B. Stockinger and P. P. Tak, The resolution of inflammation, *Nat. Rev. Immunol.*, 2013, **13**, 59–66.
98. U. Dirnagl and M. Endres, Found in translation preclinical stroke research predicts human pathophysiology, clinical phenotypes, and therapeutic outcomes, *Stroke*, 2014, **45**, 1510–1518.
99. C. Iadecola and J. Anrather, Stroke research at a crossroad: asking the brain for directions, *Nat. Neurosci.*, 2011, **14**, 1363–1368.
100. H. Kono and K. L. Rock, How dying cells alert the immune system to danger, *Nat. Rev. Immunol.*, 2008, **8**, 279–289.
101. T. Shichita, M. Ito, R. Morita, K. Komai, Y. Noguchi, H. Ooboshi, R. Koshida, S. Takahashi, T. Kodama and A. Yoshimura, MAFB prevents excess inflammation after ischemic stroke by accelerating clearance of damage signals through MSR1, *Nat. Med.*, 2017, **23**, 723–732.

CHAPTER 3

Biomarkers of Acute Brain Injury and Surrogate Endpoints in Traumatic Brain Injury and Stroke Translational Studies

LINDA PAPA[a,b]

[a] Department of Emergency Medicine, Orlando Regional Medical Center, Orlando, Florida, USA; [b] Department of Neurology and Neurosurgery, McGill University, 86 W. Underwood (S-200), Orlando, Florida 32806, USA
Email: lpstat@aol.com

3.1 Epidemiology of Acute Traumatic Brain Injury and Stroke

There are an estimated 10 million people affected annually by traumatic brain injury (TBI) across the globe.[1] In the United States, TBI is a major cause of death and disability,[2] with about 52 000 annual deaths and 5.3 million Americans impaired by its effects. European TBI prevalence data is not consistently reported, but it has been estimated that 1.6 million head-injured patients are hospitalized annually in Europe with an incidence rate of about 235 per 100 000.[3] According to the World Health Organization, TBI will surpass many diseases as the major cause of death and disability by the year 2020.[1]

Drug Discovery Series No. 62
Acute Brain Impairment: Scientific Discoveries and Translational Research
Edited by Philip V. Peplow, Svetlana A. Dambinova, Thomas A. Gennarelli and Bridget Martinez

Published by the Royal Society of Chemistry, www.rsc.org

TBI results from external blunt force trauma to the head leading to an alteration in cognitive functioning. Traditionally, TBI has been separated into three very broad categories: mild, moderate, and severe. The severity of TBI is initially based on the Glasgow Coma Scale (GCS) score which uses ocular, verbal, and motor responses to determine central nervous system impairment. A GCS score of less than 8 is considered severe, a GCS score of 9–12 is considered moderate, and a GCS score of 13–15 is considered mild. The accuracy of the GCS score can be impeded by cerebral perfusion, intoxication from drugs or alcohol, sedative medications, and other distracting injuries. Therefore, this classification scheme fails to capture the spectrum of TBI and the different types of injuries associated with it. The presence of a severe TBI is usually evident on initial examination, but the severity of milder injury is more difficult to determine, particularly within the first few hours after injury when the neurological examination is of restricted value. Individuals who incur a mild TBI are acutely at risk for intracranial bleeding and diffuse axonal injury (DAI). Structural brain abnormalities from trauma can be detected using techniques such as computed tomography (CT) or standard magnetic resonance imaging (MRI). CT scans are readily available in more settings than MRI.[4] MRI, in comparison, can be more sensitive than CT scans in determining both the presence of intracerebral hemorrhage and the degree of ischemia.[5,6]

Globally, in 2013, there were 6.5 million stroke deaths, making stroke the second-leading cause of death behind ischemic heart disease, particularly in those above the age of 60 years.[7] Every year, 15 million people worldwide suffer a stroke. Nearly 6 million die and another 5 million are left permanently disabled.[8] The prevalence of stroke events is expected to increase across the globe as the global population aged over 65 increases.[9] The number of stroke events in Europe is projected to rise from 1.1 million per year in 2000 to 1.5 million per year by 2025, largely due to the aging population.[10] In the United States, stroke is the fifth-leading cause of death, behind diseases of the heart, cancer, chronic lower respiratory disease, and unintentional injuries/accidents, killing more than 130 000 Americans each year.[7] Every year, about 795 000 people in the United States have a stroke; 610 000 of these are first or new strokes and 185 000 are recurrent strokes.[7]

Acute stroke is characterized by either an ischemic insult (a sudden reduction in the blood supply to the brain caused by thrombosis or embolisms) or a hemorrhagic insult (caused mainly by rupture of blood vessel or aneurysm) resulting in complications ranging from vision impairments and aphasia to varying degrees of paralysis, short-term memory loss, dementia, and difficulty concentrating and learning novel information. Most strokes (87%) are ischemic and the rest (13%) are hemorrhagic.[11] Current management of acute stroke is based on imaging such as MRI and CT followed by vascular recanalization or repair, rehabilitation, and supportive care. A key therapeutic strategy in the management of acute stroke is the restoration or improvement of perfusion to the ischemic area. Other strategies include reducing the rate of ischemic stroke reoccurrence through anticoagulation and modifying risk factors such as high blood pressure, diabetes, smoking, and excessive alcohol use.[12]

3.2 Brain Injury Biomarkers and Surrogate Endpoints in Clinical Trials

There are numerous blood-based biomarkers being used in medicine today; however, clinicians have not yet been afforded the opportunity to use blood-based biomarkers for the management of brain injury. Acute brain injury blood-based biomarkers have considerable diagnostic and prognostic promise in the clinical setting.[13–16] Research in blood-based brain injury biomarkers has exploded over the last two decades and continues to grow at an unwavering pace.[17] Commercialization of these biomarkers is ongoing, with a number of companies seeking FDA approval to market their tests for clinical applications. Stroke biomarkers have been suggested by many investigators as an opportunity to improve our ability to diagnose and treat stroke in a timely and safe manner.[18]

The difficulty in classifying injury severity is one that has made clinical trials in the field of brain injury challenging. There have been 30 failed clinical trials for TBI since 1993, 25 of which have been in the last 15 years and 13 of which have been in the last 5 years.[19] These negative results have come at an estimated cost of $1.1 billion.[19] Many seemingly promising neuroprotective and treatment agents were supported by extensive pre-clinical studies and Phase I and II trials. Large-scale clinical trials are costly, time consuming, and require tremendous medical resources. It has been suggested that future clinical trials should employ surrogate endpoints, such as blood-based biomarkers, to inform clinical trial decisions about inclusion, treatment effects, and outcome assessment. Potentially, blood-based biomarkers could provide early, quick, and less costly measures of injury type and severity compared to conventional neurological assessments, thereby significantly reducing the risks and costs of human clinical trials.

In the spring of 2015, the FDA–NIH Joint Leadership Council identified the harmonization of terms used in translational science and medical product development as a priority need, with a focus on terms related to study endpoints and biomarkers. The two agencies developed the BEST (Biomarkers, EndpointS, and other Tools) Resource.[20] A *biomarker* is defined as a characteristic that is measured as an indicator of normal biological processes, pathogenic processes, or responses to an exposure or intervention, including therapeutic interventions. Molecular, histologic, radiographic, or physiologic characteristics are types of biomarkers. Categories of biomarkers include susceptibility/risk biomarkers, diagnostic biomarkers, monitoring biomarkers, prognostic biomarkers, predictive biomarkers, pharmacodynamic/response biomarkers, and safety biomarkers.[21] A *surrogate endpoint* is defined as an endpoint that is used in clinical trials as a substitute for a direct measure of how a patient feels, functions, or survives.

A surrogate endpoint does not measure the clinical benefit of primary interest in and of itself, but rather is expected to predict that clinical benefit or harm based on epidemiologic, therapeutic, pathophysiologic, or other scientific evidence. From a US regulatory standpoint, surrogate endpoints and potential surrogate endpoints can be characterized by the level of clinical validation such as "validated surrogate endpoint," "reasonably likely surrogate endpoint," and "candidate surrogate endpoint."

From an acute brain injury (TBI and stroke) biomarker perspective, no blood-based biomarker has been approved by the FDA for clinical use, nor has a TBI biomarker been validated as a surrogate measure for clinical trials. Although there is an abundance of papers being published, many lack the rigorous methods and reporting required to adequately evaluate these markers for clinical use or as surrogate endpoints. Too often, there are inadequate sample sizes, inappropriate control groups and outcome measures, and variable definitions of TBI and stroke. In most studies, the kinetic profiles of the biomarkers being assessed are not even known and, therefore, the optimal sampling time relative to onset of injury is obscure. Importantly, variability in how samples are processed and the performance characteristics of the assays themselves are rarely known or incompletely described.[17] It is time to raise the bar on blood-based brain injury biomarker research and to validate biomarkers as diagnostic and monitoring tools for clinical trials. In order to achieve this, researchers, laboratories, and companies will have to conduct biomarker studies that will meet FDA standards.[17]

Key features that would make a blood-based brain injury biomarker useful both clinically and as a surrogate endpoint include: a high sensitivity and specificity for brain injury; the ability to stratify patients by severity of injury; the ability to appear rapidly in blood; the ability to monitor response to treatment; and the ability to predict functional outcome.[22] A few of the most widely studied protein biomarkers for acute brain injury in humans are presented below and stratified according to neuroanatomic location.

3.3 Biomarkers of Astroglial Injury

3.3.1 S100β

S100β is found in astrocytes and is a low-affinity calcium binding protein[23] that regulates intracellular levels of calcium and is a marker of astrocyte injury. It can also be found in cells that are not neuronal such as adipocytes, chondrocytes, and melanoma cells, making it non-brain specific.[24,25] S100β is one of the most extensively studied biomarkers for brain injury.[14,15,26] Elevated S100β levels in serum have been associated with outcome measures such as post-concussive syndrome and MRI abnormalities.[27–31] However, these findings have also been negated by others.[32–35] Studies have found correlations between elevated serum levels of S100β and CT abnormalities in adults and children,[15,36,37] but S100β is also elevated in the setting of polytrauma in trauma patients with no direct head trauma.[37–39] Since the results are so conflicting, the value of S100β as a surrogate marker in TBI is still controversial.

S100β has been evaluated for acute ischemic stroke. Several studies show S100β peaking at days 2–4 following acute ischemic stroke onset, rendering it less useful in the acute setting when decisions about thrombolysis needs to be performed within hours of post-infarction onset. However, S100β concentrations have been shown to peak earlier in cases of hemorrhage, at roughly 24 hours post-infarction onset. Thus, S100β offers more promise as a prognostic tool in the evaluation of infarction severity in coordination with MRI results and, relatedly, patient outcome and functionality. S100β concentrations within the first 2–10 days after onset have been found to be predictive of infarct volume and to have good correlation with assessments of neurological functionality such as the National Institutes of Health Stroke Scale (NIHSS).[40] Lesions greater than 5 cm in volume were correlated with significantly greater concentrations of S100β at 10, 24, and 72 hours post-onset and also exhibited potential to predict hemorrhagic transformation.[40]

3.3.2 Glial Fibrillary Acidic Protein

Glial fibrillary acidic protein (GFAP) is a protein found in the astroglial skeleton and in both white and gray brain matter[41] and is strongly upregulated during astrogliosis.[42] Serum GFAP has been shown to be elevated in various types of brain damage, including neurodegenerative disorders,[43,44] stroke,[45] and severe TBI.[38,46–51] In particular, GFAP has become a very promising brain-specific glial-derived biomarker for mild TBI in adults and children.[16,32,37,39,52–55] GFAP is released into serum following a mild TBI within an hour of injury and remains elevated for several days after injury.[16,39,52] Unlike S100β, GFAP is elevated in mild TBI patients with axonal injury as evidenced by MRI at 3 months post-injury.[32] In adults and children, serum GFAP levels distinguish mild TBI patients from trauma patients without TBI and detect intracranial lesions on CT with a sensitivity of 94–100%.[16,37,39,52,54] Moreover, GFAP outperforms S100β in detecting CT lesions in the setting of multiple trauma when extracranial fractures are present.[16,39,54] GFAP also predicts the need for neurosurgical intervention in patients with mild TBI.[16,52] The temporal profile of GFAP was evaluated in a large cohort of 584 trauma patients seen at the emergency department. GFAP performed consistently over 7 days in identifying concussion and mild to moderate TBI, detecting traumatic intracranial lesions on head CT, and predicting neurosurgical intervention.[16]

GFAP may also provide the opportunity to distinguishing ischemic and hemorrhagic stroke. It is critical to expeditiously distinguish between these two subsets of stroke because the treatment paradigm is very different between them. Thrombolytics (recombinant tissue plasminogen activator [rt-PA]) offers the only FDA-approved drug treatment for ischemic stroke beyond supportive care. Studies have shown that the risk–benefit ratio of rt-PA administration is favorable for patients treated within 3 hours of stroke onset, although some institutions place the threshold at 4.5–6 hours.[6,56–59] Endovascular procedures such as thrombectomy are becoming an increasingly explored treatment method.[60–62] Even this treatment strategy for ischemic stroke, however, is not usually undertaken without having first administered rt-PA to the patient.[63] The administration of thrombolytics in the setting of a hemorrhagic stroke could be disastrous. Clinicians have the difficult task of both rapidly confirming the diagnosis of stroke and rapidly confirming that a patient is not at risk for significant hemorrhage.[6,56–59] CT and MRI scans are currently the primary diagnostic means for making this distinction. However, recent studies have shown that even MRI scans can miss roughly 17% of strokes.[64–68] CT scans are even less effective.

A number of studies have suggested that the more immediate structural damage caused by hemorrhage in comparison with ischemic stroke creates a surge in GFAP levels.[69–73] Within the first 6 hours of stroke onset, significantly higher levels of serum GFAP have been found in patients suffering from hemorrhagic stroke than in those with ischemic stroke.[71–73] Moreover, serum GFAP levels have also been associated with acute stroke severity (NIHSS) and intracranial hemorrhage (ICH) volume and been found to be inversely proportional to functionality as long as 3 months after stroke onset. GFAP appears to be a strong candidate for a surrogate endpoint given its timely elevation and discriminatory ability.

3.4 Biomarkers of Neuronal Injury

3.4.1 Neuron-specific Enolase

Neuron-specific enolase (NSE), an isozyme of the glycolytic enzyme enolase, is found in central and peripheral neuronal cell bodies and increases in serum following cell injury.[74] Notably, it is also present in erythrocytes and endocrine cells,[75] leading to false-positive results in the setting of hemolysis.[76,77] Several studies have been published examining serum NSE following mild TBI,[74,78–81] but many of these reports contained inadequate control groups. Early levels of NSE concentrations have been correlated with outcome in children, particularly those under 4 years of age.[82–85] In the setting of DAI in severe TBI, levels of NSE at 72 hours of injury have shown an association with unfavorable outcome.[86]

Data surrounding potential correlations between NSE concentration and ischemia are also somewhat mixed. Missler *et al.* found that serum concentrations of NSE in patients suffering from acute ischemia did increase, peaking at day 1.9, and were significantly correlated with infarct volume on CT.[87] NSE levels, however, were not found to correlate with outcome at either discharge or 6 months.[87] Jung *et al.* reported that while cerebrospinal fluid (CSF) NSE levels rose in response to aneurysmal subarachnoid hemorrhage (aSAH), they failed to correlate with resulting cerebral vasospasm.[88] In another study, patients who had suffered aSAH had elevated CSF levels within 8 days that predicted the GCS, cerebral infarction, and intracranial hypertension.[89]

Unfortunately, this did not hold true for serum levels. In patients with ICH, Kuroiwa *et al.* did not observe a correlation between serum NSE concentrations and neurological status at admission, but did find elevated levels in those whose hematoma was 5 cm or greater.[90] However, unlike S100B, NSE has not been found to reliably predict outcome.[91]

3.4.2 Ubiquitin C-terminal Hydrolase

A promising candidate biomarker for TBI currently under investigation is ubiquitin C-terminal hydrolase-L1 (UCH-L1). This protein is involved in the addition and removal of excessive, oxidized, or misfolded proteins in neurons.[92] Clinical studies in humans with severe TBI have confirmed that the UCH-L1 protein is significantly elevated in human CSF and is detectable very early after injury,[93–95] and remains significantly elevated for at least 1 week post-injury.[94] CSF and serum levels of UCH-L1 correlate well.[96] Serum UCH-L1 is elevated in children with moderate to severe TBI.[97] Most recently, UCH-L1 was detected in the serum of both adult and child mild TBI patients within 1 hour of injury.[16,98,99] Interestingly, serum concentrations of UCH-L1 discriminated concussion patients from uninjured and trauma control patients without head injury.[16,98,99] A number of studies have shown serum UCH-L1 levels to be significantly higher in those with intracranial lesions on CT than in those without lesions,[16,53,55,98,99] and to be much higher in those requiring a neurosurgical intervention.[16,98] The time course of UCH-L1 was evaluated in a large cohort of 584 trauma patients presenting to the emergency department of a level 1 trauma center within 4 hours of injury, and serum levels of UCH-L1 rose rapidly, peaked at 8 hours after injury, and declined quickly over 48 hours.[16]

Despite UCH-L1's capacity to identify TBI, it is perhaps best employed as a prognostic tool in the assessment of stroke. This deubiquitinating enzyme is particularly useful in assessing both brain damage severity (*i.e.* infarction volume and extent of vasospasm) and the resulting outcome in patients who have suffered hemorrhage or ischemia. Individuals who suffered acute ischemic episodes within 12–24 hours following an aortic aneurysm repair were found to have elevated CSF concentrations of UCH-L1.[100] UCH-L1 holds therapeutic promise for ischemia patients to the extent that it and similar deubiquitinating enzymes have been found to reduce infarction in cases of rapid ischemic tolerance following brief ischemia.[101] Interestingly, polyubiquitinated protein build-ups in hippocampal synapses have been reported in response to global ischemia.[102] The potential role of UCH-L1 in cases of hemorrhage is perhaps clearer. Lewis *et al.* found that individuals suffering from aSAH had consistently higher concentrations of UCH-L1 in the CSF at 2 weeks after post-aneurysmal rupture, and elevations were associated with poor recovery.[103] Similarly, Siman *et al.* found that CSF concentrations of UCH-L1 taken over a 10-day period since aneurysmal rupture rose significantly and predicted severity of infarction, vasospasm, and outcome.[100] UCH-L1 appears to have a role in both the early detection of TBI and monitoring after stroke.

3.5 Biomarkers of Axonal Injury

3.5.1 αII-spectrin Breakdown Products

αII-spectrin is a 280 kDa protein that is an important structural component of the cortical membrane cytoskeleton and is particularly abundant in axons and presynaptic terminals.[104] It serves as a key substrate for both calpain-2 and caspase-3 cysteine proteases, which cleave cytoskeletal αII-spectrin into spectrin breakdown products (SBDPs).[105,106] These SBDPs have been reported in CSF from adults with severe TBI, and they have been shown to have a significant relationship with severity of injury and clinical outcome.[95,107–112] The time courses of calpain-mediated SBDP150 and SBDP145 (markers of necrosis) differ from that of caspase-3-mediated SBDP120 (marker of apoptosis) and have been shown to correlate with severity of injury, CT scan findings, and outcome at 6 months post-injury.[95,112] These findings were similar in children with moderate to severe TBI.[97] More recently, serum levels of SBDP150 measured in mild TBI patients have shown a significant association with acute measures of injury severity, such as GCS, intracranial injuries on CT, and neurosurgical intervention.[113]

In patients suffering from aSAH, CSF concentrations of SBDPs have been found to be significantly elevated within 72 hours post-onset and up to 12 hours pre-cerebral arterial vasospasm due to the role of necrotic proteolysis in hemorrhage and vasospasm-induced neurodegeneration.[114,115] Research on the potential role of SBDPs in human ischemia is more sparse. Caspase cascade inhibitors, which have been able to arrest processes of apoptosis, have surfaced as interesting therapeutic targets.[116,117]

3.5.2 Tau Protein

Tau is an intracellular, microtubule-associated protein that is highly enriched in axons and is related to axonal disruption.[118] After release, it is proteolytically cleaved at the N- and C-terminals. Initial elevated CSF C-tau levels in severe TBI patients have been shown to predict elevations in intracranial pressure and poor clinical outcome,[119] as well as intracranial injury on head CT.[120] However, these findings were not reproducible when C-tau was measured in peripheral blood in mild TBI.[121] Comparatively, two additional studies showed that C-tau was a poor predictor of both CT lesions and post-concussive symptoms.[33,122] Total tau (T-tau) has also been found to be correlated with severity of injury in severe TBI.[123–126] Ost *et al.* found that T-tau measured in CSF on days 2–3 discriminated between TBI and controls, as well as between good and bad outcome at 1 year.[125] Unfortunately, T-tau was not detected in serum. Recently, a study of professional hockey players showed that serum T-tau outperformed S-100B and NSE in detecting concussion at 1 hour after injury, and that levels were significantly higher in post-concussion samples at all times compared with preseason levels.[127] Phosphorylated tau is also being examined following head trauma and may have some association with chronic encephalopathy.[128] It appears that enhanced tau protein phosphorylation occurs with more severe injuries.[129,130]

Hu *et al.* observed that serum tau levels in patients with intracerebral hemorrhage were significantly predictive of 3 month mortality.[131] Elevated CSF levels of tau were reported in patients following severe episodes of aSAH in comparison with more moderate episodes, correlating with those patients with fatal outcomes.[132]

3.5.3 Neurofilaments

Neurofilaments (NFs) are heteropolymeric components of the neuron cytoskeleton that consist of a 68 kDa light NF subunit (NF-L) backbone with either 160 kDa medium (NF-M) or 200 kDa heavy subunit (NF-H) sidearms.[133] They provide structural support for the axon. Phosphorylated NF-H in CSF has been found to be elevated in adult patients with severe TBI compared to controls.[93] Hyperphosphorylated NF-H has also been correlated with severity of brain injury in children,[134] with significantly higher levels on day 2–4 in patients with poor outcome and in those children with DAI on initial CT scan.[134] Vajtr *et al.* similarly found elevated serum NF-H in patients with DAI over 10 days after admission, with the highest levels from days 4–10.[135]

Serum concentrations of phosphorylated NFL-H (pNFL-H) sampled from patients with acute ischemia have been shown to correlate with ischemia on initial CT scan and CT findings at 7 days, as well as with NIHSS and GCS.[136] Sellner *et al.* also found significantly higher serum pNFL-H concentrations in ischemic patients 24 hours after onset.[137] However, the results have been mixed. While Singh *et al.* observed elevation of serum pNFL-H in ischemic patients, levels did not reach significance or predict patient outcome or infarct volume until 3 weeks post-onset.[138] In cases of hemorrhagic stroke, NFs show promise as prognostic markers, with elevated CSF levels assayed within 10–14 days of aSAH onset correlating with GCS performance at 1 year post-onset.[139] Assays of pNFL-H were also found to correlate with early neurological deterioration and survival rates at 6 months post-onset, with comparable accuracy to that of the NIHSS.[140]

3.6 Conclusion

There are a number of potential blood-based brain injury biomarkers that correlate with TBI and stroke. However, no biomarker to date has met the rigorous standards of the FDA. There is a great need to establish a regulatory pathway for brain injury biomarker use in clinical trials. This could potentially improve drug development, open new avenues for novel study methods, and accelerate patient access to promising therapies. If we do not adopt such rigorous standards in evaluating potential brain injury biomarkers, the quest for such biomarkers will never be realized, and individuals suffering a brain injury will not benefit from potential improvements in their care. It is time to raise the standards in blood-based brain injury biomarker research and to encourage researchers, laboratories, and companies to perform at standards that will meet FDA approval, as they may prove to be the key to the development of therapies.[17]

References

1. A. A. Hyder, C. A. Wunderlich, P. Puvanachandra, G. Gururaj and O. C. Kobusingye, *NeuroRehabilitation*, 2007, **22**, 341.
2. NIH Consensus Development Panel on Rehabilitation of Persons With Traumatic Brain Injury, *JAMA*, 1999, **282**, 974.
3. F. Tagliaferri, C. Compagnone, M. Korsic, F. Servadei and J. Kraus, *Acta Neurochir. (Wien)*, 2006, **148**, 255–268, discussion 268.
4. A. Barbieri, E. Giuliani, C. Carone, F. Pederzoli, G. Mascheroni, G. Greco, C. Stucchi and S. Genedani, *Minerva Anestesiol.*, 2013, **79**, 750.
5. A. P. Amar, *World Neurosurg.*, 2012, **76**, S3–S8.
6. G. Thomalla, C. Schwark, J. Sobesky, E. Bluhmki, J. B. Fiebach, J. Fiehler, O. Zaro Weber, T. Kucinski, E. Juettler, P. A. Ringleb, H. Zeumer, C. Weiller, W. Hacke, P. D. Schellinger and J. Rother, *Stroke*, 2006, **37**, 852.
7. E. J. Benjamin, M. J. Blaha, S. E. Chiuve, M. Cushman, S. R. Das, R. Deo, S. D. de Ferranti, J. Floyd, M. Fornage, C. Gillespie, C. R. Isasi, M. C. Jimenez, L. C. Jordan, S. E. Judd, D. Lackland, J. H. Lichtman, L. Lisabeth, S. Liu, C. T. Longenecker, R. H. Mackey, K. Matsushita, D. Mozaffarian, M. E. Mussolino, K. Nasir, R. W. Neumar, L. Palaniappan, D. K. Pandey, R. R. Thiagarajan, M. J. Reeves, M. Ritchey, C. J. Rodriguez, G. A. Roth, W. D. Rosamond, C. Sasson, A. Towfighi, C. W. Tsao, M. B. Turner, S. S. Virani, J. H. Voeks, J. Z. Willey, J. T. Wilkins, J. H. Wu, H. M. Alger, S. S. Wong and P. Muntner, *Circulation*, 2017, **135**, e146–e603.
8. J. Mackay and G. A. Mensah, *Atlas of Heart Disease and Stroke*, World Health Organization (WHO) in collaboration with Center for Disease Control (CDC), Geneva, Switzerland, 2004.
9. D. Mukherjee and C. G. Patil, *World Neurosurg.*, 2011, **76**, S85.
10. T. Truelsen, B. Piechowski-Jozwiak, R. Bonita, C. Mathers, J. Bogousslavsky and G. Boysen, *Eur. J. Neurol.*, 2006, **13**, 581.
11. R. A. Grysiewicz, K. Thomas and D. K. Pandey, *Neurol. Clin.*, 2008, **26**, 871, vii.
12. A. Montagu, I. P. Reckless and A. M. Buchan, *Medicine*, 2012, **40**, 490.
13. L. Papa, *Sports Med. Arthrosc.*, 2016, **24**, 108.
14. L. Papa, M. M. Ramia, D. Edwards, B. D. Johnson and S. M. Slobounov, *J. Neurotrauma*, 2015, **32**, 661.
15. L. Papa, M. M. Ramia, J. M. Kelly, S. S. Burks, A. Pawlowicz and R. P. Berger, *J. Neurotrauma*, 2013, **30**, 324.
16. L. Papa, G. M. Brophy, R. D. Welch, L. M. Lewis, C. F. Braga, C. N. Tan, N. J. Ameli, M. A. Lopez, C. A. Haeussler, D. I. Mendez Giordano,

S. Silvestri, P. Giordano, K. D. Weber, C. Hill-Pryor and D. C. Hack, *JAMA Neurol.*, 2016, **73**, 551.
17. L. Papa and K. K. Wang, *J. Neurotrauma*, 2017, **34**, 2187–2189.
18. W. J. Oliver, T. C. Shope and L. R. Kuhns, *Pediatrics*, 2003, **112**, e174.
19. U. Samadani and S. R. Daly, *AANS Neurosurgeon*, 2016, **25**(3).
20. FDA-NIH Biomarker Working Group, *BEST (Biomarkers, EndpointS, and other Tools) Resource*, https://www.ncbi.nlm.nih.gov/books/NBK326791/, Accessed June 26, 2017.
21. M. A. Robb, P. M. McInnes and R. M. Califf, *JAMA*, 2016, **315**, 1107.
22. L. Papa, in *Poteomics – Human Diseases and Protein Functions*, ed. T. K. Man and R. J. Flores, In Tech Open Access Publisher, 2012.
23. H. Xiong, W. L. Liang and X. R. Wu, *Sheng Li Ke Xue Jin Zhan*, 2000, **31**, 217.
24. D. B. Zimmer, E. H. Cornwall, A. Landar and W. Song, *Brain Res. Bull.*, 1995, **37**, 417.
25. B. Olsson, H. Zetterberg, H. Hampel and K. Blennow, *Prog. Neurobiol.*, 2011, **95**, 520.
26. S. Schulte, L. W. Podlog, J. J. Hamson-Utley, F. G. Strathmann and H. K. Struder, *J. Athl. Train.*, 2014, **49**, 830.
27. T. Ingebrigtsen and B. Romner, *Acta Neurol. Scand.*, 1997, **95**, 51.
28. K. Waterloo, T. Ingebrigtsen and B. Romner, *Acta Neurochir. (Wien)*, 1997, **139**, 26–31, discussion 31.
29. T. Ingebrigtsen and B. Romner, *J. Neurosurg.*, 1996, **85**, 945.
30. T. Ingebrigtsen, K. Waterloo, E. A. Jacobsen, B. Langbakk and B. Romner, *Neurosurgery*, 1999, **45**, 468–475, discussion 475.
31. K. Heidari, S. Asadollahi, M. Jamshidian, S. N. Abrishamchi and M. Nouroozi, *Brain Inj.*, 2015, **29**, 33.
32. Z. Metting, N. Wilczak, L. A. Rodiger, J. M. Schaaf and J. van der Naalt, *Neurology*, 2012, **78**, 1428.
33. J. J. Bazarian, F. P. Zemlan, S. Mookerjee and T. Stigbrand, *Brain Inj.*, 2006, **20**, 759.
34. D. P. Lima, C. Simao Filho, C. Abib Sde and L. F. de Figueiredo, *Injury*, 2008, **39**, 604.
35. M. Dorminy, A. Hoogeveen, R. T. Tierney, M. Higgins, J. K. McDevitt and J. Kretzschmar, *Brain Inj.*, 2015, **29**, 1158.
36. K. Heidari, A. Vafaee, A. M. Rastekenari, M. Taghizadeh, E. G. Shad, R. Eley, M. Sinnott and S. Asadollahi, *Brain Inj.*, 2015, **29**, 1146.
37. L. Papa, M. K. Mittal, J. Ramirez, M. Ramia, S. Kirby, S. Silvestri, P. Giordano, K. Weber, C. F. Braga, C. N. Tan, N. J. Ameli, M. Lopez and M. Zonfrillo, *J. Neurotrauma*, 2016, **33**, 58.
38. L. E. Pelinka, A. Kroepfl, R. Schmidhammer, M. Krenn, W. Buchinger, H. Redl and A. Raabe, *J. Trauma*, 2004, **57**, 1006.
39. L. Papa, S. Silvestri, G. M. Brophy, P. Giordano, J. L. Falk, C. F. Braga, C. N. Tan, N. J. Ameli, J. A. Demery, N. K. Dixit, M. E. Mendes, R. L. Hayes, K. K. Wang and C. S. Robertson, *J. Neurotrauma*, 2014, **31**, 1815.

40. K. Fassbender, R. Schmidt, A. Schreiner, M. Fatar, F. Muhlhauser, M. Daffertshofer and M. Hennerici, *J. Neurol. Sci.*, 1997, **148**, 101.
41. L. F. Eng, J. J. Vanderhaeghen, A. Bignami and B. Gerstl, *Brain Res.*, 1971, **28**, 351–354.
42. L. W. Duchen, in *Greenfield's Neuropathology*, ed. J. A. Adams, J. A. N. Corsellis and L. W. Duchen, Edward Arnold, London, 1984, p. 1.
43. G. Baydas, V. S. Nedzvetskii, M. Tuzcu, A. Yasar and S. V. Kirichenko, *Eur. J. Pharmacol.*, 2003, **462**, 67.
44. P. E. Mouser, E. Head, K. H. Ha and T. T. Rohn, *Am. J. Pathol.*, 2006, **168**, 936.
45. M. Herrmann, P. Vos, M. T. Wunderlich, C. H. de Bruijn and K. J. Lamers, *Stroke*, 2000, **31**, 2670.
46. U. Missler, M. Wiesmann, G. Wittmann, O. Magerkurth and H. Hagenstrom, *Clin. Chem.*, 1999, **45**, 138.
47. L. E. Pelinka, A. Kroepfl, M. Leixnering, W. Buchinger, A. Raabe and H. Redl, *J. Neurotrauma*, 2004, **21**, 1553.
48. W. J. van Geel, H. P. de Reus, H. Nijzing, M. M. Verbeek, P. E. Vos and K. J. Lamers, *Clin. Chim. Acta*, 2002, **326**, 151.
49. K. Nylen, M. Ost, L. Z. Csajbok, I. Nilsson, K. Blennow, B. Nellgard and L. Rosengren, *J. Neurol. Sci.*, 2006, **240**, 85.
50. S. Mondello, L. Papa, A. Buki, R. Bullock, E. Czeiter, F. Tortella, K. K. Wang and R. L. Hayes, *Crit. Care*, 2011, **15**, R156.
51. P. E. Vos, B. Jacobs, T. M. Andriessen, K. J. Lamers, G. F. Borm, T. Beems, M. Edwards, C. F. Rosmalen and J. L. Vissers, *Neurology*, 2010, **75**, 1786.
52. L. Papa, L. M. Lewis, J. L. Falk, Z. Zhang, S. Silvestri, P. Giordano, G. M. Brophy, J. A. Demery, N. K. Dixit, I. Ferguson, M. C. Liu, J. Mo, L. Akinyi, K. Schmid, S. Mondello, C. S. Robertson, F. C. Tortella, R. L. Hayes and K. K. Wang, *Ann. Emerg. Med.*, 2012, **59**, 471.
53. R. Diaz-Arrastia, K. K. Wang, L. Papa, M. D. Sorani, J. K. Yue, A. M. Puccio, P. J. McMahon, T. Inoue, E. L. Yuh, H. F. Lingsma, A. I. Maas, A. B. Valadka, D. O. Okonkwo, G. T. Manley, I. S. Casey, M. Cheong, S. R. Cooper, K. Dams-O'Connor, W. A. Gordon, A. J. Hricik, D. K. Menon, P. Mukherjee, D. M. Schnyer, T. K. Sinha and M. J. Vassar, *J. Neurotrauma*, 2014, **31**, 19.
54. L. Papa, M. R. Zonfrillo, J. Ramirez, S. Silvestri, P. Giordano, C. F. Braga, C. N. Tan, N. J. Ameli, M. Lopez and M. K. Mittal, *Acad. Emerg. Med.*, 2015, **22**, 1274.
55. R. D. Welch, S. I. Ayaz, L. M. Lewis, J. Unden, J. Y. Chen, V. H. Mika, B. Saville, J. A. Tyndall, M. Nash, A. Buki, P. Barzo, D. Hack, F. C. Tortella, K. Schmid, R. L. Hayes, A. Vossough, S. T. Sweriduk and J. J. Bazarian, *J. Neurotrauma*, 2016, **33**, 203.
56. M. P. Marks, J. M. Olivot, S. Kemp, M. G. Lansberg, R. Bammer, L. R. Wechsler, G. W. Albers and V. Thijs, *Radiology*, 2008, **249**, 614.
57. C. H. Chen, S. C. Tang, L. K. Tsai, M. J. Hsieh, S. J. Yeh, K. Y. Huang and J. S. Jeng, *PLoS One*, 2014, **9**, e104862.

58. M. S. Kennedy, R. S. Ferri and D. Sofer, *Am. J. Nurs.*, 2004, **104**, 18.
59. K. Uchino, A. V. Alexandrov, Z. Garami, A. El-Mitwalli, L. B. Morgenstern and J. C. Grotta, *Cerebrovasc. Dis.*, 2005, **19**, 260.
60. B. C. Campbell, P. J. Mitchell, B. Yan, M. W. Parsons, S. Christensen, L. Churilov, R. J. Dowling, H. Dewey, M. Brooks, F. Miteff, C. Levi, M. Krause, T. J. Harrington, K. C. Faulder, B. S. Steinfort, T. Kleinig, R. Scroop, S. Chryssidis, A. Barber, A. Hope, M. Moriarty, B. McGuinness, A. A. Wong, A. Coulthard, T. Wijeratne, A. Lee, J. Jannes, J. Leyden, T. G. Phan, W. Chong, M. E. Holt, R. V. Chandra, C. F. Bladin, M. Badve, H. Rice, L. de Villiers, H. Ma, P. M. Desmond, G. A. Donnan and S. M. Davis, *Int. J. Stroke*, 2014, **9**, 126.
61. M. Goyal, A. M. Demchuk, B. K. Menon, M. Eesa, J. L. Rempel, J. Thornton, D. Roy, T. G. Jovin, R. A. Willinsky, B. L. Sapkota, D. Dowlatshahi, D. F. Frei, N. R. Kamal, W. J. Montanera, A. Y. Poppe, K. J. Ryckborst, F. L. Silver, A. Shuaib, D. Tampieri, D. Williams, O. Y. Bang, B. W. Baxter, P. A. Burns, H. Choe, J. H. Heo, C. A. Holmstedt, B. Jankowitz, M. Kelly, G. Linares, J. L. Mandzia, J. Shankar, S. I. Sohn, R. H. Swartz, P. A. Barber, S. B. Coutts, E. E. Smith, W. F. Morrish, A. Weill, S. Subramaniam, A. P. Mitha, J. H. Wong, M. W. Lowerison, T. T. Sajobi and M. D. Hill, *N. Engl. J. Med.*, 2015, **372**, 1019.
62. O. A. Berkhemer, P. S. Fransen, D. Beumer, L. A. van den Berg, H. F. Lingsma, A. J. Yoo, W. J. Schonewille, J. A. Vos, P. J. Nederkoorn, M. J. Wermer, M. A. van Walderveen, J. Staals, J. Hofmeijer, J. A. van Oostayen, G. J. Lycklama À Nijeholt, J. Boiten, P. A. Brouwer, B. J. Emmer, S. F. de Bruijn, L. C. van Dijk, L. J. Kappelle, R. H. Lo, E. J. van Dijk, J. de Vries, P. L. de Kort, W. J. van Rooij, J. S. van den Berg, B. A. van Hasselt, L. A. Aerden, R. J. Dallinga, M. C. Visser, J. C. Bot, P. C. Vroomen, O. Eshghi, T. H. Schreuder, R. J. Heijboer, K. Keizer, A. V. Tielbeek, H. M. den Hertog, D. G. Gerrits, R. M. van den Berg-Vos, G. B. Karas, E. W. Steyerberg, H. Z. Flach, H. A. Marquering, M. E. Sprengers, S. F. Jenniskens, L. F. Beenen, R. van den Berg, P. J. Koudstaal, W. H. van Zwam, Y. B. Roos, A. van der Lugt, R. J. van Oostenbrugge, C. B. Majoie and D. W. Dippel, *N. Engl. J. Med.*, 2015, **372**, 11.
63. A. Yilmaz, E. Akpinar, M. A. Topcuoglu and E. M. Arsava, *Neuroradiology*, 2015, **57**, 501.
64. P. D. Schellinger, L. L. Latour, C. S. Wu, J. A. Chalela and S. Warach, *Neuroradiology*, 2006, **48**, 69.
65. C. S. Kidwell, J. A. Chalela, J. L. Saver, S. Starkman, M. D. Hill, A. M. Demchuk, J. A. Butman, N. Patronas, J. R. Alger, L. L. Latour, M. L. Luby, A. E. Baird, M. C. Leary, M. Tremwel, B. Ovbiagele, A. Fredieu, S. Suzuki, J. P. Villablanca, S. Davis, B. Dunn, J. W. Todd, M. A. Ezzeddine, J. Haymore, J. K. Lynch, L. Davis and S. Warach, *JAMA*, 2004, **292**, 1823.
66. D. W. Kang, J. A. Chalela, W. Dunn and S. Warach, *Stroke*, 2005, **36**, 1939.

67. D. W. Kang, L. L. Latour, J. A. Chalela, J. A. Dambrosia and S. Warach, *Neurology*, 2004, **63**, 2261.
68. J. A. Chalela, D. W. Kang and S. Warach, *J. Neuroimaging*, 2004, **14**, 504.
69. J. P. Mohr and J. Kejda-Scharler, *Clin. Chem.*, 2011, **58**, 319.
70. M. B. Jensen, M. R. Chacon, J. A. Sattin, A. Aleu and P. D. Lyden, *Neurologist*, 2008, **14**, 243.
71. C. Foerch, I. Curdt, B. Yan, F. Dvorak, M. Hermans, J. Berkefeld, A. Raabe, T. Neumann-Haefelin, H. Steinmetz and M. Sitzer, *J. Neurol. Neurosurg. Psychiatry*, 2006, **77**, 181.
72. C. Foerch, M. Niessner, T. Back, M. Bauerle, G. M. De Marchis, A. Ferbert, H. Grehl, G. F. Hamann, A. Jacobs, A. Kastrup, S. Klimpe, F. Palm, G. Thomalla, H. Worthmann and M. Sitzer, *Clin. Chem.*, 2012, **58**, 237.
73. L. Papa, F. Silvestri, S. Silvestri, J. Tesar, C. N. Tan, N. J. Ameli, M. A. Lopez, C. A. Haeussler, D. Mendez Giordano, H. Tran, T. Stamile, P. Katari, P. A. Ameli and P. Giordano, *Acad. Emerg. Med.*, 2017, **24**, S38.
74. I. M. Skogseid, H. K. Nordby, P. Urdal, E. Paus and F. Lilleaas, *Acta Neurochir. (Wien)*, 1992, **115**, 106.
75. D. Schmechel, P. J. Marangos and M. Brightman, *Nature*, 1978, **276**, 834.
76. P. Johnsson, S. Blomquist, C. Luhrs, G. Malmkvist, C. Alling, J. O. Solem and E. Stahl, *Ann. Thorac. Surg.*, 2000, **69**, 750.
77. L. Ramont, H. Thoannes, A. Volondat, F. Chastang, M. C. Millet and F. X. Maquart, *Clin. Chem. Lab. Med.*, 2005, **43**, 1215.
78. R. Ergun, U. Bostanci, G. Akdemir, E. Beskonakli, E. Kaptanoglu, F. Gursoy and Y. Taskin, *Neurol. Res.*, 1998, **20**, 418.
79. Y. Yamazaki, K. Yada, S. Morii, T. Kitahara and T. Ohwada, *Surg. Neurol.*, 1995, **43**, 267–270, discussion 270.
80. S. A. Ross, R. T. Cunningham, C. F. Johnston and B. J. Rowlands, *Br. J. Neurosurg.*, 1996, **10**, 471.
81. T. Fridriksson, N. Kini, C. Walsh-Kelly and H. Hennes, *Acad. Emerg. Med.*, 2000, **7**, 816.
82. R. P. Berger, P. D. Adelson, M. C. Pierce, T. Dulani, L. D. Cassidy and P. M. Kochanek, *J. Neurosurg.*, 2005, **103**, 61.
83. R. P. Berger, S. R. Beers, R. Richichi, D. Wiesman and P. D. Adelson, *J. Neurotrauma.*, 2007, **24**, 1793.
84. S. Varma, K. L. Janesko, S. R. Wisniewski, H. Bayir, P. D. Adelson, N. J. Thomas and P. M. Kochanek, *J. Neurotrauma.*, 2003, **20**, 781.
85. S. Bandyopadhyay, H. Hennes, M. H. Gorelick, R. G. Wells and C. M. Walsh-Kelly, *Acad. Emerg. Med.*, 2005, **12**, 732.
86. S. Y. Chabok, A. D. Moghadam, Z. Saneei, F. G. Amlashi, E. K. Leili and Z. M. Amiri, *J. Trauma Acute Care Surg.*, 2012, **72**, 1654.
87. U. Missler, M. Wiesmann, C. Friedrich and M. Kaps, *Stroke*, 1997, **28**, 1956.
88. C. S. Jung, B. Lange, M. Zimmermann and V. Seifert, *Stroke Res. Treat.*, 2013, **2013**, 560305.

89. S. Moritz, J. Warnat, S. Bele, B. M. Graf and C. Woertgen, *J. Neurosurg. Anesthesiol.*, 2009, **22**, 21.
90. T. Kuroiwa, H. Tanabe, M. Arai and T. Ohta, *No Shinkei Geka*, 1994, **22**, 531.
91. M. Oertel, U. Schumacher, D. L. McArthur, S. Kastner and D. K. Boker, *J. Clin. Neurosci.*, 2006, **13**, 834.
92. B. Gong and E. Leznik, *Drug News Perspect.*, 2007, **20**, 365.
93. R. Siman, N. Toraskar, A. Dang, E. McNeil, M. McGarvey, J. Plaum, E. Maloney and M. S. Grady, *J. Neurotrauma*, 2009, **26**, 1867.
94. L. Papa, L. Akinyi, M. C. Liu, J. A. Pineda, J. J. Tepas, 3rd, M. W. Oli, W. Zheng, G. Robinson, S. A. Robicsek, A. Gabrielli, S. C. Heaton, H. J. Hannay, J. A. Demery, G. M. Brophy, J. Layon, C. S. Robertson, R. L. Hayes and K. K. Wang, *Crit. Care Med.*, 2010, **38**, 138.
95. L. Papa, C. S. Robertson, K. K. Wang, G. M. Brophy, H. J. Hannay, S. Heaton, I. Schmalfuss, A. Gabrielli, R. L. Hayes and S. A. Robicsek, *Neurocrit. Care*, 2015, **22**, 52.
96. G. Brophy, S. Mondello, L. Papa, S. Robicsek, A. Gabrielli, J. Tepas Iii, A. Buki, C. Robertson, F. C. Tortella and K. K. Wang, *J. Neurotrauma*, 2011, **28**, 861.
97. R. P. Berger, S. R. Beers, L. Papa and M. Bell, *J. Neurotrauma*, 2012, **29**, 672.
98. L. Papa, L. M. Lewis, S. Silvestri, J. L. Falk, P. Giordano, G. M. Brophy, J. A. Demery, M. C. Liu, J. Mo, L. Akinyi, S. Mondello, K. Schmid, C. S. Robertson, F. C. Tortella, R. L. Hayes and K. K. Wang, *J. Trauma Acute Care Surg.*, 2012, **72**, 1335.
99. L. Papa, M. K. Mittal, J. Ramirez, S. Silvestri, P. Giordano, C. F. Braga, C. N. Tan, N. J. Ameli, M. Lopez, C. A. Haeussler, D. Mendez Giordano and M. R. Zonfrillo, *J. Neurotrauma*, 2017, **34**, 2132–2140.
100. R. Siman, N. Giovannone, N. Toraskar, S. Frangos, S. C. Stein, J. M. Levine and M. A. Kumar, *PLoS One*, 2011, **6**, e28938.
101. R. Meller, *Neuroscientist*, 2009, **15**, 243.
102. M. C. Liu, L. Akinyi, D. Scharf, J. Mo, S. F. Larner, U. Muller, M. W. Oli, W. Zheng, F. Kobeissy, L. Papa, X. C. Lu, J. R. Dave, F. C. Tortella, R. L. Hayes and K. K. Wang, *Eur. J. Neurosci.*, 2010, **31**, 722.
103. S. B. Lewis, R. Wolper, Y. Y. Chi, L. Miralia, Y. Wang, C. Yang and G. Shaw, *J. Neurosci. Res.*, 2010, **88**, 1475.
104. S. R. Goodman, W. E. Zimmer, M. B. Clark, I. S. Zagon, J. E. Barker and M. L. Bloom, *Brain Res. Bull*, 1995, **36**, 593.
105. B. R. Pike, J. Flint, J. R. Dave, X. C. Lu, K. K. Wang, F. C. Tortella and R. L. Hayes, *J. Cereb. Blood Flow Metab.*, 2004, **24**, 98.
106. N. C. Ringger, B. E. O'Steen, J. G. Brabham, X. Silver, J. Pineda, K. K. Wang, R. L. Hayes and L. Papa, *J. Neurotrauma*, 2004, **21**, 1443.
107. S. Cardali and R. Maugeri, *J. Neurosurg. Sci.*, 2006, **50**, 25.
108. J. A. Pineda, S. B. Lewis, A. B. Valadka, L. Papa, H. J. Hannay, S. C. Heaton, J. A. Demery, M. C. Liu, J. M. Aikman, V. Akle, G. M. Brophy, J. J. Tepas, K. K. Wang, C. S. Robertson and R. L. Hayes, *J. Neurotrauma*, 2007, **24**, 354.

109. L. Papa, D. D'Avella, M. Aguennouz, F. F. Angileri, O. de Divitiis, A. Germano, A. Toscano, F. Tomasello, G. Vita, J. A. Pineda, R. L. Hayes, B. Pike and K. K. W. Wang, *Acad. Emerg. Med.*, 2004, **11**, 515.
110. L. Papa, S. B. Lewis, S. Heaton, J. A. Demery, J. J. Tepas III, K. K. W. Wang, C. S. Robertson and R. L. Hayes, *Acad. Emerg. Med.*, 2006, **13**, S39.
111. L. Papa, J. Pineda, K. K. W. Wang, S. B. Lewis, J. A. Demery, S. Heaton, J. J. Tepas III and R. L. Hayes, *Acad. Emerg. Med.*, 2005, **12**, 139.
112. G. M. Brophy, J. A. Pineda, L. Papa, S. B. Lewis, A. B. Valadka, H. J. Hannay, S. C. Heaton, J. A. Demery, M. C. Liu, J. J. Tepas, 3rd, A. Gabrielli, S. Robicsek, K. K. Wang, C. S. Robertson and R. L. Hayes, *J. Neurotrauma*, 2009, **26**, 471.
113. L. Papa, K. W. Wang, B. G. Brophy, J. A. Demery, S. Silvestri, P. Giordano, J. L. Falk, K. Schmid, F. C. Tortella, R. L. Hayes and C. S. Robertson, *J. Neurotrauma*, 2012, **29**, A28.
114. S. B. Lewis, G. J. Velat, L. Miralia, L. Papa, J. M. Aikman, R. A. Wolper, C. S. Firment, M. C. Liu, J. A. Pineda, K. K. Wang and R. L. Hayes, *J. Neurosurg.*, 2007, **107**, 792.
115. R. Nath, M. Huggins, S. B. Glantz, J. S. Morrow, K. McGinnis, R. Nadimpalli and K. K. Wanga, *Neurochem. Int.*, 2000, **37**, 351.
116. L. Fasulo, G. Ugolini, M. Visintin, A. Bradbury, C. Brancolini, V. Verzillo, M. Novak and A. Cattaneo, *J. Neurochem.*, 2000, **75**, 624.
117. H. J. Lin, K. M. Shaffer, Y. H. Chang, J. L. Barker, J. J. Pancrazio, D. A. Stenger and W. Ma, *Neurosci. Lett.*, 2002, **332**, 103.
118. K. S. Kosik and E. A. Finch, *J. Neurosci.*, 1987, **7**, 3142.
119. F. P. Zemlan, E. C. Jauch, J. J. Mulchahey, S. P. Gabbita, W. S. Rosenberg, S. G. Speciale and M. Zuccarello, *Brain Res.*, 2002, **947**, 131.
120. G. J. Shaw, E. C. Jauch and F. P. Zemlan, *Ann. Emerg. Med.*, 2002, **39**, 254.
121. D. A. Chatfield, F. P. Zemlan, D. J. Day and D. K. Menon, *Br. J. Neurosurg.*, 2002, **16**, 471.
122. M. Ma, C. J. Lindsell, C. M. Rosenberry, G. J. Shaw and F. P. Zemlan, *Am. J. Emerg. Med.*, 2008, **26**, 763.
123. G. Franz, R. Beer, A. Kampfl, K. Engelhardt, E. Schmutzhard, H. Ulmer and F. Deisenhammer, *Neurology*, 2003, **60**, 1457.
124. N. Marklund, K. Blennow, H. Zetterberg, E. Ronne-Engstrom, P. Enblad and L. Hillered, *J. Neurosurg*, 2009, **110**, 1227.
125. M. Ost, K. Nylen, L. Csajbok, A. O. Ohrfelt, M. Tullberg, C. Wikkelso, P. Nellgard, L. Rosengren, K. Blennow and B. Nellgard, *Neurology*, 2006, **67**, 1600.
126. M. Sjogren, M. Blomberg, M. Jonsson, L. O. Wahlund, A. Edman, K. Lind, L. Rosengren, K. Blennow and A. Wallin, *J. Neurosci. Res.*, 2001, **66**, 510.
127. P. Shahim, Y. Tegner, D. H. Wilson, J. Randall, T. Skillback, D. Pazooki, B. Kallberg, K. Blennow and H. Zetterberg, *JAMA Neurol.*, 2014, **71**, 684.

128. V. Puvenna, M. Engeler, M. Banjara, C. Brennan, P. Schreiber, A. Dadas, A. Bahrami, J. Solanki, A. Bandyopadhyay, J. K. Morris, C. Bernick, C. Ghosh, E. Rapp, J. J. Bazarian and D. Janigro, *Brain Res.*, 2016, **1630**, 225.
129. R. Rubenstein, B. Chang, P. Davies, A. K. Wagner, C. S. Robertson and K. K. Wang, *J. Neurotrauma*, 2015, **32**, 342.
130. W. J. Yang, W. Chen, L. Chen, Y. J. Guo, J. S. Zeng, G. Y. Li and W. S. Tong, *Acta Neurol. Scand.*, 2016, **135**, 622–627.
131. H. T. Hu, F. Xiao, Y. Q. Yan, S. Q. Wen and L. Zhang, *Clin. Biochem.*, 2012, **45**, 1320.
132. E. R. Zanier, L. Longhi, M. Fiorini, L. Cracco, A. Bersano, T. Zoerle, V. Branca, S. Monaco and N. Stocchetti, *Acta Neurochir. Suppl.*, 2008, **102**, 339.
133. J. P. Julien and W. E. Mushynski, *Prog. Nucleic Acid Res. Mol. Biol.*, 1998, **61**, 1.
134. J. Zurek, L. Bartlova and M. Fedora, *Brain Inj.*, 2012, **25**, 221.
135. D. Vajtr, O. Benada, P. Linzer, F. Samal, D. Springer, P. Strejc, M. Beran, R. Prusa and T. Zima, *Soud Lek*, 2013, **57**, 7.
136. W. Radwan, A. El-Sabbagh, S. Atweh and R. Sawaya, *N. Z. Med. J.*, 2012, **125**, 75.
137. J. Sellner, A. Patel, P. Dassan, M. M. Brown and A. Petzold, *Neurochem. Res.*, 2011, **36**, 2287.
138. P. Singh, J. Yan, R. Hull, S. Read, J. O'Sullivan, R. D. Henderson, S. Rose, J. M. Greer and P. A. McCombe, *J. Neurol. Sci.*, 2011, **304**, 117.
139. K. Nylen, L. Z. Csajbok, M. Ost, A. Rashid, J. E. Karlsson, K. Blennow, B. Nellgard and L. Rosengren, *Neurosci. Lett.*, 2006, **404**, 132.
140. J. Y. Cai, C. Lu, M. H. Chen, H. J. Ba, X. D. Chen, J. H. Lin and J. Sun, *Clin. Chim. Acta*, 2013, **424**, 182.

CHAPTER 4

Difficulties of Translating Outcomes from Animal Studies to Clinical Trials

PHILIP V. PEPLOW*[a] AND BRIDGET MARTINEZ[b]

[a] Department of Anatomy, University of Otago, Dunedin, New Zealand;
[b] Department of Molecular and Cellular Biology, University of California, Merced, USA
*Email: phil.peplow@otago.ac.nz

4.1 Introduction

Animal studies of treatment intervention for transient ischemic attack (TIA), traumatic brain injury (TBI), or ischemic stroke are deemed necessary by government regulatory bodies for providing data on the safety, toxicity, and efficacy of treatment. These are a prelude to clinical testing of such therapies, with the findings from animal studies helping in the design of clinical trials. Unfortunately, translation of findings in experimental animals to human patients has proven to be difficult for a number of reasons, and negative results from animal studies are often unpublished.

The failure to translate from animals to humans is likely due in part to poor methodology and failure of the models to accurately mimic the human disease condition. The large primates (*e.g.* chimpanzee) are the closest relatives to humans, but they are expensive and difficult to obtain, not easily housed or managed, and there are ethical issues regarding their use. Consequently, most experimental studies have used common laboratory animals such as the rat, mouse, and rabbit, and only a few have used small primates,

Drug Discovery Series No. 62
Acute Brain Impairment: Scientific Discoveries and Translational Research
Edited by Philip V. Peplow, Svetlana A. Dambinova, Thomas A. Gennarelli and Bridget Martinez

Published by the Royal Society of Chemistry, www.rsc.org

including the marmoset and macaque. Crucial genetic, molecular, immunologic, and cellular differences between humans and animal species such as the rat, mouse, and rabbit prevent these models serving as effective physiological comparators by which to assess and generate successful treatments.[1] Furthermore, the laboratory environment can have a significant effect on experimental results, and stress is a common confounding variable in caged animals.

There is also variation in the recruitment and selection of human patients in terms of the extent and nature of the injury, leading to a pronounced heterogeneity in the study groups and an inability to demonstrate a significant effect of treatment. Most of the animal models of TIA and ischemic stroke have used occlusion of the middle cerebral artery (MCA). In human patients, stroke can involve occlusion of other main arteries (*e.g.* internal carotid artery, vertebral artery, basilar artery, or posterior cerebral artery), so that the territory affected will be different.

These limitations and difficulties are discussed in the following sections. It is imperative to keep in mind that experimental animal studies may provide important information on mechanism of action of a treatment or procedure, and mouse models including transgenics and knockouts have been used extensively for this purpose. Mouse models serve to replicate specific processes or sets of processes within a disease, but not the whole spectrum of physiological changes that occur in humans in the disease setting.[2]

4.2 Differences among Animal Models of Stroke, TIA, and TBI in the Type and Extent of Cerebral Injury

Over the last four decades, a variety of animal stroke models have been developed with the aim of identifying the mechanisms that underlie cerebral ischemia and developing new agents and treatments for stroke therapy. Animal stroke models are essential for several reasons: (1) ischemic stroke in humans is extremely diverse in its manifestations, causes, and anatomical locations, whereas an experimental ischemic stroke is highly reproducible, well controllable, and standardized, allowing more precise analysis of stroke pathophysiology and drug effects; (2) molecular, biochemical, genetic, and physiological investigations often require invasive direct access to brain tissue; (3) pathophysiological events occurring during the first minute of an ischemic stroke are most often not detectable by imaging techniques in human stroke patients and thus can be studied only in an animal model; (4) perfusion and vasculature are essential in the pathophysiology of stroke and cannot be modeled in *in vitro* models.

The success of stroke studies in animals depends on the experimental model chosen. An inadequate model may lead to limitations that compromise results and analyses. In addition, the extrapolation of results from animal models to humans can be unreliable. Numerous animal species have been used to study strokes. Mice and rats are the most commonly used species, with an increasing use of larger species such as rabbits and non-human primates.[3] Detailed anatomical knowledge of the encephalic arterial vessels of various animal species is essential for developing a reliable and useful stroke model. The general arrangement of encephalic arteries among various species is maintained, with some particularities. In humans, the circle of Willis receives blood by both the paired internal carotid arteries laterally and the basilar artery caudally, so that blood supply to the forebrain is mainly from the internal carotid artery, and the caudal areas of the brain are typically supplied with vertebral arterial blood. In contrast, in both dogs and cats, the internal carotid artery is much less important, and most of the blood supply to the brain is from the maxillary artery.[4,5] Extensive extra-cranial arterial anastomoses make it difficult to render the brain ischemic in the dog and cat.[3,5]

The success of preclinical stroke research in developing new therapies might depend in part on the selection of appropriate animal stroke models to use. Most of the stroke studies are performed in small animal species such as the mouse, rat and rabbit. The rat is one of the most commonly used animals in stroke investigations for several key reasons: (1) the cerebral vasculature and physiology is similar to that of humans; (2) its moderate body size enables easy monitoring of physiologic parameters; (3) its small brain size is suited to fixation or freezing procedures (*e.g.* for biochemical analyses); (4) there is a relative homogeneity among strains; and (5) it is easy to conduct reproducible studies. The mouse is the most appropriate animal in which genetic modifications can be created, and transgenic and knockout studies are widely used to assess the molecular pathophysiology of stroke. In accord with the Stroke Therapy Academic Industry Roundtable (STAIR) recommendations, a positive result from a study in a species should always be verified in another species and should be reproducible in a second stroke model (*i.e.* permanent and then transient occlusion model).[6]

The MCA and its branches are the cerebral vessels most often affected in human ischemic stroke, accounting for approximately 70% of infarcts.[7] Hence, techniques that occlude this artery are closest to human ischemic stroke, and intra-arterial suture occlusion of the MCA is the most common method in rodents. The technique involves temporarily occluding the common carotid artery, introducing a suture directly into the internal carotid artery, and advancing the suture until it interrupts the blood supply to the MCA. A modification involves inserting the suture into the transected external carotid artery, using this artery as a path to advance a suture through the internal carotid artery. This method enables permanent MCA occlusion or transient ischemia with reperfusion. The external carotid artery approach is a better procedure for transient MCA occlusion because it maintains the anatomical integrity required for reperfusion.[8]

The most common durations of MCA occlusion using an intraluminal suture in rats are 60 minutes, 90 minutes, and 120 minutes, or permanent occlusion; the success rate of inducing an infarction is 88–100%, and that of a subarachnoid hemorrhage is 4%.[8,9] In mouse models of 60 minute transient MCA occlusion, the success rate is 96%, and the success rate for subarachnoid hemorrhage is 0%.[8,10] Depending on the duration of MCA occlusion, this technique leads to territorial infarction in rats, involving the striatum and the overlying frontoparietal and temporal cortices, as well as some portion of the occipital cortex, but also induces infarction in the thalamus and hypothalamus.[11] Infarction after MCA occlusion follows a progression from early ischemia in the striatum to delayed infarction in the cortex overlying the striatum. In mice, infarction involves a substantial portion of the hemisphere including most of the cortex, striatum, thalamus, hippocampus, and subventricular zone and is similar to that of rats undergoing MCA occlusion.[12,13] The extent and distribution of ischemic injury is highly sensitive to increasing occlusion duration in mice; the difference between 15 minutes and 30 minutes of MCA occlusion results in a five-fold increase in infarct volume and also includes regions outside the MCA occlusion territory.[13] No ischemic lesions are observed in mice subjected to MCA occlusion for ≤10 minutes, while an MCA occlusion of 15 minutes leads to a detectable infarction.[14]

Different factors affect reproducibility of infarction and include suture diameter (3–0 or 4–0 sutures for rats, 6–0 sutures for mice), insertion length of suture (which may vary from 18 to 22 mm in rats depending upon the age of the animal), animal strain, position of suture, and coating of suture. When the suture is positioned 15–16 mm from the bifurcation of the common carotid artery, only the hypothalamic and anterior choroidal arteries are occluded, leading to small subcortical infarcts that might be a potential model of lacunar stroke.[15] Sutures coated with silicone or poly-L-lysine have a greater adherence to the adjacent vascular endothelium than uncoated sutures, cause larger infarcts, and reduce inter-animal variability compared to uncoated sutures.[16] Evidence suggests that infarct size is affected by different rat and mouse strains,[17,18] whereas the extension of the penumbra is independent of rodent strain.[19] Sprague–Dawley rats exhibit small infarct volumes with considerable variability, whereas Wistar rats have the lowest variability of infarct size.

The MCA occlusion animal model is considered to be suitable for reproducing ischemic stroke and subsequent neuronal cell death, cerebral inflammation, and blood–brain barrier damage.[20] However, transient MCA occlusion lasting ≥2 hours in rats induces spontaneous hyperthermia that is associated with hypothalamic damage.[21] Hypothalamic damage rarely occurs with human stroke. Mild hypothermia occurs in mice after transient MCA occlusion.[22] These changes in core body temperature may confound experimental results. Also, there are concerns with regard to the different pathophysiologies of the transient and permanent MCA occlusion models. In promptly reversed transient ischemia, primary core injury may recover, but a secondary delayed injury occurs after a free interval of as much as 6–12 hours. This is a long therapeutic window that is not seen in human stroke.[23] In permanent MCA occlusion, primary core injury expands in peripheral brain regions and achieves a maximum at approximately 3 hours after MCA occlusion. Thus, due to the two different pathophysiologies, unequal results may be expected from the transient and permanent MCA occlusion models. The transient vascular occlusion by intraluminal suture insertion with prompt reperfusion does not replicate naturally occurring stroke and reperfusion, and therefore is considered to be less suited for clinical translation.[23]

As mentioned earlier, a high incidence of lesions (assessed by magnetic resonance imaging [MRI] at 24 hours after reperfusion) is observed with MCA occlusion ≥15 minutes in mice, but no permanent neurological deficits or MRI lesions are seen in animals with MCA occlusion ≤10 minutes. However, histopathological examination reveals minor ischemic changes even after MCA occlusion of a 2.5 minute duration. A 10 minute or shorter duration of focal cerebral ischemia provides an optimized and suitable mouse model of TIA.[14]

Several animal models of TBI have been developed and tested. They can be broadly classified as impact acceleration models, inertial (non-impact) acceleration models, and direct brain deformation models. Although no single animal model can express the great diversity of neural damage constituting human TBI, multiple animal models are capable of replicating specific features of TBI that can be examined.

Impact acceleration models involve direct head impacts using a piston, humane stunner or captive bolt pistol, calibrated pendulum, or weight drop on to the skull. These models resemble closed head injuries in motor vehicle accidents or falls where there is rapid acceleration/deceleration of the head after impact to an intact skull. These models have reproduced several important features of TBI in primates, cats, and sheep, including contusions, subarachnoid hemorrhage, and widespread axonal injury, and skull fractures are often found.[24,25]

Inertial acceleration models involve acceleration of the head without impact. Angular acceleration, especially in the coronal plane, has been shown to be particularly injurious in non-human primates.[26] Angular or translational acceleration can be generated in these animals and, by controlling head motion, acute subdural hematoma, temporary unconsciousness or prolonged coma, and diffuse axonal injury are induced.

Direct brain deformation models include both fluid percussion and rigid indentation types and use either a fluid pulse or a mechanically driven piston, respectively, to rapidly compress the exposed dura and cortex through a craniotomy site.[27] These models produce well-controlled levels of localized injury rather than diffuse damage and have been used in rats and cats.

Primary TBI is the result of mechanical forces producing tissue deformation at the moment of injury. The blood vessels, axons, neurons, and glia may be damaged in a focal, multifocal, or diffuse pattern of involvement, resulting in complex cellular, neurochemical, and metabolic alterations. Secondary TBI occurs as a complication of the different types of primary brain damage and includes ischemic and hypoxic damage and cerebral swelling, the consequences of raised intracranial pressure, hydrocephalus, and infection.[28]

4.3 Differences between Humans and Animal Stroke Models

Two critical issues that are common in experimental studies are the age and gender of the animals. Most preclinical studies typically use healthy animals of similar ages, usually young adult male animals. However, the response to stroke may be different in young and older animals in terms of mechanisms, ischemic consequences, histological changes, and functional recovery.[29] Although in humans stroke mainly affects the elderly, stroke patients are of widely varying ages and non-uniform medical comorbidity profiles. In regard to gender, its overall incidence is higher in male than in female patients, although it increases in middle-aged women compared to men of the same age. Also, the differences in mechanisms and outcomes of stroke are influenced by both hormone-dependent[30] and hormone-independent factors.[31] The non-uniformity among patients complicates clinical trials by introducing confounding variables.[32]

Small animal models may have a higher tolerance for ischemia than the human brain. The human brain is highly organized and lacks the collateralization of cerebral vessels found in small mammals, so its vulnerability to brain trauma or ischemia is greater than that in small mammals. Also, in part, the prominent vulnerability of brain tissue in humans to ischemic damage reflects the high metabolic rate, which is 3.5-times higher than even those of the brains of other primate species.[33] Central neurons have a near-exclusive dependence on glucose as an energy substrate, and brain stores of glucose or glycogen are limited. However, energetics considerations and energy substrate limitations may not be solely responsible for the brain's heightened vulnerability to ischemia. The brain's intrinsic cell-to-cell and intracellular signaling mechanisms may become harmful under ischemic conditions, hastening energy failure and enhancing the final pathways underlying ischemic cell death in all tissues, including free radical production, activation of catabolic enzymes, membrane degradation, apoptosis, and inflammation.[33]

Animal models typically involve infarction of one vascular territory through a controlled ligation or occlusion of a parent vessel. Strokes in human subjects vary widely, leading to intrinsically widely differing outcomes. Many of the clinical studies have included patients with both hemorrhagic and ischemic stroke, whereas animal studies assess one particular type of stroke (*e.g.* MCA occlusion). Hemorrhagic stroke must be kept separate from ischemic stroke as the mechanisms and pathways for damage are very different.[34]

There are general anatomical differences between human and animal brains. Rodents such as the rat and mouse are lissencephalic species and have very little white matter as compared to humans. Thus, the administration of drugs into these animal models usually assesses the rescue of gray matter. This is an important difference, as many human strokes are lacunar or within other white matter structures.[35] Animal species with gyrencephalic brains include humans, non-human primates such as the chimpanzee and rhesus macaque, pigs, dogs, cats, and ferrets. Smaller primates such as the marmoset have a lissencephalic brain. Not only are there anatomical differences between a gyrencephalic and lissencephalic brain, but evidence suggests that there are functional differences as well. The positive effects observed in the rat or mouse may be lost in other larger animals such as the dog, which has an anatomically distinct gyrencephalic brain.[36] Also, while large animal species such as the cat, dog, and rabbit have greater similarity to the human in terms of similar brain size and structure, such animals show more individual anatomical variation. Thus, reproducibility of infarction is very low for these animals.

As mentioned previously, the outcomes of animal and human studies may be affected by confounding factors that could include strain/race variability, age variability, gender variability, and comorbidities.

4.4 Differences in the Ways That Studies Are Conducted in Humans and Animal Models

In most animal studies, the animals are either pre-treated with a neuroprotective agent or treatment is started within a very short time period from a surgical vessel occlusion. The time of inducing the infarct is distinct and the time of treatment is easily measurable. The best measurement of the time of infarct in the human population is the last time point that the patient was considered to be neurologically at baseline. If a patient wakes with a neurological deficit, the exact time of their deficit is impossible to record clinically. This uncertainty coupled with delays in obtaining medical treatment lead to a wide range of treatment onset times. Given the extremely time-sensitive nature of ischemic injury, patients treated with the same neuroprotective agent may have widely varying outcomes. The last time point at which the patient was considered to be neurologically at baseline is a critical factor in determining whether a patient is a suitable candidate for acute thrombolysis treatment (intravenous intervention of tissue plasminogen activator within 4.5 hours).

The failure of previous clinical studies may be related to the late time of first administration (5.2–48.0 hours).[32] Moreover, the optimal time windows for administration of a neuroprotective drug or treatment are generally extended when the trial is moved into treating human subjects in order to best recruit patients. However, this may negate any effect seen, as many neuroprotective mechanisms must be implemented within a very short timeframe following ischemic stroke.[35]

4.5 Differences in Primary Functional Outcomes Measured in Clinical Trials and Animal Studies

There are critical discrepancies between experimental studies and clinical trials in regard to outcome measurements. In clinical trials, besides mortality and symptomatic intracranial hemorrhage rates, the primary functional outcomes are long term (usually at 90 days) and include assessment of the degree of the patient's disability using the modified Rankin scale, activities of daily life and dependence using the Barthel index, and various aspects of neurological function such as consciousness, cognitive, speech, motor, and sensory function using the National Institutes of Health Stroke Scale, as well as mortality at 30 and 90 days and development of intracranial hemorrahage.[37,38] By contrast, in animal studies, measures are short term (*e.g.* 24 hours), the primary outcome is usually the infarct size, and the neurological secondary outcome is limited.

Neurological deficits in rodent stroke models may be examined using behavioral tests, of which there are many, but no one test has stood out as being superior to others in characterizing the various deficits that occur after stroke. Different tests are sensitive to measuring deficits associated with particular areas of damage, and there are behavioral tests that assess both acute and chronic impairments in rodent models.[39] Due to loss of limb function after stroke, many tests focus on motor and sensory deficits. However, since learning and memory impairments are also common after stroke, cognitive testing is also a crucial component in elucidating the full scope of deficits. Stress and anxiety in animals during testing are confounding factors that lead to unreliable data.[39]

4.6 Strategies to Improve the Quality of Animal and Clinical Studies of Stroke Therapies

The updated and amended STAIR recommendations for preclinical studies included the following: (1) the fundamentals of good scientific inquiry should be followed by implementing randomization and eliminating assessment bias, defining inclusion/exclusion criteria *a priori* and reporting the reasons for excluding animals from the final data analysis, performing appropriate power and sample size calculations, and disclosing potential conflicts of interest; (2) after initial studies in younger healthy animals, further studies should be performed in aged animals and animals with comorbid conditions such as hypertension, diabetes, and hypercholesterolemia; (3) efficacy studies should be performed in both male and female animals; (4) interaction studies with medications commonly used in stroke patients should be performed for advanced preclinical drug development candidates; and (5) relevant biomarker endpoints such as diffusion/perfusion MRI mismatch and serum markers of tissue injury should be included that can also be obtained in human trials to indicate that the therapeutic target has been modified. It was suggested that adherence to these recommendations might enhance the chances for success in clinical trials.[6]

In animal studies, the minimum effective and maximum tolerated doses of a drug should be defined. There should be a target concentration and a tissue level of effect identified from animal histology, with behavioral studies giving some indication that when the drug is administered to humans there is a reasonable prospect of achieving clinical benefit. Rodent studies appear to be relevant to identifying a therapeutic time window for thrombolytic and neuroprotective drugs in acute stroke. Penumbral imaging using diffusion/perfusion MRI mismatch may assist in the identification of the therapeutic window in a particular animal model. Multiple time endpoints are essential, and both histological and behavioral outcomes should be assessed. Studies need to be performed at least 2–3 weeks or longer after stroke onset to demonstrate a sustained benefit on behavioral outcomes in delayed survival studies. The model of occlusion of the MCA using an intraluminal suture is imperfect, and in some situations occlusion may occur but spontaneous reperfusion may ensue, leading to variability in infarct size. Cerebral blood flow should be monitored using Doppler scanning or perfusion imaging to demonstrate adequate sustained occlusion and to monitor reperfusion in transient ischemic models. Treatment efficacy should be shown in at least two animal species using both histological and behavioral outcomes. Rodents or rabbits are suitable for initial testing, and gyrencephalic animals including primates are preferred as a second animal model species. Furthermore, positive results obtained in one laboratory need to be replicated in at least one independent laboratory before clinical studies can be initiated.[6]

In clinical trials, patient selection needs careful consideration, particularly in regard to the following: (1) inclusion/exclusion criteria that are determined by the aim of the trial. Inappropriate choice of these criteria, and having too many exclusion criteria, can contribute to the failure of the trial; (2) intervention has the most benefit when used at the most appropriate time (*e.g.* the efficacy of intravenous thrombolysis for reducing poor functional outcome declines rapidly over a few hours after stroke onset,[40] aspirin should be given within 48 hours to prevent early recurrence,[41] and hemicraniotomy needs to be performed within 48 hours to benefit patients with malignant MCA ischemic stroke).[42] Also, hazards may be maximized by treating at the wrong time (*e.g.* by delaying intravenous thrombolysis beyond 6 hours); (3) patients with mild stroke usually have a good outcome, and those with severe stroke usually have a poor outcome. Hence, recruiting patients with very mild or severe stroke will dilute the trial with patients who are less likely to respond significantly to the intervention; (4) the aim of prevention trials is to reduce or delay the occurrence of further vascular or stroke events, and if patients are recruited with a low risk of recurrence, then they will dilute the trial; (5) having a minimum age for recruitment (*e.g.* >50) will help increase stroke recurrence rates in prevention trials, but the absolute benefit of an intervention may be greatest in older people; (6) both MRI, perfusion/diffusion mismatch, and computed tomography perfusion mismatch can identify penumbral presence and extent in real time. Effective reperfusion depends on salvage of vulnerable, hypoperfused tissue in the ischemic penumbra. Reperfusion of large cores is known to be hazardous with a high risk of symptomatic hemorrhagic transformation. Advanced multimodal imaging will also identify the relevant occluded artery and avoid unnecessary catheter angiography.[43]

In addition to the issues mentioned above that have an important bearing on the outcomes of clinical trials, other major design flaws include: (1) side effects that precluded achieving adequate plasma drug concentrations; (2) inclusion of a substantial percentage of lacunar stroke patients when testing a drug with no preclinical evidence of protection of the white matter; (3) inclusion of a substantial percentage of patients who received thrombolysis in whom an additional treatment effect is difficult to measure or where the study agent adversely affected outcome with concomitant thrombolysis.[44]

Despite the difficulties with patient heterogeneity, varying time of presentation, and differences in large/small vessel occlusion, it is felt that there is still potential for the use of neuroprotective agents in acute ischemic stroke. One group of patients of greatest potential would be those with large vessel occlusions, who increasingly are treated by intra-arterial endovascular therapies in which a microcatheter is advanced into the occluded vessel and the vessel reopened directly. This group would represent a cohort of patients in whom one or a cocktail of potential neuroprotective agents could be infused immediately following reperfusion. This would combine reperfusion with neuroprotection, and these patients would provide a more homogeneous and controlled group for clinical evaluation of neuroprotective agents.[32]

4.7 Conclusion

There is an increasing global incidence of acute ischemic stroke, and very few patients are able to be examined and treated within the very narrow time window that exists for thrombolysis. Consequently, it is imperative that effective treatments be developed for these patients to minimize potential damage and aid recovery. Many animal studies of acute ischemic stroke have been performed, mainly in rodents, and they have demonstrated the beneficial effects of several different classes of neuroprotective agents, but with the exception of two or three of these agents, it has not been possible to transfer this benefit to human patients. Occlusion of the MCA by an intraluminal suture has been the most used stroke model in animal studies, allowing testing of agents for effect in animals subjected to temporary or permanent occlusion of the MCA. The transient occlusion of the MCA with prompt reperfusion does not mimic the situation in human patients, and the permanent occlusion model is considered to be more appropriate.

While rodents and rabbits can be chosen for conducting initial animal studies, it is necessary to test the findings in a second animal model with a gyrencephalic brain, preferably a non-human primate.[45] Interestingly, intravenous infusion of the peptide Tat-NR2B9c (NA-1) in male *Macaca fascicularis* at 1 hour after a 90 minute MCA occlusion reduced the infarct volume by 55% at 24 hours and by 70% at 30 days as measured by MRI and histologically. There was improved performance in behavioral tasks and in neurological tests according to the Nonhuman Primate Stroke Scale (NHPSS). In addition, the treatment also preserved cellular functionality in the ischemic penumbra.[45] The beneficial effects on infarct volume and NHPSS scores were replicated in a situation similar to the classical human one by administering Tat-NR2B9c at 1 hour after onset of a 4.5 hour MCA occlusion or 3 hours after a 3.5 hour MCA occlusion.[46–48] Tat-NR2B9c acts by inhibiting post-synaptic density-95 (PSD-95), a synaptic scaffolding protein that links N-methyl-D-aspartate (NMDA) receptors to neurotoxic signaling in neurons, and as a consequence, glutamate excitotoxicity is blocked. These studies showed that neuroprotection in primates is feasible in a clinically relevant situation.

The findings from most of the reported clinical trials have been affected by a marked heterogeneity in patients in terms of age, gender, and existing comorbidities, and the inclusion of patients with different types of stroke (ischemic *vs.* hemorrhagic). In addition, some patients who have very mild or severe stroke are less likely to respond significantly to treatment. Patient selection is therefore a critical factor contributing to the possible success or failure of the clinical trial.

References

1. A. Gawrylewski. *The Scientist*, July 1, 2007.
2. B. Fingleton, *Curr. Pharm. Des.*, 2007, **13**, 333.
3. J. B. Casals, N. C. Pieri, M. L. Feitosa, A. C. Ercolin, K. C. Roballo, R. S. Barreto and F. F. Bressan, *et al.*, Comp. Med., 2011, **61**, 305.
4. C. N. S. Vasculature - Anatomy & Physiology. https://en.wikivet.net/CNS_Vasculature_-_Anatomy_%26_Physiology (accessed 7 May 2017).
5. L. A. Gillilan, *Am. J. Anat.*, 1976, **146**, 237.
6. M. Fisher, G. Feuerstein, D. W. Howells, P. D. Hurn, T. A. Kent, S. I. Savitz and E. H. Lo, Stroke, 2009, **40**, 2244.
7. J. Bogousslavsky, G. Van Melle and F. Regli, *Stroke*, 1988, **19**, 1083.
8. S. Liu, G. Zhen, B. P. Meloni, K. Campbell and H. R. Winn, *J. Exp. Stroke Transl. Med.*, 2009, **2**, 2.
9. R. Schnid-Elsaesser, S. Zausinger, E. Hungerhuber, A. Baethmann and H. J. Reulen, *Stroke*, 1998, **29**, 2162.
10. D. Tsuchiya, S. Hong, T. Kayama, S. S. Panter and P. R. Weinstein, *Brain Res.*, 2003, **970**, 131.
11. J. H. Garcia, K. F. Liu and K. L. Ho, *Stroke*, 1995, **26**, 636.
12. K. Maeda, R. Hata and K. A. Hossmann, *Neurobiol. Dis.*, 1999, **6**, 101.

13. B. W. McColl, H. V. Carswell, J. McCulloch and K. Horsburgh, *Brain Res.*, 2004, **997**, 15.
14. E. Pedrono, A. Durukan, D. Strbian, I. Marinkovic, S. Shekhar, M. Pitkonen, U. Abo-Ramadan and T. Tatlisumak, *J. Neuropathol. Exp. Neurol.*, 2010, **69**, 188.
15. Z. He, T. Yamawaki, S. Yang, A. L. Day, J. W. Simpkins and H. Naritomi, *Stroke*, 1999, **30**, 2743.
16. J. Bouley, M. Fisher and N. Henninger, *Neurosci. Lett.*, 2007, **412**, 185.
17. M. Walberer, E. Stolz, C. Muller, C. Friedrich, C. Rottger, F. Blaes and M. Kaps, *et al.*, *Lab. Anim.*, 2006, **40**, 1.
18. M. H. Cheng, L. L. Lin, J. Y. Liu and A. J. Liu, *CNS Neurosci. Ther.*, 2012, **18**, 794.
19. E. Reid, D. Graham, M. R. Lopez-Gonzalez, W. M. Holmes, I. M. Macrae and C. McCabe, *J. Cereb. Blood Flow Metab.*, 2012, **32**, 1765.
20. L. Belayev, R. Busto, W. Zhao and M. D. Ginsberg, *Brain Res.*, 1996, **739**, 88.
21. F. Li, T. Omae and M. Fisher, *Stroke*, 1999, **30**, 2464.
22. P. A. Barber, L. Hoyte, F. Colbourne and A. M. Buchan, *Stroke*, 2004, **35**, 1720.
23. K. A. Hossmann, *J. Cereb. Blood Flow Metab.*, 2012, **32**, 1310.
24. J. W. Finnie and P. C. Blumbergs, *Vet. Pathol.*, 2002, **39**, 679.
25. W. T. O'Connor, A. Smyth and M. D. Gilchrist, *Pharmacol. Ther.*, 2011, **130**, 106.
26. T. A. Gennarelli, *Acta Neurochir. Suppl.*, 1983, **32**, 1.
27. Y. Xiong, A. Mahmood and M. Chopp, *Nature Rev.*, 2013, **14**, 128.
28. P. C. Blumbergs, N. R. Jones and J. B. North, *J. Neurol. Neurosurg. Psychiatry*, 1989, **52**, 838.
29. C. L. Rosen, V. A. DiNapoli, T. Nagamine and T. Crocco, *J. Neurosurg.*, 2005, **103**, 687.
30. B. Shao, Y. Cheng and K. Jin, *Curr. Drug Targets*, 2012, **13**, 188.
31. J. T. Lang and L. D. McCullough, *J. Transl. Med.*, 2008, **6**, 33.
32. S. Grupke, J. Hall, M. Dobbs, G. J. Bix and J. F. Fraser, *Clin. Neurol. Neurosurg.*, 2015, **129**, 1.
33. J. M. Lee, M. C. Grabb, G. J. Zipfel and D. W. Choi, *J. Clin. Invest.*, 2000, **106**, 723.
34. S. R. Robinson, T. N. Dang, R. Dringen and G. M. Bishop, *Redox Rep.*, 2009, **14**, 228.
35. L. Hoyte, J. Kaur and A. M. Buchan, *Exp. Neurol.*, 2004, **188**, 200.
36. L. Belayev. In: Translation Stroke Research: From Target Selection to Clinical Trials, eds. P. A. Lapchak and J. H. Zhang, Springer 2012, p. 471.
37. M. Kelly-Hayes, J. T. Robertson, J. P. Broderick, P. W. Duncan, L. A. Hershey, E. J. Roth, W. H. Thies and C. A. Trombly, *Stroke*, 1998, **29**, 1274.
38. Positive Stroke Clinical Trial, ClinicalTrials.gov, US National Institutes of Health https://clinicaltrials.gov/ct2/show/NCT01852201 (accessed 7 May 2017).

39. K. L. Schaar, M. M. Brennerman and S. I. Savitz, *Exp. Transl. Stroke Med.*, 2010, **2**, 13.
40. K. R. Lees, E. Bluhmki, R. von Kummer, T. G. Brott, D. Toni, J. C. Grotta and G. W. Albers, *et al.*, *Lancet*, 2010, **375**, 1695.
41. Z. M. Chen, P. Sandercock, H. C. Pan, C. Counsell, R. Collins, L. S. Liu, J. X. Xie, C. Warlow and R. Peto, *Stroke*, 2000, **31**, 1240.
42. K. Vahedi, J. Hofmeijer, E. Juettler, E. Vicaut, B. George, A. Algra and G. J. Amelink, *et al.*, *Lancet Neurol.*, 2007, **6**, 215.
43. P. M. Bath, M. Brainin, C. Brown, B. Campbell, S. M. Davis, G. A. Donnan and G. A. Ford, *et al.*, *Int. J. Stroke*, 2014, **9**, 683.
44. M. Fisher, *Stroke*, 2011, **42**(suppl. 1), S24.
45. D. J. Cook and M. Tymianski, *Neurotherapeutics*, 2012, **9**, 371.
46. D. J. Cook and M. Tymianski, *Expert Rev. Cardiovasc. Ther.*, 2011, **9**, 433.
47. D. J. Cook, L. Teves and M. Tymianski, *Nature*, 2012, **483**, 213.
48. D. J. Cook, L. Teves and M. Tymianski, *Sci. Transl. Med.*, 2012, **4**, 154ra133.

CHAPTER 5

Neuroprotective Therapies for Ischemic Stroke

YOLANDA CRUZ,[a] KARLA CANTU-SALDAÑA[a] AND ANTONIO IBARRA*[a,b]

[a] Universidad Anáhuac México Campus Norte, Av. Universidad Anáhuac No. 46, Lomas Anáhuac, Huixquilucan, Estado de México 52764, México; [b] Centro de Investigación del Proyecto CAMINA, Av. Universidad Anáhuac No. 46, Lomas Anáhuac, Huixquilucan, Estado de México 52764, México
*Email: iantonio65@yahoo.mx

5.1 Introduction

According to the World Health Organization (WHO), cerebrovascular and cardiovascular diseases represented the primary causes of death worldwide from non-communicable diseases in 2012. During the same year, cerebral ischemia (or stroke) caused around 6.7 million deaths,[1] although in 2010 the American Heart Association (AHA) reported 16.9 million deaths annually, ratifying its position as one of the most important causes of death.[1,2] It is also considered a medical emergency associated with poor neurological outcomes if not treated promptly and it represents the primary cause of permanent disability in adults.[1]

The WHO and several government health agencies are focused on reducing the incidence and recurrence of ischemic events through preventive activities aimed at reducing specific risk factors. Nonetheless, neuroprotective treatments have also been developed at the same time, in some cases providing interesting future prospects for therapy.

Drug Discovery Series No. 62
Acute Brain Impairment: Scientific Discoveries and Translational Research
Edited by Philip V. Peplow, Svetlana A. Dambinova, Thomas A. Gennarelli and Bridget Martinez

Published by the Royal Society of Chemistry, www.rsc.org

5.2 Metabolic Alterations that Increase the Risk of Developing Stroke

Atherosclerosis remains the main cause of stroke,[3] either by atherothrombotic (85% of cases) or atheroembolic (15% of cases) mechanisms, and several risk factors that are associated with its development can be classified as non-modifiable (age, gender, and ethnicity) and potentially modifiable (dyslipidemia, diabetes mellitus and poor glycemic control,[4] insufficient physical activity or a sedentary lifestyle,[5] alcoholism, smoking,[1] low levels of vitamin D,[6,7] and increased levels of homocysteine[8]) factors. Also, having already suffered an ischemic event increases the risk for recurrence,[1] and it needs to be recognized that 50% of the population in need of pharmacological treatment has no access to it,[1] making them more prone to recurrence.

The development of atherosclerotic plaque is characterized by the accumulation and oxidation of low-density lipoprotein (LDL) inside the vascular endothelium that stimulates inflammation through the increased expression of adhesion molecules, chemotactic proteins, growth factors, and immune cell migration (*e.g.* monocytes).[9] LDL molecules are later phagocytosed by macrophages that at this point turn into foam cells[10] that increase growth factor and cytokine expression and stimulate smooth muscle cell migration, leading to fibrous layer formation.[11] Foam cells undergo apoptosis, releasing cholesterol microcrystals that increase and lead to chronic inflammation, recognized by the presence of interleukin (IL)-1β, IL-6, and tumor necrosis factor (TNF) and the arrival of more inflammatory cells, including T lymphocytes.[12] Atherosclerotic plaque formation is silent and takes months or even years to occur, but sudden rupture and release can cause ischemic symptoms immediately.[12]

Cell death processes begin immediately after stroke onset. These have been widely studied and are quite important, since each mechanism of damage caused by stroke could be considered a target for treatment. Today's therapies focus on preventing and avoiding new occlusions; however, new neuroprotective therapies should be explored. Neuroprotection, can be achieved through many different ways. According to Vajda, the effects of different treatments should be aimed at preserving, recovering, or regenerating neural tissue cells after a lesion that could be irreversible,[13] all with the objective of providing a better quality of life for patients.[13]

5.3 Pathophysiology and Possible Therapeutic Strategies

In stroke, a rapid inhibition in oxygen and glucose delivery causes alterations in all ATP-dependent cellular activities, including Na^+/K^+ pumps,[14] which will quickly induce membrane depolarization.[15] Therefore, intracellular Ca^{2+} rises and subsequent excitatory neurotransmitter and cell death proteins are released, and free radicals are also produced.[16]

Oxygen and glucose deprivation is not homogenous: the "ischemic core" is a portion of tissue that gets irreversibly damaged because it is directly irrigated by the deep circulation, which gets occluded.[16] The "ischemic penumbra" refers to the surrounding tissue that is supported by collateral irrigation and experiences moderate levels of ischemia, allowing enough oxygen and glucose to be supplied in order to perform basal functions, and although no electrical activity occurs, these cells have a chance to survive, but only if treated promptly.[16]

Two main therapeutic strategies are used in the clinical setting to allow reperfusion: recanalization through thrombolytic agents and mechanical thrombectomy.[17] The most used, FDA-approved thrombolytic therapy is recombinant tissue plasminogen activator,[18] or alteplase, which has a therapeutic window of no more than 4.5 hours from onset of symptoms, and since it is very difficult to treat inside this timeframe, very few patients are eligible. It has been shown to be less effective in large vessel occlusion and/or if the clot is too large.[19] Also, this therapy increases the risk for a hemorrhagic transformation and an alteration in blood–brain barrier (BBB) integrity.[18] Nonetheless, other studies have shown that recombinant tissue plasminogen activator is well tolerated by patients and causes no adverse effects, even in those simultaneously treated with anticoagulants such as warfarin.[20]

The FDA also approved endovascular treatments such as mechanical thrombectomy with or without clot aspiration by means of a stent retriever.[21] This kind of therapy has a much larger time window of up to 8 hours from onset of symptoms.[22] The stent retriever has shown great benefits in large vessel occlusion, although it also increases the risk of hemorrhagic transformation and neurological deficit.[23] Recently, authors have suggested that there is no need to use thrombolytics before a mechanical thrombectomy.[24]

Unfortunately, endovascular treatments are costly and complex since they require imaging equipment such as magnetic resonance imaging and computerized axial tomography[25] scans to make sure that the benefits of treatment outweigh the risks. Consequently, this therapy needs highly trained personnel, making the costs even more expensive, thereby rendering the use of such treatments even more complex. Finally, endovascular treatment has several contraindications, including metabolic diseases, so patient access to it is low. For this reason, new pharmacological alternatives aimed at different damage mechanisms should continue to be explored.

5.3.1 Energy Failure

Metabolic alterations begin as soon as oxygen and glucose deprivation occur. Adenosine monophosphate-activated protein kinase (AMPK) is an enzyme that gets activated during this stage. It scans for energy and works to restore ATP levels through the stimulation of glycolysis and the adequate use of energy stores (inhibition of glycogen synthesis by astrocytes and an increase in lipid oxidation). Nevertheless, its sustained activation is associated with an increase in tissue damage due to the promotion of apoptosis and/or necrosis.[14]

Several AMPK inhibitors have been tested; for instance, in a middle cerebral artery occlusion (MCAO) model, it was demonstrated that compound C (6-[4-{2-piperidin-1-yl-ethoxy}-phenyl]-3-pyridin-4-yl-pyrrazolo[1,5-a]-pyrimidine) reduces infarct size and improves neurological deficit.[26] On the other hand, it has been recently observed that acute metformin preconditioning confers neuroprotection by pre-activation of AMPK-dependent autophagy, promoting reductions in infarct size, neurological deficit, and apoptosis.[27]

After long periods of ischemia and during reperfusion, the penumbra region experiences a hyperglycemic state mediated by Na^+/glucose cotransport,[28] which causes an increase in lactic acidosis and reactive oxygen species (ROS), leading to an increase of the infarct volume.[29] Regarding this matter, it has been observed that diabetic patients with poor glycemic control have a worse prognosis after stroke.[4] The mechanisms by which stroke induces glucose intolerance and insufficient insulin secretion are not well understood, but the neuroprotective effects of hypoglycemiants have been under scrutiny, some of which are shown in Table 5.1.

In an attempt to maintain glucose levels in the absence of oxygen, anaerobic glycolysis activates, causing an increase in lactate production and an acidic medium that drives protein denaturalization, enzymatic alterations, and pH-sensitive ionic channel disturbance,[30] intensifying neural tissue death.

Several neuroprotective compounds have been investigated for lowering lactic acid and/or lactate dehydrogenase (LDH). Melatonin, a hormone produced in the pineal gland, has been shown to induce neuroprotection through its antioxidant activities after being administered to rats lacking a

Table 5.1 Hypoglycemiant agents currently being studied for stroke therapy.

Substance/compound	Mechanism	Observed effect	Author(s)
Insulin	↓ Inflammatory transcription factors Activates vasodilation through ↑ endothelial nitric-oxide syntase (eNOS) Maintains energy homeostasis Regenerates neurites	↓ Infarcts ↓ Neurological deficits	Lioutas and Novak[95]
Glucagon-like peptide-1	Maintains glucose homeostasis ↑ GLUT-1 expression	↓ Infarcts ↓ Neurological deficits	Zhang *et al.*[96]
Liraglutide	↑ Insulin secretion ↓ Apoptosis ↓ ROS	↓ Infarcts ↓ Neurological deficits	Zhu *et al.*[97]
	↑ Neurotrophic factor expression	↓ Infarcts	Sato *et al.*[98]
Linagliptin	Stops endogenous glucagon-like peptide-1 degradation	↑ Neuronal survival	Darsalia *et al.*[99]
Sulfonylureas	↓ ATP-dependent K^+ channel activity	↑ Infarct size	Liu *et al.*[100]

pineal gland. In this case, smaller infarcts and reduced inflammatory activity, edema, and BBB permeability[31] were observed; likewise, melatonin significantly reduced LDH levels,[32] as did ethanol, which is also considered neuroprotective.[29]

Recent studies questioned the damaging effect of increased lactate after stroke. Castillo *et al.* proved in 2015[33] that lactate has neuroprotective activities if applied intravenously during reperfusion in a MCAO model. Other studies showed that lactate induces neuronal HCA1 receptor overexpression in the ischemic cortex and striated body. This receptor binds to G proteins and inhibits cAMP intracellular signaling, especially when lactate is present at high concentrations;[34] this effect reduces the release of Ca^{2+} and depolarization.

On the other hand, studies performed on patients that have suffered subarachnoid hemorrhage indicate that lactate has neuroprotective activities if it is the product of an increased glucose metabolism, but will cause greater neural tissue loss it if originates from a hypoxic state.[35]

5.3.2 Barrier Dysfunction

Cellular and molecular mechanisms that typify BBB permeability after stroke have been widely studied. All of them are characterized by tight junction protein (*e.g.* occludin) degradation, which is considered to be a biomarker for damage since its plasma levels are correlated to BBB and parenchymal alterations.[36] In the same way, it correlates with loss of junction proteins in astrocytes, actin polymerization,[37] free radical production, recruitment and activation of immune cells, and overexpression of matrix metalloproteinases (MMPs).[38] These events collaborate strongly with edema development, immune cell diapedesis, and thus damage exacerbation, providing reasons to study BBB stabilizers for neuroprotection. In line with this, the use of intravenous heat shock protein HSP27 in a mice stroke model was shown to induce neurological recovery as a result of the inhibition of actin polymerization, neutrophil reduction, and macrophage recruitment.[37] Tetramethylpyrazine-2′-O-sodium ferulate,[39] a new analog of tetramethylpyrazine, is capable of reducing edema through the preservation of occludins and junctional adhesion molecule 1 (JAM-1), as well as by reducing the expression of MMPs, especially MMP-9, and aquaporin 4 in an MCAO model in mice.[39] Inhibition of MMPs has also been studied, since activation of inflammation is diminished by reducing MMP expression. Cordycepin,[40] dipyridamole,[41] and melatonin[42] are some compounds that have been studied in this regard.

Numerous molecules possess osmotic properties that allow them to decrease edema. Hypertonic saline has been proven to reduce edema by diminishing endothelial growth factor and its receptor's expression. This effect lowers BBB rupture and thus reduces cellular extravasation to the brain parenchyma.[43] Although this therapy has not shown better survival rates in the clinical setting, it is still widely used.[44] Mannitol is the most used hyperosmotic agent because it can significantly improve blood flow and reduce edema and intracranial pressure in stroke patients.[45] Nevertheless, this compound can cause dehydration after some time[46] and increase the risk of acute renal failure;[47] consequently, its use needs to be further studied.

5.3.3 Excitotoxicity

Glutamate is the main excitatory neurotransmitter in the central nervous system. This molecule is strongly implicated in neuronal and axonal growth and communication. During stroke, large amounts of glutamate and other excitatory neurotransmitters[48] are released. The anoxic membrane depolarizations result from the inability to regulate ionic transport across membranes, and subsequent calcium-mediated glutamate membrane receptor N-methyl-D-aspartate (NMDA) release will cause an increase in intracellular Ca^{2+} that causes excitotoxicity and cell death.[48] Glutamate also interferes with the anti-apoptotic Bcl-2 protein, which causes an increase in cytochrome C release and caspase pathway activation, leading to apoptosis.[49]

In this regard, several neuroprotective agents impacting at this level of the ischemic cascade have been studied. Some of them have been directed to the blockage of calcium channels, antagonizing NMDA receptors, or decreasing neurotransmitters, among other approaches. For example, piroxicam is a non-steroidal anti-inflammatory drug indicated for rheumatoid arthritis and other inflammatory diseases, and when examining its effect in rats with cerebral ischemia, it led to significant reduction of glutamate and oxidant molecule release, which produced smaller infarcts.[50]

Pertussis toxin is also considered to be a good neuroprotective candidate, as it has been shown to be able to significantly reduce calcium levels and thereby increase the survival of neuronal cells in *in vitro* studies, in addition to leading to smaller infarcts in *in vivo* models.[51] Table 5.2 shows some of the most recently studied drugs in this area, and although they have shown results at basic levels of research, no clinical setting effects have been promising so far.

5.3.4 Oxidative Stress

Once blood flow is restored, glucose and oxygen become available for cells in order to restart their functions. This will dramatically increase ROS and nitrogen reactive species release. Excessive ROS levels will overload the intrinsic antioxidant capacity of cells.[25,52] Every part of the brain is vulnerable to ROS-induced damage, and the cells' ineffectiveness at neutralizing the increased amounts of free radicals will activate apoptosis and necrosis pathways. Research is focused on searching for treatments that will stimulate antioxidant pathways, as well as exogenously providing antioxidants in order to protect tissues.

Table 5.2 Neuroprotective agents aimed at reducing glutamate-induced neural damage.

Mechanism of action/compound	Research level	Observed effect	Author(s)
Glutamate inhibitors and grabbers			
Peritoneal dialysis	Clinical	Reduces infarct size and improves neurological deficits	Godino *et al.*[101]
Phaseic acid	Basic	Reduces ischemic injury	Hou *et al.*[102]
Channel blockers and reducers			
Arginine-rich peptides	Basic	Reduces infarct size	Meloni *et al.*[103]
Nimodipine	Clinical	No published results—ongoing trial	ClinicalTrials.gov: NCT02248233[104]
Dexmedetomidine	Basic	Protects hippocampal cells	Apkinar *et al.*[105]
NMDA inhibitors and reducers			
ACEA-1021	Clinical	No effect	Cai *et al.*[106]
Ginsenoside Rb1	Basic	Reduces neural damage	Wang *et al.*[107]

One of the most studied antioxidants is resveratrol, which is a phenol and phytoalexin with neuroprotective characteristics. This compound was shown to modulate the heme oxygenase-1 (HO-1) enzyme and to reduce apoptosis in hippocampal neurons, as well as to increase neural progenitor cell survival and proliferation in stroke.[53]

Galangin, a potent free radical scavenger, has neuroprotective effects since it improves neurological deficit and reduces the size of the infarction in a MCAO model through the regulation of four stroke-related enzyme signaling pathways.[54] Table 5.3 presents some other treatments that have shown antioxidant activities.

5.3.5 Inflammation

The release of free radicals, Ca^{2+}, cytokines, and damage-associated molecular pattern molecules (DAMPs) induces the activation of the immune system just hours after stroke.[55] The resulting microenvironment is characterized by microglial activation, neutrophil and monocyte/macrophage migration that will release proinflammatory mediators like IL-1β, IL-6, TNF-α,[56,57] nitric oxide (NO), ROS,[25] chemokines (CXCL10 and CCL12), monocyte chemotactic protein-1 (MCP-1), and CXCL-1,[58,59] as well as inducible NO synthase (iNOS) and cyclo-oxygenase-2 (COX-2) activation.[55] Under these conditions, the final outcome is not encouraging at all. Several studies have shown that activation of the immune system exacerbates neural tissue damage. That is why some therapeutic strategies have been proposed for inhibiting inflammation, as shown in Table 5.4. For instance, atorvastatin is commonly used to lower blood cholesterol, and it was shown to improve function and clinical prognosis in patients who have

Table 5.3 Antioxidant agents currently under study for stroke therapy.

Compound	Mechanism	Observed effect	Author(s)
Liraglutide	↑ Insulin production ↓ ROS ↑ Neurotrophic factor expression	↓ Infarcts ↓ Neurological deficits	Sato *et al.*[98]
Ethanol	↓ LDH ↓ NADPH oxidase	↓ Infarcts Better neurofunctional recovery	Kochanski *et al.*[29]
Pinocembrin	↓ LDH activity ↓ Oxidative stress ↓ Glutamate activity ↑ Anti-inflammatory molecule expression	↓ Infarcts	Saad *et al.*[108]
Daphnetin	↓ Glutamate Maintains glutathione and superoxide dismutase (SOD) ↓ ROS	↓ Infarcts ↓ Neural deficits	Du *et al.*[109]
Dexmedetomidine	Regulates TRPM2 and TRPV1 channels, thus ↓ ROS, ↓ SOD, ↓ Ca^{2+} ↑ Anti-apoptotic proteins	↓ Oxidative toxicity	Apkinar *et al.*[105]
	↑ BDNF expression ↓ IL-6 and TNF-α ↑ Anti-apoptotic proteins	↓ Brain cell necrosis	Rodríguez-González *et al.*[110]
Melatonin	↑ ROS ↑ Lipoperoxidation ↓ BBB permeability ↓ Inflammatory activity ↓ LDH	↓ Infarcts ↓ Edema	Watson *et al.*, Patiño *et al.*[31,32]
Uric acid	↓ Glucose-mediated oxidative stress	↓ Infarct Better outcomes	Amaro *et al.*[111]
Omega-3	↓ ROS Upregulation of Nrf2 and HO-1		Zhang *et al.*[112]

Table 5.4 Anti-inflammatory agents currently under study for stroke therapy.

Substance	Mechanism	Observed effect	Level of research	Author(s)
Sanguinarine	↓ TNF-α, IL-1β, IL-6	↓ Infarct size	Basic	Wang *et al.*[113]
Daphnetin	↓ TNF-α, IL-1β, IL-6 ↓ TLR4 expression	↑ Neurological recovery ↓ Infarct size	Basic	Liu *et al.*[114]
Indomethacin	↓ Microglial activation	↑ Neural preservation ↑ Neuroblast migration	Basic	Lopes *et al.*[115]
Uncaria sinensis	↓ iNOS, COX-2 ↓ TNF-α, IL-1β, IL-6	↓ Infarct size ↑ Neurological recovery	Basic	Kang *et al.*[116]
Co-ultramicronized *N*-palmitoylethanolamine (PEA)/luteolin composite	Inhibits infiltration/ degranulation of mast cells ↓ iNOS, COX-2	↑ Neurological recovery	Clinical, promising results	Caltagirone *et al.*[117]
Dapsone 4,4′-diamino-diphenylsulfone	Inhibits neutrophil MMPs and their recruitment	↑ Neurological recovery	Clinical, no recent updates	Nader-Kawachi *et al.*[118]
Ginsenoside Rd	Inhibits microglial activation	↑ Neurological recovery	Clinical, promising results	Zhang *et al.*[119]
Minocycline	Inhibits microglial activation, COX-2, prostaglandins and IL-1β	↑ Neurological recovery	Clinical, promising results	Fagan *et al.*[120]

suffered a cerebrovascular accident through the lowering of TNF-α and IL-6 cytokines.[60] Curcumin is a polyphenol with anti-inflammatory and antioxidant characteristics that has also been shown to decrease TNF-α and IL-6 levels. Curcumin also improves mitochondrial activity, leading to smaller infarct sizes and major neurological recovery in stroke.[61] Lercanidipine has also demonstrated positive effects on infarct size through the reduction of MMP-9.[62]

Although a great number of anti-inflammatory drugs have been developed, successful results have been scarce, and a controversy about their use has emerged. Some basic and clinical investigations have shown that the immune system also participates in the recovery and protection of injured tissue.[63,64]

A different perspective has recently been considered regarding the evolution of inflammatory events, since evidence suggests that the first response is of an anti-inflammatory rather than inflammatory nature, characterized by the activation of microglia toward an anti-inflammatory M2 phenotype[63,65] and the release of cytokines including IL-4,[66] IL-10, and transforming growth factor-β (TGF-β)[58] and molecules like lipocain-2.[67] Later on, this response is modified as free radicals, Ca^{2+}, glutamate, and DAMPs cause a polarization toward an M1 microglial activation[58] that leads to greater tissue damage.

5.3.6 Immunomodulation

The knowledge of the type of molecules released by the immune cells and their functions has driven research into trying to modulate the immune response in order to generate a benefit for the patient. Different immunomodulators have been proposed, such as platonin, a cyanine photosensitizing dye and potent antioxidant capable of modulating the immune response through the reduction of IL-1β, iNOS, COX-2, and ionized calcium binding adaptor molecule 1 (Iba-1) in a mouse MCAO model. The effect of this modulation causes a reduction of infarcts and edema and improves neurological recovery, with the only disadvantage being that it has to be intraperitoneally applied before ischemia.[68]

Thiamet G is a β-N-acetylglucosaminidase inhibitor that has displayed a powerful immunomodulatory effect in ischemic mice by reducing Iba-1, inducing an M2 microglial phenotype that reduces iNOS and COX-2 expression and results in smaller infarcts and better neurological recovery.[69]

Interferon-β (IFN-β) is an immunomodulatory cytokine approved by the FDA for multiple sclerosis treatment. In stroke, it was shown to reduce monocyte/macrophage and neutrophil infiltration and to reduce inflammatory mediators as well as microglial activation, which leads to a smaller infarct size and better neurological recovery.[70]

The intravenous application of IL-10 *via* the jugular vein 60 minutes after stroke was shown to induce neuroprotection through the Jak–Stat3, PI3K–Akt and NF-κB pathways. This therapy also induces inhibition of Th1 immune responses, causing a diminished infarct size and higher cellular survival in response to Ca^{2+} in a hippocampal cell culture.[71]

T lymphocytes have often been linked to neural tissue damage, and several experiments have shown that in the absence of these cells, ischemic lesions are smaller[72–74] Therefore, different treatments have focused on obstructing their entrance into the injured brain. Curcumin induces neuroprotection by reducing T lymphocyte infiltration and thus lowering interferon-γ (IFN-γ) and IL-17 levels.[75] Nevertheless, several reports have also suggested that the presence of these cells, especially T regulatory lymphocytes (Tregs), is necessary for promoting neuroprotection.[76,77] Brea and co-workers showed that an increase of endogenous Tregs reduced the infarct size as compared to animals with a lower number of these cells.[78] Atorvastatin was also shown to increase the Treg count in the ischemic brain, therefore reducing the infarct size and neurological deficit.[79]

During stroke, there is an autoimmune response against neural constituents.[80] This reaction is conceived under the terms of a new paradigm now called "protective autoimmunity" and is considered a normal physiological mechanism to be potentiated through active immunization with neural-derived peptides. On such grounds, several peptides have been studied.

Cop-1 is a polymeric antagonist obtained from the 82–100 region of myelin basic protein (MBP). This copolymer induces an anti-inflammatory response in other degenerative conditions like experimental allergic encephalomyelitis[81] and spinal cord injury.[82] Active immunization with Cop-1 was shown to have a neuroprotective effect since it reduces infarct size and improves neurological recovery at day 7 post-stroke.[83] Cop-1 increases neurotrophin-3 production and neurogenesis at 7 days and 2 months post-stroke.[84]

Human leucocyte antigen DR-α1 bound to oligodendrocyte glycoprotein-35-55 (DR-α1–MOG-35-55) is capable of inhibiting T lymphocyte-specific neuroantigens and hindering the binding of macrophage migration inhibitory factor[85] to its receptor in monocytes and macrophages. This strategy significantly reduces infarct volume, $CD4^+$ and $CD8^+$ T lymphocytes, and proinflammatory cytokine accumulation, accompanied by an improvement in neurological recovery.[85] MOG had previously been shown to increase IL-10 production by Tregs, in addition to reducing CD11b expression.[86]

Poly-YE is a copolymer that exerts immunomodulatory effects and facilitates effector T lymphocyte responses, promoting the development of smaller infarcts and better neurological recovery, as well as an increase in neurogenesis during the acute phase of ischemia.[87]

Some authors also consider stem cell transplant to be an immunomodulatory mechanism due to its ability to modify its surrounding microenvironment.[88] Stem cell transplant is a widely studied field of basic research in which adult human multipotent progenitor cells,[88] menstrual blood-derived stem cells,[89] mesenchymal stem cells,[90] adipocyte-derived stem cells,[91] and umbilical cord blood mesenchymal stem cells,[92] among others, have been studied for their effects. All of the aforementioned cells induce an M2 microglial response by means of increasing growth factor (such as brain-derived neurotrophic factor (BDNF) and insulin-like growth factor-1 (IGF-1)) and anti-inflammatory immunoregulatory cytokine (IL-10 and TGF-β) production. Transplanted mesenchymal stem cells are specifically involved in angiogenesis and neurogenesis mechanisms,[90,93] all of which lead to better neurological recovery and smaller infarct sizes. Basic research with stem cells has shown such success and there are currently 115 registered clinical trials on ClinicalTrials.gov, some of which have concluded, but most of which are still in the experimental phase.

Lee *et al.* evaluated the effect of autologous bone marrow stem cells and observed an improvement in neurological recovery using the Modified Rankin Scale. Clinical recovery was associated with plasma levels of stromal cell growth factor (SDF-1).[94] The latest stem cell clinical trials have not yielded the expected results, and drugs have presented limitations after reaching the clinical research stage.

5.4 Conclusion

Unfortunately, very few of the experimentally studied therapies have shown promising results and even fewer have been tested at clinical phases. Likewise, most of the therapeutic strategies tested in clinical settings have not been approved. It is likely that the failure of these therapies lies in the large number of molecular and biochemical mechanisms that are involved during the development of stroke. This creates a need to correctly recognize every mechanism of damage in order to get closer to a complete understanding of stroke and to be able to propose new and more efficient neuroprotective treatments that are able to improve patients' lives.

References

1. Global Status Report on Noncommunicable Diseases World Health Organitaztion; 2014. http://apps.who.int/iris/bitstream/10665/148114/1/9789241564854_eng.pdf.
2. D. Mozaffarian, E. J. Benjamin, A. S. Go, D. K. Arnett, M. J. Blaha, M. Cushman, S. R. Das, S. de Ferranti, J. P. Despres, H. J. Fullerton, V. J. Howard, M. D. Huffman, C. R. Isasi, M. C. Jimenez, S. E. Judd, B. M. Kissela, J. H. Lichtman, L. D. Lisabeth, S. Liu, R. H. Mackey, D. J. Magid, D. K. McGuire, E. R. Mohler, 3rd, C. S. Moy, P. Muntner, M. E. Mussolino, K. Nasir, R. W. Neumar, G. Nichol, L. Palaniappan, D. K. Pandey, M. J. Reeves, C. J. Rodriguez, W. Rosamond, P. D. J. Sorlie, J. Stein, A. Towfighi, T. N. Turan, S. S. Virani, D. Woo, R. W. Yeh, M. B. Turner and American Heart Association Statistics Committee and Stroke Statistics Subcommittee, *Circulation*, 2016, **133**, e38, DOI: 10.1161/CIR.0000000000000350.
3. R. Magge, B. C. Lau, B. P. Soares, S. Fischette, S. Arora, E. Tong, S. Cheng and M. Wintermark, *Am. J. Neuroradiol.*, 2013, **34**(2), 402.
4. C. Hjalmarsson, K. Manhem, L. Bokemark and B. Andersson, *Stroke Res. Treat.*, 2014, **2014**, 1.

5. E. P. Soler and V. C. Ruiz, *Curr. Cardiol. Rev.*, 2010, **6**(3), 138.
6. W. Ren, Y. Gu, L. Zhu, L. Wang, Y. Chang, M. Yan, B. Han and J. He, *Compr. Psychiatry*, 2016, **65**, 9.
7. D. F. Alfieri, M. F. Lehmann, S. R. Oliveira, T. Flauzino, F. Delongui, M. C. de Araújo, I. Dichi, V. D. Delfino, L. Mezzaroba, A. N. Simão and E. M. Reiche, *Metab. Brain Dis.*, 2017, **32**(2), 493.
8. H. M. Kwon, Y. S. Lee, H. J. Bae and D. W. Kang, *Stroke*, 2014, **45**(3), 871.
9. M. J. Meens, A. Pfenniger and R. B. Kwak, *Swiss. Med. Wkly.*, 2012, **142**, 1.
10. M. Koch and A. Zernecke, *IUBMB Life*, 2014, **66**(11), 735.
11. L. Badimon and G. Vilahur, *J. Intern. Med.*, 2014, **276**(6), 618.
12. G. K. Hansson, P. Libby and I. Tabas, *J. Intern. Med.*, 2015, **278**(5), 483.
13. F. J. Vajda, *J. Clin. Neurosci.*, 2002, **9**(1), 4.
14. B. Manwani and L. D. McCullough, *J. Neurosci. Res.*, 2013, **91**(8), 1018.
15. P. Kalaivani, M. Ganesh, R. Sathiya, V. Ranju, V. Gayathiri and C. Saravana Babu, *J Stroke Cerebrovasc Dis.*, 2014, **23**(10), 2753.
16. M. A. Moskowitz, E. H. Lo and C. Iadecola, *Neuron*, 2010, **67**(2), 181.
17. B. C. Campbell, G. A. Donnan and S. M. Davis, *Front. Neurol.*, 2014, **5**, 194.
18. A. D. Horsch, J. W. Dankbaar, Y. Van Der Graaf, J. M. Niesten, T. Van Seeters, I. C. Van Der Schaaf, L. J. Kappelle, B. K. Velthuis and DUST Investigators, *Neuroradiology*, 2015, **57**(12), 1219.
19. M. Alonso de Leciñana, J. Díaz-Guzmán, J. A. Egido, A. García Pastor, P. Martínez-Sánchez, J. Vivancos, E. Díez-tejedor and Ad Hoc Committee of the Madrid Association of Neurology's Stroke Forum, *Neurología*, 2013, **28**(7), 425.
20. Y. Xian, J. J. Federspiel, A. F. Hernandez, D. T. Laskowitz, L. H. Schwamm, D. L. Bhatt, E. E. Smith, G. C. Fonarow and E. D. Peterson. *Circulation*, 2017, **135**(11), 1024.
21. M. Mokin, A. A. Khalessi, J. Mocco, G. Lanzino, T. M. Dumont, R. A. Hanel, D. K. Lopes, R. D. Fessler, 2nd, A. J. Ringer, B. R. Bendok, E. Veznedaroglu, A. H. Siddiqui, L. N. Hopkins and E. I. Levy, *Neurosurg. Focus.*, 2014, **36**(1), E5.
22. T. G. Jovin, A. Chamorro, E. Cobo, M. A. de Miguel, C. Molina, A. Rovira, L. San Roman, J. Serena, S. Abilleira, M. Ribó, M. Millán, X. Urra, P. Cardona, E. López-Cancio, A. Tomasello, C. Castaño, J. Blasco, L. Aja, L. Dorado, H. Quesada, M. Rubiera, M. Hernández-Pérez, M. Goyal, A. M. Demchuk, R. Von Kummer, M. Gallofré, A. Dávalos and REVASCAT Trial Investigators, *N. Engl. J. Med.*, 2015, **372**, 2296.
23. P. Widimsky and L. N. Hopkins, *Eur. Heart J.*, 2016, **37**(40), 3081.
24. J. M. Coutinho, D. S. Liebeskind, L. A. Slater, R. G. Nogueira, W. Clark, A. Dávalos, A. Bonafé, R. Jahan, U. Fischer, J. Gralla, J. L. Saber and V. M. Pereira, *JAMA Neurol.*, 2017, **74**(3), 268.
25. D. Paspalj, P. Nikic, M. Savic, D. Djuric, I. Simanic, V. Zivkovic, N. Jeremic, I. Srejovic and V. Jakovljevic, *Mol. Cell. Biochem.*, 2015, **406**(1–2), 75.

26. J. Li, S. E. Benashski, C. Siegel, F. Liu and L. H. McCullough, *Neurosci. Lett.*, 2010, **482**(1), 62.
27. T. Jiang, J. T. Yu, X. C. Zhu, H. F. Wang, M. S. Tan, L. Cao, Q. Q. Zhang, L. Gao, J. Q. Shi, Y. D. Zhang and L. Tan, *Br. J. Pharmacol.*, 2014, **171**(13), 3146.
28. Y. Yamazaki, S. Harada and S. Tokuyama, *Neurosci. Lett.*, 2015, **604**, 134.
29. R. Kochanski, C. Peng, T. Higashida, X. Geng, M. Hüttemann, M. Guthikonda and Y. Ding, *J. Neurochem.*, 2013, **126**(1), 113.
30. J. L. Cross, B. P. Meloni, A. J. Baker, S. Lee and N. W. Knuckey, *Stroke Res. Treatment.*, 2010, **1**(9), 1.
31. N. Watson, T. Diamandis, C. Gonzales-Portillo, S. Reyes and C. V. Borlongan, *Cell Transplant.*, 2016, **25**(5), 883.
32. P. Patiño, E. Parada, V. Farré-Alins, S. Molz, R. Cacabelos, J. Marco-Contelles, M. G. López, C. I. Tasca, E. Ramos, A. Romero and J. Egea, *Neurotoxicology*, 2016, **57**, 61.
33. X. Castillo, K. Rosafio, M. T. Wyss, K. Drandarov, A. Buck, L. Pellerin, L. B. Weber and L. Hirt, *J. Cereb. Blood. Flow Metab.*, 2015, **35**(10), 1561.
34. V. Mosienko, A. G. Teschemacher and S. Kasparov, *J. Cereb. Blood Flow Metab.*, 2015, **35**(7), 1069.
35. M. Oddo, J. M. Levine, S. Frangos, E. Maloney-Wilensky, R. T. Carrera, M. Daniel, P. J. Levivier, P. D. Magistretti and P. D. LeRoux, *Stroke*, 2012, **43**(5), 1418.
36. R. Pan, K. Yu, T. Weatherwax, H. Zheng, W. Liu and K. J. Liu, *Sci Rep.*, 2017, **7**, 40331.
37. Y. Shi, X. Jiang, L. Zhang, H. Pu, X. Hu, W. Zhang, W. Cai, Y. Gao, R. K. Leak, R. F. Keep, M. V. Bennett and J. Chen, *Proc. Natl. Acad. Sci. U. S. A.*, 2017, **114**(7), E1243.
38. H. T. Zhang, P. Zhang, Y. Gao, C. L. Li, H. J. Wang, L. C. Chen, Y. Feng, R. Y. Li, Y. L. Li and C. L. Jiang, *Mol. Med. Rep.*, 2017, **15**(1), 57.
39. S. H. Xu, M. S. Yin, B. Liu, M. L. Chen, G. W. He, P. P. Zhou, Y. J. Cui, D. Yang and Y. L. Wu, *Hum. Exp. Toxicol.*, 2016, DOI: 10.1177/0960327116657401.
40. J. Yuan, A. Wang, Y. He, Z. Si, S. Xu, S. Zhang, K. Wang, D. Wang and Y. Liu, *Brain Res. Bull.*, 2016, **127**, 171.
41. M. Massaro, E. Scoditti, M. A. Carluccio, M. Pellegrino, N. Calabriso, C. Storelli, G. Martines and R. De Caterina, *Thromb. Haemost.*, 2013, **109**(2), 280.
42. J. W. Jang, J. K. Lee, M. C. Lee, M. S. Piao and S. H. Kim, *J. Neurol. Sci.*, 2012, **323**(1–2), 221.
43. L. Huang, W. Cao, Y. Deng, G. Zhu, Y. Han and H. Zeng, *BMC Neurosci.*, 2016, **17**(1), 64.
44. C. J. Ong, S. G. Keyrouz and M. N. Diringe, *Neurocrit. Care.*, 2015, **23**(2), 285.
45. Q. Wang, H. Ye and Y. Su, *Turk. Neurosurg.*, 2014, **24**(3), 333.
46. H. S. Mangat and R. Härtl, *Ann. N. Y. Acad. Sci.*, 2015, **1345**, 83.

47. S. Y. Lin, S. C. Tang, L. K. Tsai, S. J. Yeh, L. J. Shen, F. L. Wu and J. S. Jeng, *Medicine*, 2015, **94**(47), e2032.
48. Y. Liao, A. M. Kristiansen, C. P. Oksvold, F. A. Tuvnes, N. Gu, E. Rundén-Pran, P. Ruth, M. Sausbier and J. F. Storm, *PLoS One*, 2010, **5**(12), e15601.
49. U. Anilkumar and J. H. Prehn, *Front. Cell Neurosci.*, 2014, **8**, 281.
50. P. Bhattacharya, A. K. Pandey, S. Paul and R. Patnaik, *J. Physiol. Biochem.*, 2014, **70**(4), 901.
51. Z. Tang, S. Li, P. Han, J. Yin, Y. Gan, Q. Liu, J. Wang, C. Wang, Y. Li and J. Shi, *J Neurochem.*, 2015, **135**(5), 998.
52. S. Manzanero, T. Santro and T. V. Arumugam, *Neurochem. Int.*, 2013, **62**(5), 712.
53. G. P. Dias, G. Cocks, M. C. do Nascimento Bevilaqua, A. E. Nardi and S. Thuret, *Oxid. Med. Cell Longev.*, 2016, **2016**, 1.
54. R. Yang, K. Chen, Y. Zhao, P. Tian, F. Duan, W. Sun, Y. Liu, Z. Yan and S. Li, *Molecules*, 2016, **21**(4), 438.
55. J. Montaner, Fisiopatología de la Isquemia cerebral, ed. España: ICG Marge SL. 2008, capítulo 8, 117.
56. A. Tuttolomondo, D. Di Raimondo, R. di Sciacca and G. L. Pinto, *Curr. Pharm. Des.*, 2008, **14**(33), 3574.
57. K. L. Lambertsen, K. Biber and B. Finsen, *J. Cerebral Blood Flow Metabol.*, 2012, **32**, 1677.
58. A. Patel, R. Ritzel and F. L. McCullough, *Int. J. Physiol. Pathophysiol. Pharmacol.*, 2013, **5**(2), 73.
59. P. Liu, J. W. Xiang and S. X. Jin, *Neurol. Res.*, 2015, **37**(10), 853.
60. A. Tuttolomondo, D. di Raimondo, R. Pecoraro, C. Maida, V. Arnao, V. Della Corte, I. Somonetta, F. Corpora, D. Di Bona, R. Maugeri, D. G. Iacopino and A. Pinto, *Medicine*, 2016, **95**(13), e3186.
61. Y. Miao, S. Zhao, Y. Gao, R. Wang, Q. Wu, H. Wu and T. Luo, *Brain Res. Bull.*, 2016, **121**, 9.
62. S. Gupta, U. Sharma, N. R. Jagannathan and Y. K. Gupta, *Exp. Neurol.*, 2017, **288**, 25.
63. A. Rayasam, M. Hsu, G. Hernández, J. Kijak, A. Lindstedt, C. Gerhart, M. Sandor and Z. Fabry, *Neurochem. Int.*, 2017, DOI: 10.1016/j.neuint.2017.02.009.
64. W. N. Jin, S. X. Shi, Z. Li, M. Li, K. Wood, R. J. Gonzales and Q. Liu, *J. Cereb. Blood Flow. Metab.*, 2017, DOI: 10.1177/0271678×17694185.
65. C. Y. Xia, S. Zhang, Y. Gao, Z. Z. Wang and N. H. Chen, *Int. Immunopharmacol.*, 2015, **25**(2), 377.
66. X. Zhao, H. Wang, G. Sun, J. Zhang and J. A. Edwards, *J. Neurosci.*, 2015, **35**(32), 11281.
67. C. Xing, X. Wang, C. Cheng, J. Montaner, E. Mandeville, W. Leung, K. van Leyen, J. Lok, X. Wang and E. H. Lo, *Stroke*, 2014, **45**(7), 2085.
68. J. R. Sheu, Z. C. Chen, T. Jayakumar, D. S. Chou, T. L. Yen, H. N. Lee, S. H. Pan, C. H. Hsia, C. H. Yang and C. Y. Hsieh, *Sci. Rep.*, 2017, **7**, 42277.

69. Y. He, X. Ma, D. Li and J. Hao, *J. Cereb. Blood Flow Metab.*, 2016, DOI: 10.1177/0271678×16679671.
70. P. C. Kuo, B. A. Scofield, I. C. Yu, F. L. Chang, D. Ganea and J. H. Yen, *J Am Heart Assoc.*, 2016, **5**, 1.
71. E. A. Tukhovskaya, E. A. Turovsky, M. V. Turovskaya, S. G. Levin, A. N. Murashev, V. P. Zinchenko and O. V. Godukhin, *Neurosci Lett.*, 2014, **571**, 55.
72. G. Yilmaz, T. Arumugam, K. Stokes and D. Granger, *Circulation*, 2006, **113**, 2105.
73. A. Liesz, W. Zhou, E. Mracskó, S. Karcher, H. Bauer, S. Schwarting, L. Sun, D. Bruder, S. Stegemann, A. Cerwenka, C. Sommer, A. H. Dalpke and R. Veltkamp, *Brain*, 2011, **134**(Pt.3), 704.
74. S. Wang, H. Zhang and Y. Xu, *Neurol. Res.*, 2016, **38**(6), 495.
75. N. W. Liu, C. C. Ke, Y. Zhao, Y. A. Chen, K. C. Chan, D. T. Tan, J. S. Lee, Y. Y. Chen, T. W. Hsu, Y. J. Hsieh, C. W. Chang, B. H. Yang, W. S. Huang and R. S. Liu, *Transl. Stroke Res.*, 2016, DOI: 10.1007/s12975-016-0512-4.
76. J. Kipnis, T. Mizrahi, E. Hauben, I. Shaked, E. Shevach and M. Schwartz, *Proc. Natl. Acad. Sci. U. S. A.*, 2002, **99**(24), 15620.
77. M. Schwartz and J. Kipnis, *Ann. N. Y. Acad. Sci.*, 2005, **1051**, 701.
78. D. Brea, J. Agulla, M. Rodrıguez-Yañez, D. Barral, P. Ramos-Cabrer, F. Campos, A. Almeida, A. Dávalos and J. Castillo, *J. Cell Mol. Med.*, 2014, **18**(8), 1571.
79. A. L. Rodríguez-Perea, J. Gutierrez-Vargas, G. P. Cardona-Gómez, C. J. Guarin, M. Rojas and P. A. Hernández, *J. Neuroimmune Pharmacol.*, 2017, **12**(1), 152.
80. C. Iadecola and J. Anrather, *Nat. Med.*, 2011, **17**(7), 796.
81. R. Aharoni, B. Kayhan, R. Eilam, M. Sela and R. Arnon, *Proc. Natl. Acad. Sci. U. S. A.*, 2003, **100**(24), 14157.
82. E. Hauben, O. Butovsky, U. Nevo, E. Yoles, G. Moalem, E. Agranov, F. Mor, R. Leibowitz-Amit, E. Pevsner, S. Akselrod, M. Neeman, I. R. Cohen and M. Schwartz, *J Neurosci.*, 2000, **20**(17), 6421.
83. A. Ibarra, H. Avendaño and Y. Cruz, *Neurosci. lett.*, 2007, **425**, 110.
84. Y. Cruz, P. Suárez-Meade and A. Ibarra, *Neural. Regener. Res.*, 2015, **10**(11), 1733.
85. J. Wang, Q. Ye, J. Xu, G. Benedek, H. Zhang, Y. Yang, H. Lui, R. Meza-Romero, A. A. Vandenbark, H. Offner and Y. Gao, *Transl. Stroke Res.*, 2016, 1, DOI: 10.1007/s12975-016-0514-2.
86. D. Frenkel, Z. Huang, R. Maron, D. N. Koldzic, W. W. Hancock, M. A. Moskowitz and H. L. Weiner, *J. Immunol.*, 2004, **172**, 6549.
87. Y. Ziv, A. Finkelstein, Y. Geffen, J. Kipnis, I. Smirnov, S. Shpilman, I. Vertkin, M. Kimron, A. Lange, T. Hecht, K. G. Reyman, J. B. Marder, M. Schwartz and E. Yoles, *Stroke*, 2007, **38**, 774.
88. B. Yang, J. A. Hamilton, K. S. Valenzuela, A. Bogaerts, X. Xi, J. Aronowski, R. W. Mays and S. I. Savitz, *Stem Cells*, 2017, DOI: 10.1002/stem.2600.

89. M. C. Rodrigues, T. Lippert, H. Nguyen, S. Kaelber, P. R. Sanberg and C. V. Borlongan, *Adv. Exp. Med. Biol.*, 2016, **951**, 111.
90. Y. C. Hsuan, C. H. Lin, C. P. Chang and M. T. Lin, *Brain Behav.*, 2016, **6**(10), e00526.
91. F. Zhou, S. Gao, L. Wang, C. Sun, L. Chen, P. Yuan, H. Zhao, Y. Yi, Y. Qin, Z. Dong, L. Cao, H. Ren, L. Zhu, Q. Li, B. Lu, A. Liang, G. T. Xu, H. Zhu, Z. Gao, J. Ma, J. Xu and X. Chen, *Stem Cell Res. Ther.*, 2015, **6**, 92.
92. Q. Cheng, Z. Zhang, S. Zhang, H. Yang, X. Zhang, J. Pan, L. Weng, D. Sha, M. Zhu, X. Hu and Y. Xu, *Brain Res.*, 2015, **1594**, 293.
93. J. Xiang, J. Hu, T. Shen, B. Liu, F. Hua, K. Zan, J. Zu, G. Cui and X. Ye, *Neurosci. Lett.*, 2017, **644**, 62.
94. J. S. Lee, J. M. Hong, G. J. Moon, P. H. Lee, Y. H. Ahn, O. Y. Bang and STARTING Collaborators, *Stem Cells*, 2010, **28**(6), 1099.
95. V. A. Lioutas and V. Novak, *Neural. Regen. Res.*, 2016, **11**(3), 400–401.
96. S. Zhang, W. Zuo, X. F. Guo, W. B. He and N. H. Chen, *Neurochem. Int.*, 2014, **70**, 22.
97. H. Zhu, Y. Zhang, Z. Shi, D. Lu, T. Li, Y. Ding, Y. Ruan and A. Xu, *Sci. Rep.*, 2016, **6**, 26859.
98. K. Sato, M. Kameda, T. Yasuhara, T. Agari, T. Baba, F. Wang, A. Shinko, T. Wakamori, A. Toyoshima, H. Takeuchi, T. Sasaki, S. Sasada, A. Kondo, C. V. Borlongan, M. Matsumae and I. Date, *Int. J. Mol. Sci.*, 2013, **14**(11), 21513.
99. V. Darsalia, M. Larsson, G. Lietzau, D. Nathanson, T. Nyström, T. Klein and C. Patrone, *Diabetes Obes. Metab.*, 2016, **18**(5), 537.
100. R. Liu, H. Wang, B. Xu, W. Chen, E. Turlova, N. Dong, C. L. Sun, Y. Lu, H. Fu, R. Shi, A. Barszczyk, D. Yang, T. Jin, E. Mannucci, Z. P. Feng and H. S. Sun, *Diabetes*, 2016, **65**(9), 2795.
101. M. C. Godino, V. Romera, J. A. Sánchez-Tomero, J. Pacheco, S. Canals, J. Lerma, J. Vivancos, M. A. Moro, M. Torres, I. Lizasoain and J. Sánchez-Prieto, *J.* Clin. Invest., 2013, **123**(10), 4359.
102. S. T. Hou, S. X. Jiang, L. I. Zaharia, X. Han, C. L. Benson, J. Slinn and S. R. Abrams, *J. Biol. Chem.*, 2016, **291**(53), 27007.
103. B. P. Meloni, L. M. Brookes, V. W. Clark, J. L. Cross, A. B. Edwards, R. S. Anderton, R. M. Hopkins, K. Hoffmann and N. W. Knuckey, *J. Cereb. Blood Flow Metab.*, 2015, **35**(6), 993.
104. L. I. Runhui, 2014, in Shenyang Medical College, Liaoning, China. Sponsored by the Fengtian Hospital. https://clinicaltrials.gov/ct2/show/NCT02248233.
105. H. Apkinar, M. Nazıroğlu, I. S. Övey, B. Çiğ and O. Apkinar, *Sci. Rep.*, 2016, **6**, 37196.
106. S. X. Cai, *Curr. Top. Med. Chem.*, 2006, **6**(7), 651.
107. S. Wang, M. Li, Y. Guo, C. Li, L. Wu, X. F. Zhou, Y. Luo, D. An, S. Li, H. Luo and L. Pu, *J. Ethnopharmacol.*, 2017, **202**, 138.
108. M. A. Saad, R. M. Abdel Salam, S. A. Kenawy and A. S. Attia, *Pharmacol. Rep.*, 2015, **67**(1), 115.

109. G. Du, H. Tu, X. Li, A. Pei, J. Chen, Z. Miao, J. Li, C. Wang, H. Xie, X. Xu and H. Zhao, *Neurochem. Res.*, 2014, **39**(2), 269.
110. R. Rodríguez-González, T. Sobrino, S. Veiga, P. López, J. Rodríguez-García, S. Veiras del Rio, A. Baluja, J. Castillo and J. Álvarez, *Life Sci.*, 2016, **144**, 162–169.
111. S. Amaro, L. Llull, A. Renú, C. Laredo, B. Perez, E. Vila, F. Torres, A. M. Planas and A. Chamorro, *Ann. Neurol.*, 2015, **77**(5), 775.
112. M. Zhang, S. Wang, L. Mao, R. Leak, Y. Shi, W. Zhang, X. Hu, B. Sun, G. Cao, Y. Gao, Y. Xu, J. Chen and F. Zhang, *J. Neurosci.*, 2014, **34**(5), 1903.
113. Q. Wang, P. Dai, H. Bao, P. Liang, W. Wang, A. Xing and J. Sun, *Exp. Ther. Med.*, 2017, **13**(1), 263.
114. J. Liu, Q. Chen, Z. Jian, X. Xiong, L. Shao, T. Jin, X. Zhu and L. Wang, *Biomed. Res. Int.*, 2016, **2016**, 2816056.
115. R. S. Lopes, M. M. Cardoso, A. O. Sampaio, M. S. Barbosa, C. C. Souza, M. C. Da Silva, E. M. Ferreira, M. A. Freire, R. Rodriguez Lima and W. Gomes-Leal, *J. Biosci.*, 2016, **41**(3), 381.
116. B. K. Kang, M. K. Kim, S. Y. Kim, S. J. Lee, Y. W. Choi, B. T. Choi and H. K. Shin, *Am. J. Chin. Med.*, 2015, **43**(6), 1099.
117. C. Caltagirone, C. Cisari, C. Schievano, R. Di Paola, M. Cordaro, G. Bruschetta, E. Esposito, S. Cuzzocrea and Stroke Study Group, *Transl. Stroke Res.*, 2016, **7**(1), 54.
118. J. Nader-Kawachi, F. Góngora-Rivera, J. Santos-Zambrano, P. Calzada and C. Ríos, *Neurol. Res.*, 2007, **29**(3), 331.
119. G. Zhang, F. Xia, Y. Zhang, X. Zhang, Y. Cao, L. Wang, X. Liu, G. Zhao and M. Shi, *Mol. Neurobiol.*, 2016, **53**(4), 2529.
120. S. C. Fagan, J. L. Waller, F. T. Nichols, D. J. Edwards, L. C. Pettigrew, W. M. Clark, C. E. Hall, J. A. Switzer, A. Ergul and D. C. Hess, *Stroke.*, 2010, **41**(10), 2283.

CHAPTER 6

The Molecular Neuroprotective Strategies in Cerebral Ischemia: An Insight into Emerging Treatments for Oxidative Stress

SREEKALA S. NAMPOOTHIRI,[a] SAURABH A. JAIN,[b] DAVID BURROWS,[b] OK-NAM BAE,[c] ARSHAD MAJID[b] AND G. K. RAJANIKANT*[a]

[a] School of Biotechnology, National Institute of Technology Calicut, Calicut 673601, Kerala, India; [b] Sheffield Institute for Translational Neuroscience, University of Sheffield, 385a Glossop Road, Sheffield S10 2HQ, UK; [c] College of Pharmacy, Hanyang University, Ansan 15588, Republic of Korea

*Email: rajanikant@nitc.ac.in

6.1 Introduction

A myriad of interrelated pathophysiological events following cerebral ischemia culminate in neuronal death. Neurons are highly vulnerable to ischemic changes as they require an interminable supply of oxygen and glucose to maintain ionic gradients across the membrane. With the onset of ischemic stroke, there is a concomitant depolarization of neurons, energy failure and subsequent generation of excessive free radicals. This results in

Drug Discovery Series No. 62
Acute Brain Impairment: Scientific Discoveries and Translational Research
Edited by Philip V. Peplow, Svetlana A. Dambinova, Thomas A. Gennarelli and Bridget Martinez

Published by the Royal Society of Chemistry, www.rsc.org

oxidative stress and disruption of cell ion homoeostasis, leading to glutamate excitotoxicity and a profuse influx of calcium (Ca^{2+}) ions. Activation of the inflammatory and immune response ultimately activates apoptotic pathways, leading to neuronal loss. The tolerance threshold and the spectrum of severity of ischemic stroke outcome are dependent on the brain region and the population of neurons subjected to ischemic insult. The core brain tissue in the vicinity of ischemic injury undergoes rapid degradation and elicits necrotic cell death, whereas the infarction evolves slowly at the penumbral site, acting as a frontier region for employing post-stroke therapeutic strategies. The use of thrombolytic therapy, within a limited time window of 3.0–4.5 hours, reduces the mortality and morbidity associated with ischemic stroke by restoring the blood flow in the brain. Nonetheless, it is an admissive fact that the development of single-target drug therapies has frequently failed in clinical trials, and developing treatment strategies for ischemic stroke is innately onerous. Despite these challenges, pleiotropic treatment approaches are under investigation to provide more practical and efficacious clinical therapeutics. Research continues to elucidate the interrelationship between the molecular mechanisms of ischemic stroke and neuroprotection. This chapter focuses on oxidative stress-induced cerebral ischemia and various neuroprotective strategies employed to ameliorate neuronal damage.

6.2 Free Radical Release and Oxidative Stress in Cerebral Ischemia

Free radicals were first hypothesized to be the by-products of enzymatic reactions and were later chronicled to be 'Pandora's Box of evils', resulting in severe cellular damage, cancer, mutagenesis and senescence. Research has highlighted that moderate amounts of free radicals and free radical derivatives are crucial for various physiological cellular functions, such as maintaining response to oxygen tension, regulation of vascular tone, signal transduction from membrane receptors and oxidative stress.[1] Redox regulation is, therefore, pertinent for maintaining a balance between the propitious and pernicious effects of free radicals during pathological conditions.

Free radicals contain unpaired valence electron/electrons that can pair with DNA, lipids or protein molecules, leading to oxidation. There are primarily three phases of reactive oxygen species (ROS) generation during cerebral ischemia/reperfusion (I/R) injury. There is an initial outbreak of ROS from the mitochondria soon after hypoxia-ischemia (HI). The second phase of ROS production occurs 25–30 minutes following HI due to intracellular ATP depletion and conversion of hypoxanthine to xanthine *via* xanthine oxidase (XO). Restoration of blood flow/reperfusion triggers the third phase of ROS generation, causing severe neuronal injury.[2] ROS and their derivatives include superoxide anions, hydrogen peroxide and hydroxyl radicals.[3] Approximately 2–5% of the electrons in the electron transport chain contribute to superoxide anion production in the mitochondria. In addition, the pro-oxidant enzymes such as nicotinamide adenine dinucleotide phosphate oxidase (NADPH oxidase [NOX]) and XO contribute to the formation of superoxide ions. These radicals are constantly scavenged by superoxide dismutase (SOD) to produce hydrogen peroxide, which is subsequently detoxified to water and oxygen by catalase (CAT) or glutathione peroxidase (GSH-Px). Nitric oxide (NO) is another free radical known to be constantly produced in the brain *via* NO synthase (NOS). NO production leads to cellular dysfunction by the nitrosylation of tyrosines in protein molecules. NO also reacts with superoxide anions to contribute to the formation of hydroxyl radicals, with peroxynitrite ($ONOO^-$) as the reaction product that degrades rapidly to $^\bullet OH$ and nitrogen dioxide (NO_2).[3] The presence of low concentrations of these oxidants forms an essential part in maintaining oxygen tension; however, a surfeit of oxidants can damage the cell irreversibly. Thus, it is evident that endogenous anti-oxidants such as SOD, GSH-Px and CAT help in maintaining redox homoeostasis.

6.3 Endogenous Anti-oxidant Defense Mechanisms and Therapeutic Targets

Since the oxidative stress induced by brain injury transgresses multiple pathways and targets, the potential influences of the most important endogenous anti-oxidants and their functions in neuroprotection are discussed in the following sections.

6.3.1 SODs in Cerebral Ischemia

Endogenous anti-oxidants, particularly SODs, are crucial in scavenging the surplus production of free radicals following ischemic brain damage. SODs in brain cells can be categorized into three types depending on the metal ion requirement and anatomic distribution, namely copper/zinc SOD (SOD1), manganese SOD (SOD2) and extracellular SOD (ECSOD). Transgenic and mutant studies have shown that overexpression of SODs confers neuroprotection following cerebral ischemia.[4]

Copper/zinc SOD is a dimeric cytosolic protein encoded by the *SOD1* gene and requires copper and zinc as the metal ion cofactors. SOD1 does not induce a direct effect following ischemic damage, owing to its extremely short half-life of 6 minutes in the circulating blood and inability to cross the blood–brain barrier (BBB). However, lecithin- or liposome-bound SOD1 alters the enzymatic properties of SOD1 and leads to a prolonged half-life and increased BBB permeability. Treatment with liposome-entrapped copper/zinc SOD results in the attenuation of oxidative damage and mitigation of cerebral infarction in rodents subjected to focal cerebral ischemia.[2] Manganese SOD (SOD2) is a tetrameric protein expressed in mitochondria that requires Mn^{2+} for its activity, and extracellular SOD (SOD3) is a copper-containing SOD confined to the extracellular spaces, cerebral fluid and vessels.[4]

The effects of SODs in ischemia and traumatic brain injury have been identified through studies using transgenic animals. Transgenic mice with homozygous SOD1 overexpression results in a five-fold increase in the SOD1 activity in comparison with the heterozygous ($SOD1^{+/-}$) mice that exhibit a three-fold SOD1 elevation.[5] SOD1 activity reduces infarct volume by differing amounts depending on the localization of ischemia. For example, SOD1 overexpression decreases infarct volume by 35% 24 hours following permanent focal cerebral ischemia, induced by occluding the distal middle cerebral artery and the bilateral carotid artery,[6] whereas increased SOD1 activity in global ischemia reduces hippocampal CA1 cell death by 50%, which is thought to be partly due to hampered mitochondria-associated apoptosis.[7] Regulation of neuronal viability by SOD1 is further corroborated in SOD1-deficient mice that show 100% mortality 24 hours following ischemia, whereas $SOD1^{+/-}$ mice exhibited 30% mortality.[8] Furthermore, the introduction of a SOD mimetic, manganese tetrakis (4-benzoic acid) porphyrin, inhibits global ischemia-induced neurological deficits.[9] However, SOD1 overexpression fails to offer neuroprotection in the absence of reperfusion following permanent focal ischemia.[10,11]

SOD2 overexpression thwarts apoptosis induced by Fe^{2+}, amyloid-β peptide or NO-generating agents. SOD2 is also downregulated following 3 hours of reperfusion in an ischemic brain.[12] SOD2 transcription is regulated by signal transducer and activator of transcription 3 (STAT3), which binds to the promoter region of SOD2. STAT3 levels decrease as early as at 1 hour of reperfusion following transient focal cerebral ischemia (tFCI), curbing its recruitment to the SOD2 promoter. SOD2 levels in the brain are lowered, which coincide with a subsequent increase in oxygen radical generation, further aggravating the oxidative damage following ischemic reperfusion.[13]

SODs are also known to target redox-sensitive signaling pathways following ischemia, including the Akt, p38 and nuclear factor-κB (NF-κB) pathways. $SOD1^{+/-}$ transgenic mice with less oxidative damage alleviate NF-κB activation compared to the wild-type controls following cerebral ischemia. In addition, SOD1 promotes neuroprotection by the activation of NF-κB1-related rapid immune responses and anti-apoptotic factors (BCL2 and X chromosome-linked inhibitor of apoptosis protein) following tFCI.[14] $SOD1^{+/-}$ transgenic mice suppress the excess production of cytokines and chemokines such as interleukin (IL)-1β, tumor necrosis factor-α (TNF-α), IL-6, monocyte chemoattractant protein 1 and macrophage inflammatory protein 1α.[15] p38, a member of the mitogen-activated protein kinase (MAPK) family, is stimulated by ROS and leads to neuronal apoptosis and inflammatory cytokine production post-ischemia. $SOD1^{+/-}$ transgenic mice impede the activation of p38 and cytosolic phospholipase A2 (a p38 MAPK pathway substrate), resulting in a remarkable 70% reduction in infarct volume compared to wild-type mice following 1 and 3 days of reperfusion.[16]

SOD1 expression also modulates cell survival pathways (Akt) in contrast to p38 and NF-κB pathways. Activated Akt phosphorylates BCL2-associated agonist of cell death (BAD), a critical component of the intrinsic cell death machinery, to mitigate apoptosis and enhance cell survival.[17] Cerebral ischemia inhibits the Akt pathway *via* post-ischemic oxidant production, thus reducing the generation of phosphorylated proline-rich Akt substrate (pPRAS), subsequently shifting the cell fate to death signaling pathways. SOD1 overexpression reduces the direct effect of oxidative stress after reperfusion injury by activating the Akt pathway and blocking the inhibition of pPRAS.[18]

In the central nervous system (CNS), the BBB distinctly separates the brain extracellular fluid from the circulating blood. BBB disruption as a result of ischemic injury is detrimental, causing severe brain damage due to amassed edema and subsequent hemorrhage. BBB impairment following ischemia is associated with either vascular endothelial cell death or the degradation of basal laminal extracellular matrix components. The evident deleterious role of vascular endothelial cells following ischemia has been highlighted using *in vitro* and *in vivo* models. Oxygen glucose deprivation (OGD) induces DNA damage in vascular endothelial cells, and following tFCI, vascular endothelial cells generate excess oxygen free radicals, further contributing to the oxidative damage. This is further corroborated in $SOD1^{+/-}$ transgenic endothelial cells that show attenuation of OGD-induced endothelial cell death.[19]

Matrix metalloproteinases (MMPs) are a group of proteolytic enzymes that degrade extracellular matrices. Generation of free radicals post-ischemia activates MMPs that degrade the neurovascular matrix, resulting in BBB injury. A pronounced increase in pro-MMP-9 levels occur following 3 hours of reperfusion, and these levels are significantly higher in $SOD1^{-/-}$ mice compared to wild-type controls. MMP-2 levels are also elevated in the ischemic region of $SOD1^{-/-}$ mice. Hence, active MMPs are identified to co-localize with superoxide ions in the ischemic brain regions at the capillary wall and astrocytic processes. There is substantial evidence suggesting a spatial relationship between MMP-9 activity and superoxide production.[20] SOD1 overexpression reduces lesion size, and this effect is directly associated with decreased MMP-9 activity. Conversely, SOD1 deficiency activates MMP-9 and aggravates BBB damage. Long-term incubation of endothelial cells in a high-glucose medium enhances MMP-9 promoter activity, with an increased generation of ROS. This enhanced MMP-9 activity could be controlled with an anti-oxidant treatment, including PEG-SOD and *N*-acetylcysteine.[20] Furthermore, IL-1β treatment triggers superoxide anion release in vascular smooth muscle cells associated with MMP-9 and extracellular signal-regulated kinase (ERK) activation. This effect can be arrested by pre-treating the cells with *N*-acetylcysteine or overexpressing SOD2.[21] Therefore, it is evident that SODs play a considerable role in alleviating the oxidative stress-induced neuronal injury following cerebral ischemia and are fundamental targets for neuroprotection.

6.3.2 NOX Regulates ROS Generation in Ischemic Stroke

One of the remarkable therapeutic approaches against oxidative stress is to inhibit the enzymatic source of ROS generation. NOXs are the only known group of enzymes that function solely on ROS production, whereas other sources, such as XO, NOS and cyclooxygenase, produce ROS as a by-product. NOX is composed of multiple protein subunits with an active NOX catalytic subunit to transfer electrons from NADPH to oxygen, thereby forming ROS. There are five NOX isoforms (NOX-1–5), and each homolog differs in its cellular localization, requirement of protein subunits and the nature of ROS produced. Out of the five NOX isoforms, NOX-1, NOX-2 and NOX-4 are known to be expressed in the CNS. NOX-3 is not pertinent to stroke pathology because it is expressed in the inner ear and NOX-5 is less characterized.[22]

NOX-4 is abundantly expressed in the cerebral arteries compared to the peripheral vasculature. Ischemic stroke increases NOX-2 and NOX-4 protein expression within 24 hours; however, their expression is confined to specific cell types. NOX-2 expression increases in the microglia, whereas NOX-4 expression is observed in the neurons and endothelial cells.[23] The pertinence of NOX-2 in ischemic stroke pathology is highly conflicting, and reports suggest that NOX-2 deletion might reduce infarct size in aged mice *via* inflammatory mechanisms.[24] By contrast, NOX-4 deletion in an *in vivo* ischemic stroke model and the intrathecal administration of VAS2870 (a NOX inhibitor) in wild-type mice reduces brain infarctions drastically by 75%. However, VAS2870 has off-target effects such as alkylating cysteine thiol residues, and its pharmacokinetic limitations restrict intravenous administration.[25] NOX-4 overexpression increases infarct size, which further supports the relevance of NOX-4 as a key therapeutic target in ischemic stroke.[25] NOX isoform-specific inhibitors such as gp91ds-tat (for NOX-2) and pyrazolopyridine derivate GKT136901 (for NOX1/4) act as efficient NOX inhibitors. Other popular NOX inhibitors include apocyanin and diphenylene iodonium (DPI); however, the therapeutic efficiency of both of these inhibitors is questionable. Apocyanin loses its therapeutic relevance as it provides neuroprotection when administered pre-stroke, with no effect observed if applied post-ischemic injury. DPI is not a specific NOX inhibitor as it also interacts with NOS and flavoproteins.[26] Nevertheless, development of novel target-specific inhibitors for NOX homologs, especially NOX-2 and NOX-4, is crucial in regulating the generation of ROS and preventing oxidative stress-induced ischemic stroke.

6.3.3 Activation of Nrf2/ARE Signaling Alleviates Oxidative Stress

Nuclear factor E2-related factor 2 (Nrf2) belongs to the Cap'n'Collar (CNC) family of proteins that regulate the expression of genes containing the antioxidant response element (ARE) or the MAF recognition element (MARE).

Nrf2 is normally repressed by Kelch-like erythroid cell-derived protein (Keap1), containing more than 25 cysteine residues. The Nrf2–Keap1 complex mechanism was recently clarified by the identification of distinct cysteine residues that recognize different electrophiles making Keap1 highly sensitive.[27] Under oxidative stress, these cysteine residues are blocked and cause the translocation of Nrf2 from the cytoplasm into the nucleus to dimerize with ARE or MARE.[28] This promotes the transcription of downstream anti-oxidant target genes and restores the activity of endogenous anti-oxidant enzymes (SODs, CATs, GSH-Px and GSH) essential for redox homeostasis.[29]

Nrf2 expression is specific to oxidative stress-induced ischemic stroke and is insensitive to glutamate/*N*-methyl-D-aspartate (NMDA)-induced neurotoxicity.[30] Activation of Nrf2 during the acute phase of ischemic stroke varies depending on the site of infarct, the time of ischemic insult and the type of cells. Nrf2 expression increases following an ischemic insult, with the upregulation beginning at 3 hours and the maximum expression at 24 hours, whereas 72 hours of reperfusion reduces Nrf2 expression in a transient middle cerebral artery occlusion (MCAO) model. The peri-infarct region exhibits a greater Nrf2 expression than the core ischemic zone because the core experiences comparatively less oxidative stress than the penumbra. Furthermore, Nrf2 activation is observed only in neurons during the early hours of reperfusion (8 hours), whereas 24 hours of reperfusion activates Nrf2 in neurons, astrocytes and microglia. Several reports suggest that the loss of Nrf2 worsens infarct volume and the neurological outcome in *in vivo* ischemic stroke models.[31]

Nrf2 exerts protection against mitochondrial dysfunction by inhibiting mitochondrial uncoupler carbonyl cyanide-4-(trifluoromethoxy) phenylhydrazone (FCCP).[32] This inhibits ROS production and elevates the levels of anti-oxidant enzymes. Nrf2 activation augments BCL2 expression and attenuates cytochrome-c release from mitochondria, inhibiting apoptosis.[33] A phytoestrogen, notoginsenoside R1, exerts neuroprotection against oxidative stress in a cerebral I/R *in vivo* model through the estrogen receptor-dependent activation of Nrf2/Akt pathways. This effect is primarily associated with the inhibition of NOX and mitochondrial dysfunction, which in turn attenuates capase-3 activation and DNA fragmentation.[34]

The overwhelming rise in ROS also causes endoplasmic reticulum (ER) stress in cerebral ischemia due to slow folding of proteins and high accumulation of aberrantly folded/unfolded proteins in the ER lumen. It triggers the interaction between p62 (found in protein aggregates) and Keap1 to activate Nrf2 with subsequent autophagic degradation of Keap1. Therefore, it is speculated that p62–Nrf2–Keap1 interaction regulates autophagy and might be beneficial in the removal of ROS, resulting in oxidative stress prevention.[35] The activation of Nrf2 by different physical mechanisms (calorie restriction, ozone therapy, hyperbaric oxygen therapy and physical exercise) might be beneficial to combat oxidative stress-induced neuronal injury in cerebral ischemia.[36,37] Other Nrf2 activators with roles in neuroprotection against cerebral ischemia are discussed later in this chapter.

6.3.4 Other Endogenous Oxidative Stress Targets in Ischemic Stroke

Anti-oxidant defense following cerebral ischemia is multifaceted, involving several possible therapeutic targets. PARK-7/DJ-1, silent information regulator factor 2-related enzyme 1 (sirtuin 1 [Sirt1]), cofilin and transglutaminases (TGs) are a few other oxidative stress regulators in ischemic stroke. PARK-7/DJ-1 is a redox-sensitive protein that acts as a biomarker in neurodegenerative diseases and ischemic stroke. DJ-1 functions against intracellular oxidative stress and also uses extracellular paracrine/autocrine cues between neighboring neuronal cells to render neuroprotection. DJ-1 expression increases in the early phases of experimental stroke and is translocated into the mitochondria, with the subsequent secretion of DJ-1 protein in primary rat cortical neurons exposed to HI.[38] Sirt1 is a NAD-dependent protein deacetylase that regulates oxidative stress and cell death. Sirt1 overexpression deacetylates endothelial NOS following ischemic stroke and restores cerebral blood flow, whereas knockdown of Sirt1 exacerbates neuronal damage. Its downstream targets include NF-κB, Foxos, p53 and peroxisome proliferator-activated receptor-γ (PPAR-γ) coactivator-1α (PGC-1α), which are aberrantly expressed following HI. Calorie restriction, RS treatment and ischemic preconditioning are some of the known strategies to enhance Sirt1 expression following cerebral ischemia.[39] TGs are calcium-dependent enzymes that are translocated into the nucleus in response to oxidative stress. TG inhibitors are reported to reduce neuronal injury and have a wider therapeutic window.[40] Another target, cofilin, exhibits multicellular function in oxidative stress, and its potential as a therapeutic target in ischemic stroke is rapidly expanding. It is primarily involved in actin cytoskeleton reorganization; however, cofilin oxidation leads to apoptotic and necrotic neuronal cell death.[41]

6.4 Sequestering Oxidative Stress-induced Apoptosis and Necrosis Post-ischemia

Oxidative stress leads to neuronal death and there are primarily two mechanisms by which neurons die: apoptosis (programmed and targeted cell death) or necrosis (unplanned cell death). Apoptosis dismantles cellular function in a rather organized fashion compared to necrosis. The understanding of neuronal death in cerebral ischemia is conventionally limited to necrosis in the core brain tissue within minutes of tFCI due to the reduced blood flow. However, the ischemic penumbra or peri-infarct zone remains less affected even after hours or days of ischemic stroke, acting as a region of salvage following cerebral infarction. Neurons in these regions undergo apoptosis after several hours of ischemic insult and are potential targets for recovery.

6.4.1 Apoptosis Regulation and the Mechanistic Targets

Apoptosis is a highly conserved genetic mechanism that is activated by two general pathways: intrinsic and extrinsic pathways. Intrinsic apoptotic mechanisms post-ischemia involve the excessive influx of calcium ions, whereas extrinsic apoptotic mechanisms involve the Fas death receptor and TNF. Both pathways cause mitochondrial dysfunction and the release of pro-apoptotic proteins and excessive ROS, the activation of caspase-dependent apoptosis/caspase-independent apoptosis and DNA damage. Apoptosis is largely regulated by BCL2 and the caspase family of proteins. BCL2 overexpression prevents mitochondrial damage and protects from ischemic injury. BCL2 inhibition with a small molecule mimetic specific for the BH-3 domain (pro-apoptotic member of BCL2 family) activates apoptosis.[42] This strategy is under clinical trial in cancer; however, the activation of BCL2 for neuroprotection is still controversial due to its expression in multiple cancer types, such as leukemia and lymphoma. Instead, repressing a pro-apoptotic BH-3-only member of the BCL2 family, Bid, is considered a better replacement to target the anti-apoptotic features of BCL2. This has been shown to reduce mitochondrial damage and provide neuroprotection in a MCAO model.[43]

Mammalian caspases belong to a large family of cysteine proteases that consists of 11 members and is of cardinal importance in apoptotic mechanisms. Caspase-1 and caspase-3 were first identified for their roles in neuronal death following cerebral ischemia. Caspase-3 mediates direct intracellular neuronal death, whereas caspase-1 initiates apoptosis through the extracellular secretion of cytokines such as IL-1β and IL-1α. During the early stages of infarction and degradation of the necrotic core, caspase-8 and caspase-1 initiate early apoptosis, resulting in cytoplasmic and nuclear condensation. During the secondary stage of injury in the penumbral region, caspase-9 mediates mitochondria-associated cell death.[42] Inhibition of caspases for neuroprotection is highly selective and depends on the brain tissues undergoing apoptosis and the severity of stroke. Caspase inhibitors z-VAD-FMK and z-DEVD-FMK alleviate brain injury during tFCI, but are ineffective when the injury is severe following global ischemia. In addition, caspase inhibitors produce futile results after permanent ischemia, which demonstrates that caspase activation requires at least a low amount of residual blood flow and reperfusion. Research has confirmed this speculation where caspase-3 expression increases within 24 hours of reperfusion following transient MCAO, and the levels remain significantly lower even after the same duration of reperfusion in permanent MCAO.[44] Furthermore, caspase inhibitors are incapable of targeting a specific site of injury and are degraded *in situ*, which may hamper their success in clinical trials. Despite these challenges, it is a ubiquitous fact that ischemic stroke initiates apoptotic mechanisms involving caspase activation; thus, blocking caspase-dependent pathways is a viable neuroprotective strategy.

c-Jun N-terminal kinase (JNK), a stress-inducible kinase, is activated in permanent, transient focal and global cerebral ischemia. JNK signaling is triggered *via* the expression of c-Jun, a downstream target of JNK in the nuclei of dying neurons. Systemic delivery of JNK inhibitor (SP600125) reduces the infarct volume in tFCI as well as in permanent global ischemia. Out of the three JNK isoforms (JNK1, JNK2 and JNK3), JNK3 deficiency attenuates ischemic brain injury while JNK1 contributes towards most of the JNK activity in the brain. JNK also induces the activation of various pro-apoptotic genes, such as *Bim* and *Fas*. JNK3-deficient mice have deficient *Bim* and *Fas* expression and JNK inhibitor SP600125 inhibits *Bim* and *Fas* expression, condemning apoptosis following ischemia.[45,46] Therefore, designing specific inhibitors for JNK3 could be a promising therapeutic strategy to render neuroprotection post-ischemia.

6.4.2 Targeting Necrosis in Cerebral Ischemia

Necrosis is an immediate effect following brain injury leading to cellular disintegration. Analogous to apoptosis, post-ischemic necrosis is also accompanied by the activation of various death mechanisms, but is distinguished by necrotic morphology. Poly-ADP-ribose-1 (PARP-1) is activated upon traumatic brain or neuronal injury. PARPs play a dual functional role in neuronal death mediated by apoptosis and necrosis. While PARP-1 cleavage by caspase forms a part of the apoptotic neuronal death, PARP-1 mediates poly-ADP-ribosylation by consuming NAD^+, resulting in PARP-1 over-activation. This over-activation eventually leads to necrosis due to exorbitant energy loss. PARP-1$^{-/-}$ primary neuronal cultures combat neurotoxicity induced by OGD, NMDA or NO. PARP-1-deficient mice exhibit a reduced infarct volume following MCAO, and furthermore, PARP-1 inhibitors elicit neuroprotection.[47] Concomitantly, studies prove that apoptotic markers remain unaltered in PARP-1-deficient mice following ischemic injury, which further emphasizes that PARP-1$^{-/-}$ mice are resilient to injury, mainly due to necrotic cell death following energy loss by poly-ADP-ribosylation.

Calpains are calcium-dependent proteases with two major isozymes: μ- and m-calpains. Calpains generally regulate homeostatic processes under normal conditions, and the increase in intracellular Ca^{2+} concentration following ischemia leads to its activation and neuronal dysfunction. Calpain over-stimulation causes the breakdown of cytoskeletal protein, loss of structural integrity, axonal transport discord and, ultimately, necrotic cell death. Cbz-Val-Phe-H is a calpain inhibitor known to reduce infarction volume in the MCAO model; however, this has not been tested in clinical trials so far.[48] One of the anticipated restrictions might be its bioavailability, being a cysteine protease.

Necroptosis is a programmed caspase-independent form of necrosis that shares common pathways with apoptosis. This neuronal death process is initiated with the activation of death receptors by their own ligands, namely Fas and TNF-α, forming death-inducing signaling complex (DISC). DISC comprises the adaptor protein Fas-associated death domain (FADD), caspase-8 and the death domain-containing kinase RIPK1. Under activated apoptotic conditions, caspase-8 and caspase-3 elicit mitochondrial damage, whereas apoptotic-deficient conditions stimulate the death receptors, leading to the activation of RIPK1 and eventually necroptosis. Necrostatins, a family of specific small molecule inhibitors, arrest necroptosis in cells co-treated with FasL or TNF-α and the pan-caspase inhibitor.[49] Necrostatin-1 (Nec-1) confers neuroprotection with an extended time window and alleviates ischemic brain injury, leaving the family of necrostatins as a promising therapeutic target for ischemic stroke.

6.5 Exogenous Anti-oxidants to Decrease ROS Generation

A large number of natural anti-oxidants aid in providing post-ischemic neuroprotection by repressing multiple oxidative stress targets and thereby enhancing the activity of endogenous anti-oxidants (Table 6.1). One of the broadly studied naturally occurring anti-oxidants is resveratrol (RS), a polyphenol found in grape skin, pines and pomegranate. RS promotes neuroprotection in cerebral ischemia by acting on several oxidative stress cell death mechanisms. It inhibits XO, prevents the formation of free radicals, increases malondialdehyde, reduces glutathione levels and improves the brain metabolism by restoring ATP.[50] RS pretreatment exerts neuroprotection against oxidative stress in focal cerebral I/R injury by activating Nrf2/ARE signaling mechanism, with a consequent reduction in infarct volume and neuronal recovery.[51] Neuroprotection by RS differs according to its prophylactic, acute and post-acute delivery (50 $mg\,kg^{-1}$) following focal cerebral ischemia in mice. A sustained neurological recovery with enhanced neurogenesis and angiogenesis occurs in the prolonged post-acute RS delivery following 4 weeks of ischemic insult, whereas prophylactic and acute RS delivery reduces infarct volume and enhances BBB integrity after 2 days of ischemic insult, with no further recovery after 4 weeks. RS neuroprotection is accompanied by the elevation of its downstream target Sirt1, an increase in cell survival signals such as p-Akt, hemeoxygensae-1 (HO-1) and BCL2 and a decrease in cell death proteins (BAX and caspase-3).[52] It also reduces astroglial response and preserves myelination in HI neonatal rats, with improved long-term behavioral outcomes lasting until adulthood. Furthermore, oral RS treatment prevents stroke recurrence by protecting endothelial cerebrovasculature *via* Sirt1.[53] Despite substantial experimental evidence showing RS-induced neuroprotection, there are no studies proving the clinical significance of RS in ischemic stroke.

XO mediates ROS formation by converting hypoxanthine into xanthine following HI injury, and XO inhibition by allopurinol (AP) in the early phase of cerebral ischemia is identified to be neuroprotective. AP administration 15 minutes following HI injury reduces atrophy and the water content in

Table 6.1 Advantages and limitations of selected exogenous anti-oxidants on oxidative stress induced by cerebral ischemia.

Exogenous anti-oxidant	Advantages	Limitations	Ref.
RS	• Inhibits xanthine oxidase and prevents the formation of free radicals • Improves brain metabolism by restoring ATP • Activates Nrf2/ARE signaling mechanism • Reduces infarct volume and promotes neuronal recovery • Oral RS treatment prevents stroke recurrence *via* Sirt1 expression	• There are no studies proving the clinical significance of RS in ischemic stroke	50–53
AP	• Provides neuroprotection during the early phase of cerebral ischemia • Prevents long-term cerebral injury • Potentiates intrinsic adenosine neuroprotective mechanisms • Tested in infants and human populations • Reduces the overall stroke risk in the elderly population	• Lack of substantial results proving neuroprotection in animal models • High dose of AP required to attain neuroprotection; however, it did not impose any side effects	54–60
Omega-3 fatty acids (ALA, EPA and DHA)	• Promotes neuroprotection in adult and neonatal animal models • Oral and intravenous ALA supplementation increases the survival of hippocampal neurons • ALA provides cognitive improvement with enhanced spatial learning and memory post-stroke • Both pre- and post-treatment with DHA is beneficial against experimental stroke • DHA complexed with human serum albumin provides high-grade neuroprotection with a therapeutic window of 7 hours • Combination of DHA and delayed hypothermia produces sensorimotor improvement	• Lack of human trials • Finding an appropriate dosage to be used in post-ischemic treatment • EPA improves neurological outcome only when pretreated, and no effect on post-stroke treatment	61–68

Table 6.1 (*Continued*)

Exogenous anti-oxidant	Advantages	Limitations	Ref.
NAC	• Increases the expression of HIF-1α to provide neuroprotection in a MCAO model • Combination of NAC treatment and systemic hypothermia improves infarct volume in asphyxiated neonatal rats	• No pilot studies have been carried out revealing its neuroprotective efficacy	69, 70
Melatonin	• Does not have a pro-oxidant effect and functions independently to scavenge ROS • Randomized controlled pilot study confirms that the combination of melatonin/hypothermia therapy improves neurological outcome in 6 month old asphyxiated neonates	• Free melatonin degrades easily and is readily cleared from the circulation	71–73
UA	• Acts as a PPAR-γ agonist and reduces MMP activity • UA treatment after cerebral I/R injury reduces infarct volume in a dose-dependent manner • Activates the Nrf2/ARE pathway	• Warrants further studies to elucidate the neuroprotective mechanism of UA	74–76

ipsilateral hemisphere of 7-day-old Wistar rat pups, preventing long-term cerebral injury.[54] AP treatment prior to hypoxia increases the adenosine and inosine levels in the brain tissue of newborn piglets and potentiates intrinsic adenosine neuroprotective mechanisms.[55] Despite the lack of substantial results on the neuroprotective effect of AP in animal models, it was tested in infant and human populations. AP reduces the NO serum levels in asphyxiated infants and provides long-term neurodevelopmental benefits.[56] A high dose of AP (500 mg) administration to mothers with fetal hypoxia reduces the cord blood levels of S100B, a biomarker for neuronal damage.[57] A recent study found that the use of AP reduces the overall stroke risk in the elderly population and promotes long-term protection against ischemic stroke health hazards.[58] A follow-up of two randomized controlled trials put forth the long-term neuroprotective effects and the negative side effects of using a high dose of AP (40 $mg\,kg^{-1}$) in moderately asphyxiated infants.[59] A phase IV trial (XILO-FIST) to study the long-term outcome of AP treatment on the recovery and recurrence frequency of ischemic stroke by the inhibition of XO is currently ongoing.[60] Therefore, the majority of completed and ongoing trials using AP indicate persistent neuroprotection, especially in infants, by reducing ROS generation following a HI insult.

Omega-3 fatty acids are natural anti-oxidants with three distinct types, namely α-linolenic acid (ALA), eicosapentaenoic acid (EPA) and docosahexaenoic acid (DHA), which promote neuroprotection in adult and neonatal animal models. Oral and intravenous ALA supplementation increases the survival of hippocampal neurons and leads to cognitive improvement with enhanced spatial learning and memory post-stroke.[61] Pretreatment with ethyl-EPA for 5 and 7 days results in a significant reduction in infarct volume and neurological outcome, whereas post-stroke ethyl-EPA treatment does not have any significant neuroprotective effect.[62] EPA pretreatment exerts neuroprotection in gerbils with bilateral common carotid artery occlusion *via* anti-oxidant and anti-inflammatory properties.[63] Unlike EPA, both pre- and post-treatment with DHA is beneficial against experimental stroke and is one of the most studied omega-3 fatty acids in adult ischemic stroke. DHA complexed with human serum albumin is considered to provide high-grade neuroprotection with a broad therapeutic window of 7 hours after stroke in young and aged rats.[64–66] A recent study shed light on the treatment of triglyceride DHA post-HI injury, where the accumulation of DHA and DHA derivatives after injury causes the interaction of DHA with mitochondria, improving Ca^{2+} permeabilization and reducing oxidative brain damage.[67] Furthermore, a combination of DHA and delayed hypothermia produces a remarkable sensorimotor improvement in neonatal asphyxiated rats;[68] however, the mechanism is not completely elucidated yet.

N-Acetyl-L-cysteine (NAC) is a thiol-containing anti-oxidant that prevents the excessive generation of free radicals and protects the neuronal cells from apoptosis in experimental stroke models. NAC pretreatment increases the expression of hypoxia-inducible factor-1α (HIF-1α) and its downstream target genes (erythropoietin and GLUT-3) to provide neuroprotection in a MCAO model. This effect is associated with the upregulation of heat shock protein

90 (Hsp90), which maintains an increased HIF-1α stability.[69] The combination of NAC treatment and systemic hypothermia improves infarct volume and attenuates HI injury in neonatal rats.[70]

Melatonin does not have a pro-oxidant effect and functions independently to scavenge ROS. Melatonin reduces neuronal injury, restores GSH and diminishes oxidative stress in hippocampal slice cultures subjected to OGD rexoxygentaion.[71] A randomized controlled pilot study confirms that the combination of melatonin/hypothermia therapy improves neurological outcomes even after 6 months in asphyxiated term neonates.[72] A recent study focuses on nanoencapsulated melatonin delivery, as free melatonin degrades easily and is rapidly cleared by the circulating blood. Nanoencapsulated melatonin shows greater neuroprotection and rescues neuronal cells even at lower concentrations by restoring anti-oxidative enzyme activity in an experimental cerebral I/R injury model for aged rats.[73]

Ursolic acid (UA), a pentacyclic triterpenoid compound, is an anti-oxidant that acts as a PPAR-γ agonist and reduces MMP activity with the subsequent activation of the MAPK pathway. Intragastric UA treatment after cerebral I/R injury reduces infarct volume in a dose-dependent manner.[74] The neuroprotective effect of UA in traumatic brain injury is also shown to be associated with the activation of the Nrf2/ARE pathway.[75] Other ROS scavengers that activate the Nrf2/ARE pathway include nobiletin (a polymethoxylated flavone) and myricetin (a naturally occurring flavonol found in berries, tea and wine).[76,77] Pre-administration of nobiletin 3 days prior to and immediately after the ischemic insult significantly activates Nrf2, HO-1, SOD1 and GSH, with a reduction of NF-κB expression.[76] Ruscogenin (a steroid sapogenin derived from *Ophiopogon japonicus*), phlorofucofuroeckol (a phlorotannin derived from marine brown algae) and ConBr (a lectin purified from the seeds of *Canavalia brasiliensis*) are a few recently studied natural anti-oxidants that protect against ischemic neuronal injury.[78–80]

6.6 Neuronal Restoration Strategies Following Cerebral Ischemia

Due to the constant failure of drug and small molecule therapies in clinical trials, neuronal repair and regeneration strategies were developed to restore neuronal loss following cerebral ischemia. Neuronal repair and regeneration strategies offer a greater therapeutic window compared to neuroprotective drugs and range from a few days to months, compared to 0–8 hours. These strategies involve cell and cytokine therapies, stimulation of endogenous neurogenesis and ischemic preconditioning (IP) to promote neuronal repair.

6.6.1 Cellular Therapy

Transplantation of neural progenitor cells (NPCs), neural cell lines, bone marrow/umbilical cord blood-derived mesenchymal stem cells (MSCs), stromal cells or hematopoietic progenitor/endothelial cells promotes neuronal regeneration in CNS disorders, including ischemic stroke. Induction of neural stem cells (NSCs) isolated from the subventricular zone (SVZ) of the adult rat brain augments sensory and motor recovery in an *in vivo* stroke model.[81] Furthermore, human neurospheres that are transplanted into the striatum of stroke-induced rats differentiate into functional mature neurons and express neuronal markers, such as calbindin, HuD and parvalbumin.[82] Despite the direct influence of NSC transplantation in neuronal regeneration, other cellular therapies were sought after, as NSCs have some limitations, including a poor survival rate *in vivo*, a high rate of immune rejection and ethical concerns. To overcome these limitations, a number of specific neurotrophic, growth and differentiation factors have been identified that can promote the differentiation of multipotent MSCs into functional mature neurons.[83] MSCs readily migrate to the lesion sites to reduce scar thickness and increase the number of oligodendrocyte precursor and proliferating cells.[84] Extensive efforts have been made to evaluate the therapeutic potential of MSCs *via* the intravenous infusion of autologous MSCs in ischemic stroke patients. MSCs are less susceptible to immune rejection and overcome the ethical issues associated with human embryonic stem cells. However, MSCs do not overcome the low *in* vivo cell survival rate.

6.6.2 Endogenous Neurogenesis

Exogenous administration of stem cells to promote neuron regeneration is often accompanied by poor *in vivo* cell viability. In such cases, stimulating endogenous neurogenesis after ischemic stroke might be a beneficial approach to restoring neuronal loss. The SVZ and subgranular layer are both neurogenesis active sites in adult brains and show a dramatic increase in NPC proliferation following ischemia. Endogenous neurogenesis is a multistep process involving the migration of NPCs to the site of damage and their differentiation into the various neural cell lineages to produce functional mature neural cells.[85] Ischemic stroke does not hinder NSC proliferation; rather, it impairs the survival of these newly generated progenitors. Unlike mature cells, neural progenitor stem cells continue to undergo asymmetric cell division owing to their post-mitotic nature and maintain a pool of progenitor cells, thus allowing continued brain repair. Stroke causes these NSCs to switch from asymmetric to symmetric division, which further leads to the expansion of the progenitor pool. Stroke-induced NSC proliferation is identified to be regulated by several extracellular neurogenic signals, including fibroblast growth factor-2 (FGF-2), insulin-like growth factor-1 (IGF-1), brain-derived neurotrophic factor (BDNF) and vascular endothelial growth factor (VEGF).[86]

Another crucial parameter in endogenous neurorepair is the successful migration of NSCs from the SVZ to the region of infarction. Ischemic stroke initiates neuroblast migration, with the upregulation of chemotactic factors. However, the consequent peri-infarct scar formation and barrier molecules such as chondroitin sulfate proteoglycans obstruct neuroblast migration. NSC migration in stroke-induced endogenous neurogenesis is modulated *via* receptor–ligand signaling pathways, chemokines and proteinases such as stromal cell-derived factor-1 (SDF-1), CXC chemokine receptor 4 (CXCR4), monocyte chemoattractant protein-1 (MCP-1) and MMPs.[87]

Endogenous neurorepair is recognized to be associated with the brain vasculature and links neurogenic and angiogenic processes after ischemic stroke. Neuroblast migration is particularly localized in the infarct regions with active post-ischemic angiogenesis. Blocking angiogenesis results in a ten-fold decrease in neuroblasts within the infarct region and an overall decrease in vascular density.[88] Studies indicate that VEGF aids the coupling of angiogenesis and neurogenesis, as NSCs secrete relatively large amounts of VEGF in an ischemic brain. In addition, the activation of endothelial cells following stroke onset increases SDF-1α and MMPs to eventually guide neuroblast migration.[87] However, brain damage reduces neurogenic signaling, which makes the neurogenic processes slower. It is therefore important that the development of any prospective endogenous neurorepair therapy aims at enhancing progenitor cell survival and migration to successfully promote functional recovery of an injured brain.

6.6.3 Ischemic Preconditioning

IP refers to the activation of endogenous cell survival mechanisms triggered by HI, which is otherwise considered a noxious stimulus. Brief ischemic stimuli (rapid/delayed) primes the brain against neuronal injury, resulting in a relatively reduced infarct volume in animal stroke models. This phenomenon activates neuroprotective mechanisms by reducing excitotoxicity, balancing oxidative stress, regulating inflammatory responses and stimulating neurovascular protection as described in the previous sections.[89] Another common strategy involves remote IP, where an arm or a leg is exposed to ischemia to provide neuroprotection. This effect is attributed to the release of vasoactive substances and the activation of neural pathways.[90] IP is also associated with the activation of the HIF pathway; however, the pro-survival and pro-death effects of HIF are still under investigation. IP reduces ATP use in cells to preserve the mitochondrial function, rendering tolerance to low oxygen levels, a mechanism similar to hibernation in vertebrate species. This tolerance is partially regulated by HIF-1 transcription factor by reducing cellular metabolism.[91]

It is imperative to consider that IP is not a measure to prevent the occurrence of ischemic stroke; rather, it reduces the impact of infarction and improves post-stroke recovery. Additionally, the preconditioning stimulus intensity must be low and must not exceed a threshold to prevent aggravated brain injury. Several clinical settings have been proposed to test the effect of IP in humans. Desferrioxamine, erythropoietin and sevoflurane are approved to be safe and effective for early preconditioning and are under randomized controlled trials in patients with heart disease.[91] Adequate therapeutic guidelines need to be set for the treatment of cerebral ischemia using preconditioning for their successful implementation in clinical trials.

6.7 Conclusion

The pathophysiology of ischemic stroke is particularly capricious, rendering multiple challenges to concoct an adequate neuroprotective therapy. Attenuating oxidative stress in cerebral ischemia is a key to targeting several other downstream mechanisms, including excitotoxicity, calcium overload, inflammation and apoptotic and necrotic cell death. Oxidative damage can be curbed by inhibiting enzymes that generate ROS, scavenging ROS through exogenous anti-oxidants and upregulating endogenous anti-oxidants. Several crucial ischemic stroke molecular targets and pathways with their neuroprotective counterparts have already been discovered; however, one of the most remonstrated issues is the *in vivo* bioavailability, adequate doses and the effective delivery of small molecule inhibitors across the BBB, congenial with the acute nature of ischemic stroke.

Abbreviations

ALA	α-linolenic acid
AP	allopurinol
ARE	anti-oxidant response element
ATP	adenosine triphosphate
BAD	BCL2 associated agonist of cell death
BBB	blood–brain barrier
BCL2	BCL2, apoptosis regulator
BDNF	brain derived neurotrophic factor
CAT	catalase
DHA	docosahexaenoic acid
DPI	diphenylene iodonium
EPA	eicosapentaenoic acid
ER	endoplasmic reticulum
ERK	extracellular signal-regulated kinase
ETC	electron transport chain
FGF-2	fibroblast growth factor-2
GLUT3	glucose transporter 3
GSH	glutathione
GSH-Px	glutathione peroxidise
HI	hypoxia-ischemia
HIF-1α	hypoxia inducible factor-1α
HO-1	hemeoxygenase-1
Hsp90	heat shock protein 90
I/R injury	ischemia/reperfusion injury
IGF-1	insulin-like growth factor-1
IL	interleukin

IP	ischemic preconditioning
JNK	c-Jun N-terminal kinase
Keap1	kelch-like erythoid cell derived protein
MAPK	mitogen-activated protein kinase
MCAO	middle cerebral artery occlusion
MMP	matrix metalloproteinase
NAC	*N*-Acetyl-L-cysteine
NF-κB	nuclear factor-κB
NMDA	*N*-methyl-D-aspartate
NO	nitric oxide
NOS	nitric oxide synthase
NOX	nicotinamide adenine dinucleotide phosphate-oxidase
Nrf2	nuclear factor-E2-related factor 2
NSCs	neural stem cells
OGD	oxygen glucose deprivation
PARP-1	poly ADP-ribose-1
PPAR-γ	peroxisome proliferator-activated receptor gamma
pPRAS	phosphorylated proline-rich Akt substrate
ROS	reactive oxygen species
RS	Reseveratrol
SDF-1	stromal cell derived factor-1
Sirt1	silent information regulator factor 2-related enzyme 1/sirtuin1
SOD	superoxide dismutase
STAT3	signal transducer and activator of transcription 3
SVZ	subventricular zone
tFCI	transient focal cerebral ischemia
TG	transglutaminase
TNF-α	tumor necrosis factor-α
UA	ursolic acid
VEGF	vascular endothelial growth factor
XO	xanthine oxidase

References

1. W. Droge, *Physiol. Rev.*, 2002, **82**, 47.
2. A. Y. Abramov, A. Scorzielloand and M. R. Duchen, *J. Neurosci.*, 2007, **27**, 1129.
3. H. Chen, H. Yoshioka, G. S. Kim, J. E. Jung, N. Okami, H. Sakata, C. M. Maier, P. Narasimhan, C. E. Goeders and P. H. Chan, *Antioxid. Redox Signalling*, 2011, **14**, 1505.
4. P. H. Chan, *J.Cereb. Blood Flow Metab.*, 2001, **21**, 2.
5. C. J. Epstein, K. B. Avraham, M. Lovett, S. Smith, O. Elroy-Stein, G. Rotman, C. Bry and Y. Groner, *Proc. Natl. Acad. Sci. U. S. A.*, 1987, **84**, 8044.
6. H. Kinouchi, C. J. Epstein, T. Mizui, E. Carlson, S. F. Chen and P. H. Chan, *Proc. Natl. Acad. Sci. U. S. A.*, 1991, **88**, 11158.

7. K. Murakami, T. Kondo, C. J. Epstein and P. H. Chan, *Stroke*, 1997, **28**, 1797.
8. T. Kondo, A. G. Reaume, T.-T. Huang, E. Carlson, K. Murakami, S. F. Chen, E. K. Hoffman, R. W. Scott, C. J. Epstein and P. H. Chan, *Neuroscience*, 1997, **17**, 4180.
9. B. I. Lee, P. H. Chan and G. W. Kim, *Stroke*, 2005, **36**, 2712, [Erratum: *Stroke*, 2006, **37**, 576].
10. P. H. Chan, H. Kamii, G. Yang, J. Gafni, C. J. Epstein, E. Carlson and L. Reola, *Neuroreport*, 1993, **5**, 293.
11. G. Yang, P. H. Chan, J. Chen, E. Carlson, S. F. Chen, P. Weinstein, C. J. Epstein and H. Kamii, *Stroke*, 1994, **25**, 165.
12. J. N. Keller, M. S. Kindy, F. W. Holtsberg, D. K. St Clair, H.-C. Yen, A. Germeyer, S. M. Steiner, A. J. Bruce-Keller, J. B. Hutchins and M. P. Mattson, *J. Neurosci.*, 1998, **18**, 687.
13. J. E. Jung, G. S. Kim, P. Narasimhan, Y. S. Song and P. H. Chan, *J. Neurosci.*, 2009, **29**, 7003.
14. Y. S. Song, Y.-S. Lee and P. H. Chan, *J.Cereb. Blood Flow Metab.*, 2005, **25**, 1301.
15. T. Nishi, C. M. Maier, T. Hayashi, A. Saito and P. H. Chan, *J. Cereb. Blood Flow Metab.*, 2005, **25**, 1312.
16. C. Nito, H. Kamada, H. Endo, K. Niizuma, D. J. Myer and P. H. Chan, *J.Cereb. Blood Flow Metab.*, 2008, **28**, 1686.
17. S. R. Datta, H. Dudek, X. Tao, S. Masters, H. Fu, Y. Gotoh and M. E. Greenberg, *Cell*, 1997, **91**, 231.
18. A. Saito, T. Hayashi, S. Okuno, T. Nishi and P. H. Chan, *Stroke*, 2006, **37**, 513.
19. P. Narasimhan, J. Liu, Y. S. Song, J. L. Massengale and P. H. Chan, *Stroke*, 2009, **40**, 1467.
20. K. Jian Liu and G. A. Rosenberg, *Free Radical. Biol. Med.*, 2005, **39**, 71.
21. M. V. Gurjar, J. Deleon, R. V. Sharma and R. C. Bhalla, *Am. J. Physiol. Heart Circ. Physiol.*, 2001, **281**, H2568.
22. K. A. Radermacher, K. Wingler, P. W. Kleikers, S. A. Altenhofer, J. R. Hermans, C. Kleinschnitz and H. H. Schmidt, *Exp. Transl. Stroke Med.*, 2012, **4**, 11.
23. D. W. Infanger, R. V. Sharma and R. L. Davisson, *Antioxid. Redox Signalling*, 2006, **8**, 1583.
24. V. E. O'Collins, M. R. Macleod, G. A. Donnan, L. L. Horky, B. H. van der Worp and D. W. Howells, *Ann. Neurol.*, 2006, **59**, 467.
25. C. Kleinschnitz, H. Grund, K. Wingler, M. E. Armitage, E. Jones, M. Mittal, D. Barit, T. Schwarz, C. Geis, P. Kraft, K. Barthel, M. K. Schuhmann, A. M. Herrmann, S. G. Meuth, G. Stoll, S. Meurer, A. Schrewe, L. Becker, V. Gailus-Durner, H. Fuchs, T. Klopstock, M. H. de Angelis, K. Jandeleit-Dahm, A. M. Shah, N. Weissmann and H. H. Schmidt, *PLoS Biol.*, 2010, **8**, e1000479.
26. K. A. Radermacher, K. Wingler, F. Langhauser, S. Altenhöfer, P. Kleikers, J. J. Hermans, M. Hrabě de Angelis, C. Kleinschnitz and H. H. Schmidt, *Antioxid. Redox Signalling*, 2013, **18**, 1418.

27. S. K. Niture, R. Khatri and A. K. Jaiswal, *Free Radical Biol. Med.*, 2014, **66**, 36.
28. T. Suzuki, H. Motohashi and M. Yamamoto, *Trends Pharmacol. Sci.*, 2013, **34**, 340.
29. D. D. Zhang, *Antioxid. Redox Signalling*, 2010, **13**, 1623.
30. Z. A. Shah, R. C. Li, R. K. Thimmulappa, T. W. Kensler, M. Yamamoto, S. Biswal and S. Dore, *Neuroscience*, 2007, **147**, 53.
31. R. Zhang, M. Xu, Y. Wang, F. Xie, G. Zhang and X. Qin, *Mol. Neurobiol.*, 2016, DOI: 10.1007/s12035-016-0111-0.
32. M. Leiros, E. Alonso, J. A. Sanchez, M. E. Rateb, R. Ebel, W. E. Houssen, M. Jaspars, A. Alfonso and L. M. Botana *ACS Chem. Neurosci.*, 2014, **5**, 71.
33. C. Zhang, Z. Zhang, Q. Zhao, X. Wang, H. Ji and Y. Zhang, *Neurotoxicol. Res.*, 2014, **26**, 16.
34. X. Meng, M. Wang, X. Wang, G. Sun, J. Ye, H. Xu and X. Sun, *Free Radical Res.*, 2014, **48**, 823.
35. W. Wang, J. Kang, H. Li, J. Su, J. Wu, Y. Xu, H. Yu, X. Xiang, H. Yi, Y. Lu and L. Sun, *Brain Inj.*, 2013, **27**, 924.
36. S. Jiang, C. Deng, J. Lv, C. Fan, W. Hu, S. Di, X. Yan, Z. Ma, Z. Liang and Y. Yang, *Mol. Neurobiol.*, 2017, **54**, 1440.
37. V. Bocci and G. Valacchi, *Front. Chem.*, 2015, **3**, 4.
38. P. Pantcheva, M. Elias, K. Duncan, C. V. Borlongan, N. Tajiri and Y. Kaneko, *Neural Regen. Res.*, 2014, **9**, 1430.
39. X. Meng, J. Tan, M. Li, S. Song, Y. Miao and Q. Zhang, *Cell Mol. Neurobiol.*, 2017, **37**, 17.
40. M. Basso and R. R. Ratan, *J. Cereb. Blood Flow Metab.*, 2013, **33**, 809.
41. Q. Alhadidi, M. S. Bin Sayeed and Z. A. Shah, *Transl. Stroke Res.*, 2016, **7**, 33.
42. N. N. Danial and S. J. Korsmeyer, *Cell*, 2004, **116**, 205.
43. N. Plesnila, S. Zinkel, S. Amin-Hanjani, J. Qiu, S. J. Korsmeyer and M. A. Moskowitz, *Eur. Surg. Res.*, 2002, **34**, 37.
44. H. Li, F. Colbourne, P. Sun, Z. Zhao, A. M. Buchan and C. Iadecola, *Stroke*, 2000, **31**, 176.
45. Q. H. Guan, D. S. Pei, X. M. Liu, X. T. Wang, T. L. Xu and G. Y. Zhang, *Brain Res.*, 2006, **1092**, 36.
46. Y. Gao, A. P. Signore, W. Yin, G. Cao, X. M. Yin, F. Sun, Y. Luo, S. H. Graham and J. Chen, *J.Cereb. Blood Flow Metab.*, 2005, **25**, 694.
47. M. J. Eliasson, K. Sampei, A. S. Mandir, P. D. Hurn, R. J. Traystman, J. Bao, A. Pieper, Z. Q. Wang, T. M. Dawson, S. H. Snyder and V. L. Dawson, *Nat. Med.*, 1997, **3**, 1089.
48. A. Degterev, Z. Huang, M. Boyce, Y. Li, P. Jagtap, N. Mizushima, G. D. Cuny, T. J. Mitchison, M. A. Moskowitz and J. Yuan, *Nat. Chem. Biol.*, 2005, **1**, 112.
49. X. Teng, A. Degterev, P. Jagtap, X. Xing, S. Choi, R. Denu, J. Yuan and G. D. Cuny, *Bioorg. Med. Chem. Lett.*, 2005, **15**, 5039.

50. M. S. Lopez, R. J. Dempsey and R. Vemuganti, *Neurochem. Int.*, 2015, **89**, 75.
51. J. Ren, C. Fan, N. Chen, J. Huang and Q. Yang, *Neurochem. Res.*, 2011, **36**, 2352.
52. D. M. Hermann, A. Zechariah, B. Kaltwasser, B. Bosche, A. B. Caglayan, E. Kilic and T. R. Doeppner, *Neurobiol. Dis.*, 2015, **83**, 16.
53. D. Clark, U. I. Tuor, R. Thompson, A. Institoris, A. Kulynych, X. Zhang, D. W. Kinniburgh, F. Bari, D. W. Busija and P. A. Barber, *PLoS One*, 2012, **7**, e47792.
54. C. Palmer, J. Towfighi, R. L. Roberts and D. F. Heitjan, *Pediatr. Res.*, 1993, **33**, 405.
55. P. J. Marro, O. P. Mishra and M. Delivoria-Papadopoulos, *Brain Res.*, 2006, **1073**, 444.
56. T. Gunes, M. A. Ozturk, E. Koklu, K. Kose and I. Gunes, *Pediatr. Neurol.*, 2007, **36**, 17.
57. H. L. Torrance, M. J. Benders, J. B. Derks, C. M. Rademaker, A. F. Bos, P. Van Den Berg, M. Longini, G. Buonocore, M. Venegas, H. Baquero, G. H. Visser and F. Van Bel, *Pediatrics*, 2009, **124**, 350.
58. J. A. Singh and S. Yu, *BMC Neurol.*, 2016, **16**, 164.
59. J. J. Kaandorp, M. J. N. L. Benders and C. M. A. Rademaker *et al.*, *BMC Pregnancy Childbirth*, 2010, **10**, 8.
60. ClinicalTrials.gov Identifier: NCT02122718.
61. M. Bourourou, C. Heurteaux and N. Blondeau, *Neuropharmacology*, 2016, **108**, 60.
62. M. Ueda, T. Inaba, C. Nito, N. Kamiya and Y. Katayama, *Brain Res.*, 2013, **1519**, 95.
63. N. Okabe, T. Nakamura, T. Toyoshima, O. Miyamoto, F. Lu and T. Itano, *J. Stroke Cerebrovasc. Dis.*, 2011, **20**, 188.
64. T. N. Eady, L. Khoutorova, K. D. Atkins, N. G. Bazan and L. Belayev, *Exp. Transl. Stroke Med.*, 2012, **4**, 19.
65. T. N. Eady, L. Belayev, L. Khoutorova, K. D. Atkins, C. Zhang and N. G. Bazan, *PLoS One*, 2012, **7**, e46151.
66. T. N. Eady, L. Khoutorova, D. V. Anzola, S. H. Hon, A. Obenaus, A. Mohd-Yusof, N. G. Bazan and L. Belayev, *PLoS One*, 2013, **8**, e77237.
67. K. Mayurasakorn, Z. V. Niatsetskaya, S. A. Sosunov, J. J. Williams, H. Zirpoli, I. Vlasakov, R. J. Deckelbaum and V. S. Ten, *PLoS One*, 2016, **11**, e0160870.
68. D. R. Berman, E. Mozurkewich, Y. Liu, Y. Shangguan, J. D. Barks and F. S. Silverstein, *Neonatology*, 2013, **104**, 71.
69. Z. Zhang, J. Yan, S. Taheri, K. J. Liu and H. Shi, *Free Radical Biol. Med.*, 2014, **68**, 8.
70. M. Jatana, I. Singh, A. K. Singh and D. Jenkins, *Pediatr. Res.*, 2006, **59**, 684.
71. P. Patiño, E. Parada, V. Farré-Alins, S. Molz, R. Cacabelos, J. Marco-Contelles, M. G. López, C. I. Tasca, E. Ramos, A. Romero and J. Egea, *Neurotoxicology*, 2016, **57**, 61.

72. H. Aly, H. Elmahdy, M. El-Dib, M. Rowisha, M. Awny, T. El-Gohary, M. Elbatch, M. Hamisa and A. R. El-Mashad, *J. Perinatol.*, 2015, **35**, 186.
73. S. Sarkar, A. Mukherjee, N. Das and S. Swarnakar, *Exp. Gerontol.*, 2017, **92**, 13.
74. Y. Wang, Z. He and S. Deng, *Drug Des. Devel. Ther.*, 2016, **10**, 1663.
75. H. Ding, H. Wang, L. Zhu and W. Wei, *Neurochem. Res.*, 2017, **42**, 337.
76. L. Zhang, X. Zhang, C. Zhang, X. Bai, J. Zhang, X. Zhao, L. Chen, L. Wang, C. Zhu, L. Cui, R. Chen, T. Zhao and Y. Zhao, *Brain Res.*, 2016, **1636**, 130.
77. S. Wu, Y. Yue, A. Peng, L. Zhang, J. Xiang, X. Cao, H. Ding and S. Yin, *Food Funct.*, 2016, **7**, 2624.
78. D. K. Rieger, E. Navarro, I. Buendia, E. Parada, L. González-Lafuente, R. Leon, A. P. Costa, I. A. Heinrich, K. S. Nascimento, B. S. Cavada, M. G. Lopez, J. Egea and R. B. Leal, *Neurochem. Res.*, 2017, **42**, 347.
79. J. J. Kim, Y. J. Kang, S. A. Shin, D. H. Bak, J. W. Lee, K. B. Lee, Y. C. Yoo, D. K. Kim, B. H. Lee, D. W. Kim, J. Lee, E. K. Jo and J. M. Yuk, *PLoS One*, 2016, **11**, e0163433.
80. G. Cao, N. Jiang, Y. Hu, Y. Zhang, G. Wang, M. Yin, X. Ma, K. Zhou, J. Qi, B. Yu and J. Kou, *Int. J. Mol. Sci.*, 2016, **17**, pii: E1418.
81. F. Yu and C. M. Morshead, *Curr. Stem Cell Res. Ther.*, 2011, **6**, 190.
82. V. Darsalia, T. Kallur and Z. Kokaia, *Eur. J. Neurosci.*, 2007, **26**, 605.
83. D. Woodbury, E. J. Schwarz, D. J. Prockop and I. B. Black, *J. Neurosci. Res.*, 2000, **61**, 364.
84. L. H. Shen, Y. Li, J. Chen, A. Zacharek, Q. Gao, A. Kapke, M. Lu, K. Raginski, P. Vanguri, A. Smith and M. Chopp, *J. Cereb. Blood Flow Metab.*, 2007, **27**, 6.
85. O. Lazarov, M. P. Mattson, D. A. Peterson, S. W. Pimplikar and H. van Praag, *Trends Neurosci.*, 2010, **33**, 569.
86. R. J. Lichtenwalner and J. M. Parent, *J. Cereb. Blood Flow Metab.*, 2006, **26**, 1.
87. L. Ruan, B. Wang, Q. ZhuGe and K. Jin, *Brain Res.*, 2015, **1623**, 166.
88. P. Thored, J. Wood, A. Arvidsson, J. Cammenga, Z. Kokaia and O. Lindvall, *Stroke*, 2007, **38**, 3032.
89. U. Dirnagl, R. P. Simon and J. M. Hallenbeck, *Trends Neurosci.*, 2003, **26**, 248.
90. C. D. Hahn, C. Manlhiot, M. R. Schmidt, T. T. Nielsen and A. N. Redington, *Stroke*, 2011, **42**, 2960.
91. U. Dirnagl, K. Becker and A. Meisel, *Lancet Neurol.*, 2009, **8**, 398.

CHAPTER 7

Identification of MicroRNAs as Targets for Treatment of Ischemic Stroke

CREED M. STARY,[a] JOSH D. BELL,[a,b] JANG EUN CHO[a,c] AND RONA G. GIFFARD*[a]

[a] Stanford University School of Medicine, Dept. Anesthesiology, Perioperative & Pain Medicine, 300 Pasteur Drive, Stanford, CA 94305-5117, USA; [b] University of Toronto School of Medicine, Dept. Anesthesiology, Room 1200, 12th Floor, 123 Edward Street, Toronto, ON M5G 1E2, Canada; [c] Anam Hospital, Korea University College of Medicine, Department of Anesthesiology and Pain Medicine, 73 Inchon-ro, Seongbuk-gu, Seoul, South Korea

*Email: rgiffard@stanford.edu

7.1 Introduction

Ischemic stroke remains a leading cause of death and long-term disability worldwide.[1] Despite hundreds of promising pre-clinical trials in animal models of stroke, the only clinical treatment that remains is early restoration of blood flow with thrombolysis or clot retrieval. Cerebral ischemia research has traditionally focused on developing gene therapy strategies by evaluating and/or targeting specific proteins or the genes that encode those proteins. However, in recent years, our knowledge of the processes that determine whether a DNA protein-coding sequence becomes a functional protein has grown substantially.

Drug Discovery Series No. 62
Acute Brain Impairment: Scientific Discoveries and Translational Research
Edited by Philip V. Peplow, Svetlana A. Dambinova, Thomas A. Gennarelli and Bridget Martinez

Published by the Royal Society of Chemistry, www.rsc.org

Sequencing data from the Human Genome Project suggested that <2% of our genomic DNA codes for proteins, suggesting regulatory roles were possible for the non-protein-coding regions. Since 2003, determining the function of the other >98% of our genome has been the goal of the Encyclopedia of DNA Elements (ENCODE) project. A major ENCODE finding is the observation that ~76% of our genome is transcribed into RNA and is associated with regulation of the expression of coding genes.[2] Thus, the majority of RNA species generated fall into one of several non-translated RNA families, demonstrating that gene regulation is far more complex than originally believed. Clinical therapies targeting the non-coding regions of our genome may help overcome some of the translational hurdles that have impeded the development of effective alternative stroke therapies.

MicroRNAs (miRs) are a large family of small, generally evolutionarily conserved, single-stranded RNAs of about 22 nucleotides that act as guide molecules for the silencing complex to target messenger RNAs (mRNAs). MiRs bind to complementary sequences (typically 6–8 nt long) on the 3′-untranslated region (3′UTR) of mRNAs. This targeted binding results in degradation or translational silencing by the RNA-induced silencing complex (RISC).[3] MiR biogenesis within the nucleus involves multiple steps (Figure 7.1). Long primary miRs (pri-miRs) are first processed to 70 nt hairpin precursor miRs (pre-miRs) and then, eventually, short mature miRs.[3,4] Within the mature miR, a short sequence known as the seed sequence (5–7 nt long) determines the specificity of the miR for a given mRNA strand. However, that specificity is not exclusive, and any given miR generally has multiple mRNA targets to which it binds. Similarly, a given mRNA can be coordinately regulated by multiple miRs, which adds to the complexity of the interactions. Some estimates suggest that >30% of human genes are regulated by miRs, and over 1000 human miRs have been identified.[5] With the successful phase II trial of the first miR-targeted drug, a locked nucleic acid targeting miR-122 to treat hepatitis C,[6] enthusiasm for miR-based treatments in cerebral ischemia is increasing.

7.2 Mechanisms of Neuronal Injury in Ischemic Stroke

Neuronal cell death from cerebral ischemia/reperfusion injury has been described as occurring by several different cell death mechanisms. Necrosis is concentrated in the anoxic core and is generally rapid. In the surrounding peri-ischemic zone, or penumbra,[7,8] delayed cell death can result from apoptosis, either intrinsically, mediated by intracellular signaling and mitochondria, or extrinsically, triggered by binding of death ligands such as tumor necrosis factor-α (TNF-α) or Fas ligand to their receptors,[9] or anoikis, detachment-mediated cell death.[10] The volume of tissue that constitutes the penumbra therefore represents a cell population that could be targeted for rescue from delayed cell death and implies that a temporal window exists in which reversal or prevention of induction of specific forms of cell death can occur.

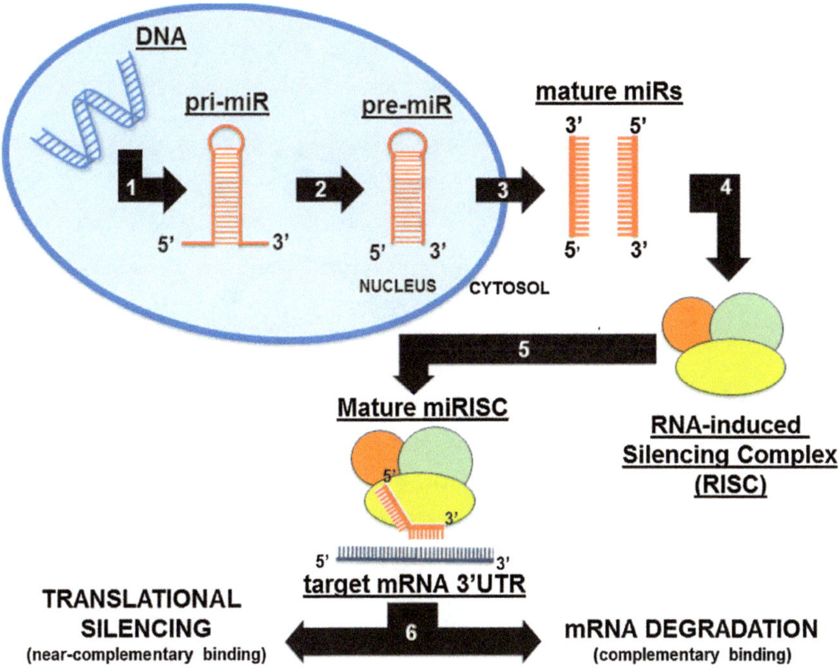

Figure 7.1 MiR biogenesis begins in the nucleus with genomic transcription of pri-miR (1). Drosha-mediated cleavage results in pre-miR (2), which is then exported to the cytosol by Exportin-5 and processed to the final mature miR forms by Dicer (3). In the cytosol, either the leading or reverse complementary mature miR strand then interacts with components of the RISC complex (4) to form the mature miRISC complex (5), which then can bind the 3′UTR of target mRNAs. MiRISC/mRNA complexes are then targeted (6) for either mRNA degradation or translational silencing, depending on the degree of miR/mRNA binding complementarity.

Recently, autophagy, which is a physiological process usually constituting a protective response, has also been found to be a mechanism of cell death.[11–13] Autophagy is an evolutionarily conserved method to eliminate damaged cellular components and to degrade them by fusion with lysosomes. Autophagy may participate in cell death during cerebral ischemia, where it participates in the destruction of accumulated unfolded proteins and stressed endoplasmic reticulum (ER) components.[11] Interestingly, the inhibition of autophagy protects against excitotoxicity and reactive oxygen species (ROS)-mediated cell death in neuronal cells, suggesting that this pathway can contribute to neuronal injury.[12]

Neuronal death after ischemia involves a complex interplay between multiple pathways including excitotoxicity, mitochondrial dysfunction, ionic imbalance, oxidative stress, and inflammation.[14–17] Ischemia leads to ATP depletion and loss of ionic homeostasis, inducing opening of the mitochondrial permeability transition pore, activation of caspases, formation of the apoptosome, and, finally, programmed cell death. As cells are unable to maintain a negative plasma membrane potential, they depolarize,

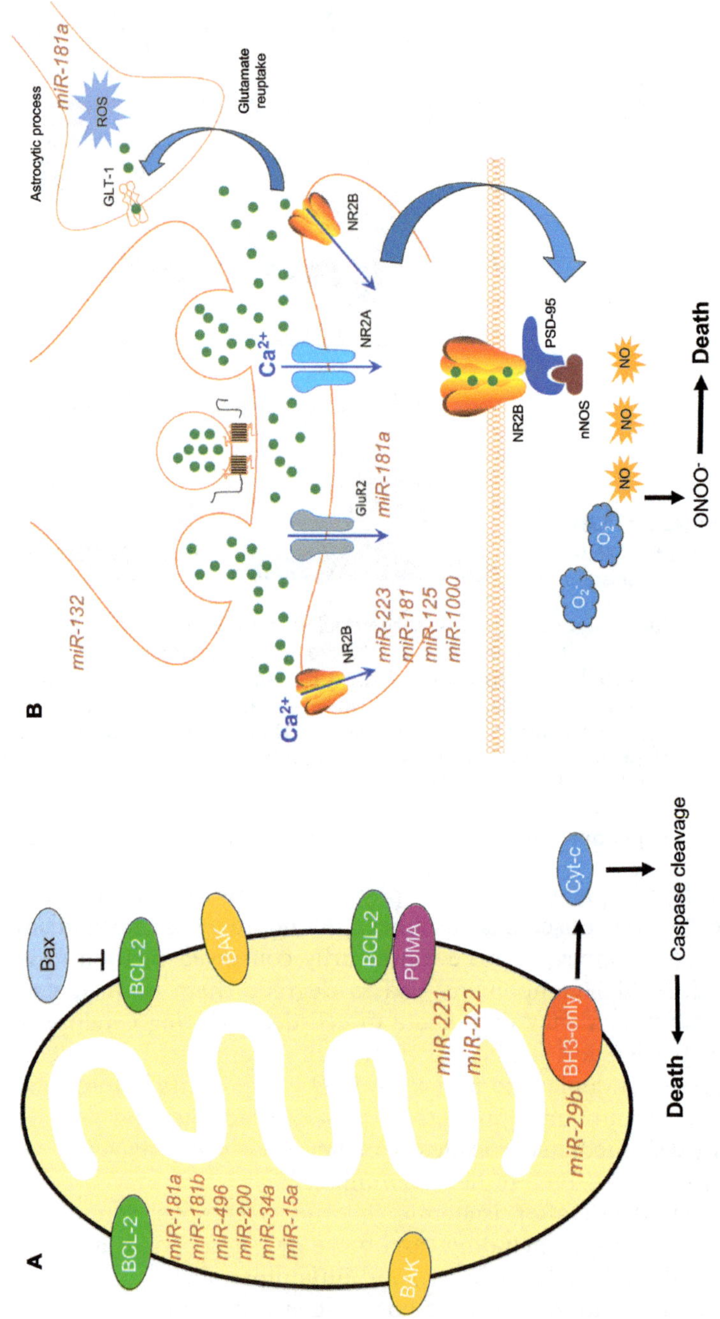
A
BCL-2
BAK
Bax
BCL-2
BAK
BCL-2
PUMA
miR-181a
miR-181b
miR-496
miR-200
miR-34a
miR-15a
miR-221
miR-222
miR-29b
BH3-only
Cyt-c
Death
Caspase cleavage
B
miR-132
Astrocytic process
GLT-1
ROS
miR-181a
Glutamate
reuptake
Ca2+
NR2B
miR-223
miR-181
miR-125
miR-1000
GluR2
miR-181a
Ca2+
NR2A
NR2B
NR2B
PSD-95
nNOS
NO
NO
NO
O2-
O2-
ONOO-
Death

opening voltage-gated calcium channels, permitting unregulated fusion of pre-synaptic vesicles and release of excitatory amino acids into the extracellular space. This cascade of events leads to a massive post-synaptic entry of calcium into the cell. Collectively, this process is known as excitotoxicity, which leads to oxidative stress, proteolysis of the cytoskeleton, calcium-mediated calcium influx, and mitochondrial membrane permeability changes that ultimately end in amplification of the apoptotic cycle.[15]

7.2.1 MiRs and the Regulation of Apoptosis

The B-cell lymphoma 2 (BCL-2) protein family is a major regulator of apoptosis through regulating mitochondrial membrane integrity, function, and apoptotic signaling.[13,18] The BCL-2 protein family consists of three subgroups: pro-survival proteins (BCL-2, BCLxL, BCLw, MCL1, and A1), multi-domain pro-apoptotic proteins BAX and BAK, and the BH3 domain-only pro-apoptotic proteins (BIM, PUMA, BID, BAD, BIK, BMF, HRK, and NOXA).[19,20] Interactions between these three groups of proteins determine whether a stressed cell will undergo apoptosis. Pro-apoptotic pathways are mediated by BH3-only proteins, which are upregulated in response to apoptotic stimuli and transduce the damage signal. BH3-only proteins inhibit anti-apoptotic proteins and activate pro-apoptotic proteins (Figure 7.2A), causing mitochondrial outer membrane permeabilization,

Figure 7.2 MiRs involved in apoptosis and excitotoxicity following cerebral ischemia. (A) Schematic of a mitochondrion. B-cell lymphoma 2 (BCL-2), a pro-survival, anti-apoptotic protein of the outer mitochondrial membrane, is regulated by a number of different miRs, including the miR-181 family, the miR-200 family, and others. As BCL-2 is anti-apoptotic, its downregulation by miRs increases cell death in numerous ischemic models. Therefore, pro-survival strategies have employed miR antagomirs to inhibit miR function, thereby upregulating BCL-2 protein and mitigating apoptotic cell death. The BH3-only pro-apoptotic protein PUMA is similarly regulated by miR-221/222. Other BH3-only proteins and Bax are pro-apoptotic, leading to the release of cytochrome c (Cyt-c), cleavage of caspases, and initiation of apoptosis. (B) Schematic of a glutamatergic synapse and the downstream generation of reactive oxygen and nitrogen species (ROS and RNS). The attenuation of glutamatergic synaptic transmission by miRs protects against excitotoxicity and ROS generation through a variety of mechanisms. Glutamate receptor expression and synaptogenesis is regulated by miR-132, while miR-181a indirectly regulates the expression of astrocytic glutamate transporter GLT-1. Particularly neurotoxic NR2B-containing NMDA receptors are downregulated by miR-223, miR-181, miR-1000, and miR-125, while the AMPA receptor GluR2 subunit, which is present in most glutamatergic receptor complexes, is regulated by miR-181a. MiR-mediated inhibition of excitotoxicity confers neuroprotection by mitigating signaling downstream of the NMDA receptor, which involves PSD-95-coupled generation of NO *via* neuronal NO synthase (nNOS). This couples with mitochondrially derived superoxide to generate the highly reactive peroxynitrite. Cell damage and death result from these ROS and RNS.

cytochrome c release, and activation of caspases to initiate apoptosis.[13,19] It is widely recognized that overexpressing pro-survival BCL-2 family members protects against cerebral ischemia *in vivo* and *in vitro*,[21–23] orchestrated through maintenance of mitochondrial function (for a review, see Ouyang and Giffard[24]). *In vivo* observations suggest decreased BCL-2 and increased BAX and BH3-only proteins in CA1 neurons after global ischemia,[25] a hippocampal sub-region that is selectively vulnerable to even short periods of ischemia. After global ischemia, p53-upregulated modulator of apoptosis (PUMA) expression increases in CA1 neurons, localizes to mitochondria, and binds BCLxL and BAX.[26]

Initial observations in several organs and models of disease indicate that the BCL-2 family is regulated by miRs[27–29] (for a review, see Pileczki *et al.*[30]). MiR-181b, miR-497, and miR-200 modulate multi-drug resistance of human cancer cell lines by targeting BCL-2.[31,32] In cardiomyocytes, experimental data indicate that miR-1 regulates apoptosis following ischemic injury through the post-transcriptional repression of BCL-2.[33] Additionally, abnormal expression of miR-34a may contribute to the pathogenesis of Alzheimer's disease by affecting the expression of BCL-2.[34]

In the setting of cerebral ischemia and related disease models, several miRs have been shown to target BCL-2 family members. MiR-15a was first observed to contribute to the pathogenesis of ischemic vascular injury through direct inhibition of the anti-apoptotic gene *Bcl-2*.[35] MiR-497 was shown to hybridize to the predicted 3′UTR target sites of *Bcl-2* and to increase stress-induced neuronal death by inhibiting BCL-2 translation.[35,36] Repression of miR-497 *in vivo* effectively lowered miR-497 levels, reduced infarct volume, and improved neurologic deficits with a corresponding increase in BCL-2 protein.[35,37] We observed that miR-181a inhibition protected against glucose deprivation (GD)-induced apoptosis, mitochondrial dysfunction, and loss of mitochondrial membrane potential in astrocytes.[38] Notably, survival was associated with increased BCL-2 and MCL1 protein levels. We used a luciferase reporter assay to confirm that BCL-2 and MCL1 are direct targets of miR-181a. A prior report demonstrated a correlation between human miR-181a levels and BCL-2 levels in glioma cells, but did not use a luciferase assay to validate the putative target sequence.[39]

Thus, several different miRs have been shown to target BCL-2 in relevant settings. That miR-181a antagomir treatment is a potentially compelling pro-survival therapy for cerebral ischemia is also indicated by the *in vivo* efficacy of reducing miR-181a levels in rodent models of focal ischemia and forebrain ischemia, with additional targets, including chaperone proteins, also implicated.[38,40] Further, post-treatment with antagomirs to reduce miR-181a levels was still protective in focal ischemia.[41]

7.2.2 Autophagy, Anoikis, and miRs

Autophagy is also regulated in part by miRs, and there is a thorough recent review on this topic.[42] Briefly, many core components of the autophagic pathway are regulated by miRs, including miR-885, miR-181a, miR-375, miR-630, and miR-34a, which target induction, vesicle elongation, vesicle nucleation, and retrieval, respectively.[42] Notably, LC3 (microtubule-associated protein 1 light chain 3/Atg8) is a marker of autophagosome formation in myocardial ischemia, and data suggest that miR-204 targets this protein to protect against autophagic cell death in ischemia–reperfusion injury of cardiac myocytes.[43]

A number of miRs promote sensitivity to anoikis. SIRT1, a histone deacetylase that degrades a critical anoikis-related protein LKB-1, is a direct target of miR-204, and data suggest that overexpression of miR-204 elevated LKB1 expression and increased anoikis sensitivity in gastric cancer cells.[44] Similarly, miR-26a also promotes anoikis in hepatocellular carcinoma cells.[45] Overexpression of miR-31 promoted anoikis in breast cancer cell lines,[46] while miR-451 induced anoikis in human glioma cells.[47] Thus, a potential therapeutic approach in ischemia might be antagomirs to one or more of these miRs, to reduce this cell death process.

7.2.3 Excitotoxicity

Excitotoxicity can be attenuated through a number of mechanisms. In the face of elevated glutamate levels, post-synaptic modifications can render neurons more vulnerable to a given quanta of glutamate. Trafficking and phosphorylation of post-synaptic glutamate receptors is deregulated during and after stroke, potentially altering receptor phenotype and composition.[48–50] Work from several laboratories indicates that specific miRs regulate glutamate receptor expression levels and potentially are neuroprotective by reducing excitotoxicity (Figure 7.2B).[51–54] Many miRs regulate glutamate receptor levels, including miR-223, miR-181, miR-125a, miR-125b, miR-1000, miR-132, and miR-124a.[55,56]

Glutamate receptor-associated trafficking proteins are also thought to undergo miR-mediated repression and regulation in ischemia.[52,55] Altering the composition of glutamate receptors might help to mitigate post-stroke excitotoxicity and its associated sequelae of oxidative stress and apoptosis initiation.

MiR-223 regulates GluR2, an α-amino-3-hydroxy-5-methyl-4-isoxazolepropionic acid (AMPA) receptor subunit and *N*-methyl-D-aspartate receptor subtype 2B (NR2B), an *N*-methyl-D-aspartate (NMDA) receptor subunit. Upregulation of miR-223 reduces vulnerability to *in vitro* excitotoxicity.[51] This study further found that miR-223 was protective against transient global ischemia *in vivo*, while absence of miR-223 conferred vulnerability to excitotoxic injury, global ischemia, and post-ischemic behavioral deficits.[51] NR2B is primarily in extra-synaptic glutamate receptors, where it couples to neuronal nitric oxide (NO) synthase *via* post-synaptic density protein 95 (PSD-95), leading to increased NO production, which combines with superoxide to make highly toxic peroxynitrite.[14,15] Attenuation of NR2B-mediated calcium influx is profoundly neuroprotective, and clinical trials are ongoing using peptides that perturb the post-synaptic interaction between NR2B and PSD-95.[57] Further evidence suggests that miR-223 is clinically relevant and circulating miR-223 is increased in post-stroke patients, and a correlation was observed between increased miR-223 expression levels and infarct volume.[54] In animal models, miR-223 is also upregulated in response to ischemia,[58] suggesting that it may be a viable therapeutic target for the reduction of excitotoxicity and stroke protection.

Other interesting data come from studies investigating the miR-181 family. In addition to their role in downregulating BCL-2 expression, miR-181 family members also appear to target glutamate receptor subunits and proteins involved in synaptic glutamate cycling. Data suggest that miR-181a targets the AMPA receptor subunit GluR2.[59,60] AMPA receptors mediate the majority of excitotoxic neurotransmission in the brain. To that end, we and others have previously demonstrated that aberrant trafficking of the AMPA receptor GluR2 subunit occurs in post-traumatic and post-ischemic neurons.[1,49,51] Now, evidence suggests that miR-181a overexpression is capable of downregulating GluR2 expression (the dominant subunit present in most heteromeric AMPA receptor complexes) and significantly reducing activity at glutamatergic synapses.[53] Post-ischemic or post-traumatic hyperexcitability is a clinical phenomenon often resulting in movement disorders and seizure occurrence, highlighting an interesting role for miRs that are capable of downregulating excitatory neurotransmission, such as miR-181a. It is therefore not surprising that antagomirs to miR-181 have consistently demonstrated neuroprotective effects,[38,40] some of which may come from a reduction of excitotoxicity, in addition to antiapoptotic effects.

Adding to the complexity of glutamate receptor regulation by miRs are data suggesting upregulation rather than downregulation of glutamate receptors by miR-132, and its own regulation by neurotrophic factors. MiR-132 is widely expressed in the brain, is neuron specific, and plays an important role in synaptogenesis and morphological development of neurons.[61,62] Indeed, miR-132 downregulation impairs neuronal development and proper synapse formation. It has previously been demonstrated that brain-derived neurotrophic factor (BDNF) upregulates glutamate receptor expression *via* decreased miR-132.[63] This effect was inhibited by the co-administration of glucocorticoids, suggesting inflammatory molecules may play a role in the downregulation of miR-132 and the upregulation of glutamate receptors. The post-ischemic environment is profoundly pro-inflammatory, and also involves upregulation of trophic factor release from neighboring glial cells. BDNF-associated inhibition of miR-132 promoting the expression of glutamate receptor subunits following ischemia would confer vulnerability to ischemia, suggesting that miR-132 antagomirs might confer neuroprotection from excitotoxicity.

7.3 Heat Shock Proteins and MiRs

A large body of evidence demonstrates that heat shock proteins (HSPs), often referred to as molecular chaperones or stress proteins, are regulated in ischemia and are critical targets of miRs. Molecular chaperones are a functionally related group of proteins that bind unfolded proteins and participate in the folding of proteins, the sequestration of denatured proteins, and the assembly or disassembly of multiprotein complexes. The aggregation of unfolded and partially unfolded proteins contributes to neurodegeneration and cellular dysfunction in a variety of central nervous system (CNS) diseases, including ischemia. The tendency for proteins to aggregate is increased by heat and oxidative stress. HSPs were first identified and named decades ago because they are induced by heat stress. HSPs are grouped into families based on size. The human approximately 70 kDa molecular weight HSP family (HSP70 family) consists of at least 11 members,[64] some of which are cytosolic, others of which are organelle specific. The best known members are the strongly heat-inducible cytosolic form, HSP72, the constitutively expressed HSC70/HSP73/HSC73, the ER form glucose-regulated protein 78 (GRP78) binding immunoglobulin protein (BIP), and the mitochondrial form GRP75/mortalin/mtHSP70.

Many of these isoforms, including HSP72, GRP75, and GRP78, have been shown to provide protection from ischemia in a variety of models, and their mechanism of protection is attributed not only to their ability to sequester unfolded proteins and carry out chaperone functions, but also their ability to inhibit neuroinflammation and programmed cell death (apoptosis).[65–67] Several studies also suggest that GRP78 plays an important role in the regulation of cell death, including both apoptotic Purkinje cell death in the cerebellum[68] and autophagy.[69] The latter process plays a critical role in degrading unnecessary or damaged proteins and organelles following ischemia. Supporting these findings, pharmacologic induction of GRP78 reduced neuronal loss in both forebrain[70] and focal cerebral ischemia.[71] We observed that GRP78 overexpression protects primary cultured astrocytes against ischemic injury *in vitro*,[72] and plays an important role in the protection seen by reducing miR-181 levels in astrocyte injury and mouse stroke.[73]

Protein–protein interactions are stabilized by chaperone proteins, highlighting a role for these proteins in maintaining the integrity of intracellular functional networks. Chaperones are part of an organelle network of particular relevance to ischemia: the interaction between the ER and mitochondria in the regulation of intracellular calcium. Indeed, several chaperones are involved in ER–mitochondrial Ca^{2+} transfer, which is a potent initiator of cell injury following ischemia *via* post-synaptic Ca^{2+} deregulation. The inositol trisphosphate receptor (IP3R) and the ryanodine receptor are both Ca^{2+} release channels in the ER, as is the voltage-dependent anion channel (VDAC) in the mitochondrial outer membrane. HSP70 family members have been shown to modulate the conductance of these Ca^{2+} channels. GRP75 directly interacts with both VDAC and IP3R, playing a central role in scaffolding this ER–mitochondrial complex.[74] Ischemia-like stress or ER stress induces the translocation of GRP78 from the ER to mitochondria and may play a role in ER–mitochondria crosstalk during cerebral ischemia.[75] Chaperone complexes at both the ER and mitochondrion orchestrate the regulation of Ca^{2+} signaling between these two organelles and control bioenergetics, cell survival, and cell death decisions.

Our knowledge of the regulation of molecular chaperones by miRs continues to grow. Levels of miR-1 are increased in ischemic myocardium,[33,76] and two of miR-1's validated targets are HSP60 and HSP70.[77] Also, in cardiac ischemic injury, miR-320 influenced injury by targeting HSP20,[78] as validated by luciferase assay and reciprocal changes in levels of expression of HSP20 with miR-320 overexpression and knockdown in cardiomyocytes. MiRs may also be important secreted factors in ischemic preconditioning. Injection of miRs extracted from the hearts of mice following ischemic preconditioning protected hearts against subsequent ischemia/reperfusion injury, possibly *via* the upregulation of HSP72 and heat shock factor-1 (HSF-1).[79] Recently, we demonstrated that miR-181 regulates GRP78 expression and outcome from cerebral ischemia in a mouse model of transient focal ischemia.[73] Most compelling was our observation of the reciprocal expression of miR-181a and GRP78 protein, found both in the necrotic ischemia core and the salvageable ischemic penumbra. We further observed *in vitro* that miR-181a mimic decreases while antagomir increases GRP78 protein expression.[73] These findings suggest that stabilization of GRP78 by miR-181a inhibition also contributes to the potential use of reducing miR-181 as a novel neuroprotective strategy in stroke.

7.4 MiRs and Glia in Ischemic Stroke

The failure to translate preclinical potential stroke therapies may be due in part to the limited attention paid thus far to glial cells in this setting. Neuron-targeted strategies ignore other cellular targets in the brain that play critical roles in the response to injury, and which may more effectively coordinate inter-cellular signaling cascades that contribute to neuronal injury. Astrocytes comprise the most numerous cell type in the brain and play a crucial role in neuronal homeostasis for both normal physiologic functioning and in response to cell stress.[80] Microglia coordinate growth and remodeling of the neural network and regulate the neuroinflammatory response to stroke.[81,82] MiR-based therapeutic strategies aimed at maintaining glial function and survival following stroke may provide a useful new approach to reducing injury and improving clinical outcomes.

7.4.1 Astrocytes

Neurogenesis, neuronal maintenance, and repair of the injured brain are coordinated by local astrocytes.[83–85] Astrocytes influence blood circulation, extracellular ion homeostasis, and release of energy substrates and growth factors. In addition to their role in neuronal housekeeping and protection, astrocytes play a significant role in neurotransmission.[84,86] Astrocytes are central to synapse formation and stabilization in development and disease[80,87,88] and modulate synaptic transmission *via* glutamate uptake.[89] Astrocytes extend fine-branching processes to contact neuronal cell bodies, dendrites, and synaptic terminals. An individual astrocyte may contact up to 100 000 neurons.[90] Moreover, astrocytes communicate with adjacent astrocytes *via* intercellular gap junctions.[91,92] As a consequence, astrocytes actively regulate and organize local and distant synaptic activity, excitability, transmission, and plasticity of the neuronal network by functioning as a coordinated syncytium.[93–96]

Following cerebral ischemia, astrocytes perform multiple functions that are beneficial for neuronal survival, but have also been suggested to exert negative effects in some settings, such as inhibiting recovery by participating in glial scar formation. Astrocytes protect neurons from glutamate excitotoxicity (see Section 7.2.3) during pathophysiologic stresses such as stroke[97] and traumatic brain injury.[98] The astrocytic syncytium may also influence neuronal survival by coordinating delivery of metabolic fuels, thereby maintaining both mitochondrial and cellular integrity. Gap junctions are permeable to both glucose and lactate,[99] regulate post-injury edema,[100] and have the potential to facilitate delivery of substrates to metabolically active neurons in local areas of decreased perfusion. Astrocytes, like microglia, can show different patterns of activation or reactivity. A novel mechanism for support of neuronal function was reported recently: the transfer of functional mitochondria from astrocytes to neurons in the setting of stroke.[101] We profiled gene expression from astrocytes isolated from brains of mice subjected to either systemic lipopolysaccharide injection or focal ischemia/reperfusion by middle cerebral artery occlusion and found that they differed.[102] A recent report details how microglia can interact with astrocytes to induce a neurodestructive phenotype.[103]

Microarray analysis of miR expression patterns in the principal cell types of the CNS delineate a preferential cellular expression pattern of individual miRs.[104] Of note, miR-181a and miR-29a appear to be more highly expressed in astrocytes, corroborating both our[105] and others'[106] observations. A review by our group[107] outlines the effects of these astrocyte-enriched miRs. MiR-181a was shown to directly target GRP78,[38] anti-apoptotic members of the BCL-2 family, BCL-2 and MCL-1,[108] and X-linked inhibitor of apoptosis (XIAP),[106] as well as additional targets involved in controlling mitochondrial function, redox state, and inflammatory pathways (for a recent review, see Ouyang *et al.*[109]). Consistent with the observation that miR-181a is expressed at greater levels in astrocytes, targeted reduction of miR-181a increased BCL-2 and increased survival of primary astrocytes,[108] while in primary neurons it failed to significantly change levels of BCL-2 and did not improve survival after ischemia-like injury.[40] The difference in effects of miR-181a suppression between different brain cell types may be the result of differences in baseline levels of expression or changes in expression in response to ischemia, but highlight the cell-specific nature of miR effects.

Luciferase assays indicate that the miR-29 family targets both pro- and anti-apoptotic BCL-2 family members.[107] While downregulation of miR-29 protected hearts against ischemia/reperfusion injury,[110] upregulation of miR-29a protected neurons from apoptosis during both global[105] and focal cerebral ischemia.[111] In the setting of *in vitro* ischemia, we observed that miR-29a mimic protected and miR-29a inhibitor aggravated astrocyte injury and mitochondrial function by targeting the BCL-2 family member PUMA[105] and the mitochondrial anion channel VDAC1.[112] Our results demonstrating targeting of several BCL-2 family members strongly suggest that the reported pro-apoptotic and anti-apoptotic effects of the miR-29 family likely reflect inhibition of different targets in different cells, dependent on the specific physiological or pathological setting. Indeed, verification of predicted miR targets by luciferase assay reflects only the *potential* to target *in vivo*; cell type-specific actions appear to play a substantial role in the physiological activity of individual miRs and also require validation. However, cell-type specificity[113] reveals an additional level of endogenous regulation that may be exploited for cell type-specific therapies aimed at reducing cerebral ischemic injury.

7.4.2 Microglia

Microglia constitute 10–15% of all cells in the brain and play an important role in neuronal migration, axonal growth, synaptic remodeling, and response to ischemic injury (for reviews, see Yenari *et al.*, Kim *et al.*, and Benakis *et al.*[81,82,114]). Microglia share a common myeloid lineage with monocytes and macrophages, and similarly act as the primary form of tissue immune defense *via* pathogen and danger-associated pattern recognition, phagocytosis of damaged cells, inactive synapses, debris, and infectious agents, regulation of T-cell responses, and induction of inflammation. Under normal physiological conditions, microglia exist in a "resting" or "surveilling" state. Cerebral ischemia induces microglial activation,[114] characterized by a change from a ramified to an amoeboid shape, loss of branching processes, and production of lysosomes and phagosomes. Microglial activation in response to stroke is a polarized process, with phenotypes lying along a spectrum from M1 or pro-inflammatory activation to M2 or anti-inflammatory or recovery and repair activation states (for a review, see Kim *et al.*[82]).

M1 or "classical" activation triggers the production of pro-inflammatory factors such as TNF-α, interleukin-1β (IL-1β), NO, and ROS, which, in excess, can exacerbate brain injury. TNF-α is a critical pro-inflammatory cytokine released from M1 microglial cells following ischemia that serves as an activator of receptor-mediated pro-apoptotic pathways within the neuron, and can further stimulate microglia *via* inducible NO synthase (iNOS) and cyclooxygenase 2.[115] Activation of the M2 phenotype by IL-4, IL-10, and/or IL-13 induces surface receptor expression of several distinct markers, including arginase, heparin-binding lectin Ym-1, CD206, and CD36. M2 polarization may also result in a greater capacity for phagocytosis.[116]

A review by Parisi *et al.*[117] outlines the role that miRs may play in determining microglial activation polarization. They identify miR-125b as an activator of M1 activation mediated by TNF-α in a mouse model of amyotrophic lateral sclerosis. In cerebral ischemia, observations suggest that the time course of polarization and the relative abundance of the two phenotypes depend on the severity, location, and duration of ischemia and reperfusion.[114] Microglial activation can depend on astrocytic release of ATP in response to local injury, suggesting that astrocyte mitochondrial function plays a direct role in microglial activation state.[97] Therefore, miR-based therapeutic interventions targeting mechanisms that mediate crosstalk between astrocytes and microglia may provide an alternative approach to simultaneously coordinating both glial mitochondrial function and microglial and astrocyte activation polarization. In this context, miR-181c, another member of the miR-181 family, was identified as directly targeting TNF-α following ischemia, thereby regulating microglial activation and microglial-mediated neuronal injury.[118] Zhang *et al.* observed that ectopic overexpression of miR-181c suppressed expression of iNOS, leading to decreased production of NO following combined oxygen GD (OGD).[118] More recently, the same group demonstrated that the microglial activator Toll-like receptor 4 (TLR4) was a target of miR-181c in microglia.[119] Overexpression of miR-181c inhibited NF-κB activation induced by OGD and the downstream production of pro-inflammatory mediators by suppressing TLR4 expression.

7.5 Clinical Utility of MiRs in Ischemic Stroke

Studies targeting miRs for therapies in cancer have established that circulating miRs are largely stable in blood,[120] indicating that miRs likely represent a sampling of cellular activity, a route of intercellular communication, and a biomarker for cerebral injury. High-throughput profiling by microarray analysis has provided an avenue to explore post-injury changes in miR expression profiles in order to identify potential therapeutic targets. Comprehensive reviews of miR profiling following cerebral ischemia have recently been presented by Chandran *et al.*[121] and Martinez and Peplow.[122] Summaries of relevant pre-clinical and clinical post-stroke miR profiling studies are presented in Tables 7.1 and 7.2, respectively.

Several principal studies have provided a foundation for the use of miRs as biomarkers for stroke. Tan *et al.*[123] conducted the first clinical analysis of circulating miRs in a cohort of post-injury stroke patients. Among the 157 miRs identified, 138 were upregulated and 19 were downregulated. Among the overexpressed miRs, 17 were associated with large-vessel atherosclerosis, small-vessel disease, and cardioembolism. The miR profile of patients with small artery stroke showed a distinctly different pattern from that of the large-artery stroke patients, suggesting for the first time that the mechanism of injury plays a role in determining the miR expression profile. A subsequent study by Zeng *et al.*[124] identified miR-210 in blood from ischemic stroke patients, increasing at 3, 7, and 14 days after stroke. In contrast to the findings by Tan *et al.*, the authors observed no changes in miR-210 among large-artery atherosclerotic stroke, cardioembolic stroke,

Table 7.1 MiR profiles in human profiling studies.

Author	Subjects and age	Time of sampling	MiRs increased	MiRs decreased
Yang *et al.*[130]	61 ± 11 years; 68% male ischemic stroke patients 56 ± 4 years; 60% male healthy volunteers	24 hours	miR-107, miR-128b, miR-153	
Sepramaniam *et al.*[131]	60 ± 1 years	1–7 days of stroke 2–24 months from stroke onset	miR-125b-2, miR-27a, miR-422a, miR-488, miR-627	
He *et al.*[132]	2 men and 3 women, 50–83 years of age	Not indicated	13 miRs, including miR-145	3 miRs, including miR-122
Tan *et al.*[123]	18–49 years	6–18 months from index stroke	138 miRs (upregulated, fold change >1.0)	19 miRs (downregulated, fold change <1.0)
Tan *et al.*[133]	18–49 years	2–24 months from stroke onset	miR-25, miR-34b, miR-483-5p, miR-498	
Zeng *et al.*[124]	Ischemic stroke patients	Within 3, 7, and 14 days after stroke	miR-210	
Li *et al.*[134]	≥50 years			miR-146a, miR-185
Gan *et al.*[135]	18–49 years	Not indicated		
Jickling *et al.*[125]	63 years, 50% female	28 hours from stroke onset	miR-363, miR-487b	miR-122, miR-148a, let-7i, miR-19a, miR-320d, miR-4429
Jia *et al.*[136]	146 acute ischemic stroke patients, 96 controls	Within 24 hours of stroke onset of control group	Serum miR-145	Serum miR-23a, miR-221

Table 7.2 MiR profiles in a rodent model of stroke.

Author	Total miRs tested	Time point	Numbers of miRs changed in more than one study: Same changes (including both)	Only opposite changes	Both	Increased miRs	Decreased miRs
Dharap *et al.*[137]	265	MCAO 6–24–72 hours	18	10	3	miR-374, miR-98, miR-340-5p, miR-21, miR-352, miR-379, miR-335, miR-181b, miR-26b, miR-15b, miR-146a	miR-466c, miR-292-5p, miR-328, miR-873, miR-494, let-7d, miR-345-5p, miR-30c-2, miR-322
						MiRs altered only at 6 hours after preconditioning (PC)	
						miR-384-3p, miR-30e, miR-153, miR-27a, miR-27b, miR-204, miR-539, miR-376b-3p, miR-376b-5p, miR-30a, miR-137, miR-382, miR-384-5p	miR-433, miR-331, miR-99b, miR-760-3p, miR-324-3p, miR-92a, miR-485, miR-129, miR-140 miR-30d, miR-425, miR-320, miR-674-3p, miR-330, miR-145
Lee *et al.*[127]		MCAO 3 hours	5	1	5	miR-200a, miR-200b, miR-200c, miR-141, miR-429, miR-182, miR-183, miR-96	
Lusardi *et al.*[138]	489	MCAO 24 hours	37	21	9	180 miRs	93 miRs
Peng *et al.*[139]		MCAO 24 hours	2		1	miR-30a	miR-181b
Sun *et al.*[140]	1067	tMCAO Sham control 1–7 days to 1–6 months	2	3	1	mmu-miR-15a-5p, sha-miR-24, oan-let-7b-3p, mmu-miR-125b-5p, mmu-miR-132-5p, mmu-miR-181c-5p, mmu-miR-378a-5p	
Shin *et al.*[141]		Ischemic brain extract, *in vitro*	2			miR-223	miR-27b
Jeyaseelan *et al.*[58]		Within 24 hours of reperfusion				miR-19b, miR-290, miR-292-5p, miR-138, miR-145	miR-103, miR-107
		48 hours				miR-150, miR-195, miR-320, let-7 family	miR-103, miR-107, miR-191
Liu *et al.*[85]		24 hours after MCAO					miR-122

and small-artery stroke. However, when stratified by the modified Rankin Scale (mRS; a commonly used scale for measuring the degree of disability or dependence in daily activities), miR-210 levels in stroke patients with good outcomes (mRS $\leq$2) were significantly higher than in patients with poorer mRS scores. A study by Jickling *et al.*[125] of 48 patients found miR-122, -148a, -19a, and -320d and let-7i were decreased while miR-363 and -487b were increased in peripheral blood cells. These findings demonstrate a high degree of complexity in detailing the clinical miR response to ischemic injury.

Ischemic stroke is most commonly modeled in rodents by occlusion of the middle cerebral artery, the cerebral blood vessel that is most frequently occluded in humans. Jeyaseelan *et al.*[58] observed in a male rat model of transient middle cerebral artery occlusion (tMCAO) an increase in miR-19b, miR-290, and miR-292-5p in peripheral blood at 24 hours of reperfusion, whereas miR-103 and miR-107 were decreased. Dharap *et al.*[126] profiled miRs in the brains of adult male spontaneously hypertensive rats after tMCAO. Of the 238 miRs evaluated, 8 were increased and 12 were decreased. They then built upon their observations and tested whether a predicted downstream target of miR-145 identified by *in silico* complementarity analysis was modulated by miR-145 levels. They found that inhibition by miR-145 antagomir led to an increased protein expression of superoxide dismutase-2 (SOD2), an endogenous antioxidant and cytoprotective protein. This observation validated the prediction that stroke-mediated miR upregulation could decrease translation of target proteins, and that miRs could indeed play a role in injury outcomes from stroke.

In a rodent model, Lee *et al.*[127] observed that even short periods of ischemia followed by reperfusion could induce miRs that permanent occlusion ischemia did not, suggesting that the reperfusion event plays a role in miR induction. The group then observed that overexpression of the reperfusion-induced miR-200c was protective in neuronal N2a cells subjected to OGD injury, a combined substrate deprivation plus oxidative stress injury model. In contrast, we demonstrated in the same N2a neuronal cell line that *decreasing* levels of miR-200c were protective against pure oxidative injury.[128] We further observed a protective effect of miR-200c inhibition in a mouse model of tMCAO. These observations highlight the consideration that the reperfusion event (as occurs during tMCAO) may also be a significant factor in the role miRs play in the evolution of injury from stroke, a factor that should be accounted for when modeling miR expression profiles following cerebral ischemia.

Liu *et al.*[129] made an additional important observation in rat MCAO. They observed decreased circulating levels of miR-122 after MCAO, and subsequently demonstrated that elevating levels of miR-122 with mimic by intravenous administration provided protection (decreased infarct volume and improved neurologic score), attenuated intracellular cell adhesion molecule-1 (ICAM-1) expression, and maintained blood vessel integrity. Treatment with intravenous miR-122 mimic also downregulated multiple genes predicted to affect cell adhesion, diapedesis, leukocyte extravasation, and eicosanoid and atherosclerosis signaling. These observations support the prediction that modulating a single miR identified in a profiling array can have a simultaneous effect on several related downstream targets and can validate miRs as a relevant clinical tool for stroke therapy. Extending the clinical utility of miR-based therapies, our group utilized a post-treatment intravenous strategy to demonstrate that miR-181a antagonism can reduce injury even when given after MCAO.[41] Future investigations optimizing both delivery and dosing strategies of miR-based therapies for cerebral ischemia are warranted.

7.6 Conclusions

The noncoding 98% of our genome represents a relevant target for efforts to overcome the translational hurdles that have impeded the development of an effective alternative stroke therapy. MiR-based strategies represent a promising new frontier. The short miR complementary sequence determines the binding specificity to target mRNAs. As noted, a single miR can bind to multiple mRNAs, and a single mRNA can be bound by multiple miRs. This arrangement creates a new and complex layer of post-transcriptional control. Identifying miRs that target multiple genes and gene families may produce a substantial effect *versus* single-gene silencing or overexpression approaches. However, identifying high-yield target miRs remains a challenge, given the variability in expression observed in different types of ischemic injury. Additionally, delineating off-target effects of miRs with high clinical utility is another important goal of pre-clinical studies. Alternative cellular targets, such as astrocytes and microglia, are also likely to provide fruitful targets for the development of an effective adjuvant stroke therapy.

Acknowledgements

This work was supported by American Heart Association grant FTF-19970029 to CMS and NIH grants NS084396 and NS080177 to RGG. The authors thank Chynna Villanueva for assistance with manuscript preparation.

References

1. V. Roger, A. Go, D. Lloyd-Jones, R. Adams, J. Berry, T. Brown, M. Carnethon, S. Dai, G. de Simone, E. Ford, C. Fox, H. Fullerton, C. Gillespie, K. Greenlund, S. Hailpern, J. Heit, P. Ho, V. Howard, B. Kissela, S. Kittner, D. Lackland, J. Lichtman, L. Lisabeth, D. Makuc, G. Marcus, A. Marelli, D. Matchar, M. McDermott, J. Meigs, C. Moy, D. Mozaffarian, M. Mussolino, G. Nichol, N. Paynter, W. Rosamond, P. Sorlie, R. Stafford, T. Turan, M. Turner, N. Wong and J. Wylie-Rosett, *Circulation*, 2011, **123**, e18–e209.
2. P. C. ENCODE, *Nature*, 2012, **489**, 57–74.
3. V. N. Kim, *Nat. Rev. Mol. Cell Biol.*, 2005, **6**, 376–385.

4. J. A. Makarova, M. U. Shkurnikov, D. Wicklein, T. Lange, T. R. Samatov, A. A. Turchinovich and A. G. Tonevitsky, *Prog. Histochem. Cytochem.*, 2016, **51**, 33–49.
5. R. C. Friedman, K. K. Farh, C. B. Burge and D. P. Bartel, *Genome Res.*, 2009, **19**, 92–105.
6. H. Janssen, S. Kauppinen and M. Hodges, *N. Engl J. Med.*, 2013, **369**(9), 878.
7. M. Chopp and Y. Li, *Acta Neurochir. Suppl.*, 1996, **66**, 21–26.
8. M. P. Mattson, C. Culmsee and Z. F. Yu, *Cell Tissue Res.*, 2000, **301**, 173–187.
9. D. A. Dawson, D. Martin and J. M. Hallenbeck, *Neurosci. Lett.*, 1996, **218**, 41–44.
10. S. R. Lee, K. Tsuji, S. R. Lee and E. H. Lo, *J. Neurosci.*, 2004, **24**, 671–678.
11. B. Gabryel, A. Kost and D. Kasprowska, *Pharmacol. Rep.*, 2012, **64**, 1–15.
12. C. Kubota, S. Torii, N. Hou, N. Saito, Y. Yoshimoto, H. Imai and T. Takeuchi, *J. Biol. Chem.*, 2010, **285**, 667–674.
13. D. R. Green and F. Llambi, *Cold Spring Harb. Perspect Biol.*, 2015, **7**, a006080.
14. M. M. Aarts and M. Tymianski, *Curr. Mol. Med.*, 2004, **4**, 137–147.
15. M. Arundine and M. Tymianski, *Cell Mol. Life Sci.*, 2004, **61**, 657–668.
16. M. A. Moskowitz, E. H. Lo and C. Iadecola, *Neuron*, 2010, **67**, 181–198.
17. K. P. Doyle, R. P. Simon and M. P. Stenzel-Poore, *Neuropharmacology*, 2008, **55**, 310–318.
18. J. M. Adams and S. Cory, *Curr. Opin. Immunol.*, 2007, **19**, 488–496.
19. D. Brenner and T. W. Mak, *Curr. Opin. Cell Biol.*, 2009, **21**, 871–877.
20. R. J. Youle and A. Strasser, *Nat. Rev. Mol. Cell Biol.*, 2008, **9**, 47–59.
21. K. Kitagawa, M. Matsumoto, Y. Tsujimoto, T. Ohtsuki, K. Kuwabara, K. Matsushita, G. Yang, H. Tanabe, J. C. Martinou, M. Hori and T. Yanagihara, *Stroke*, 1998, **29**, 2616–2621.
22. L. Xu, J. E. Lee and R. G. Giffard, *Neurosci. Lett.*, 1999, **277**, 193–197.
23. H. Zhao, M. A. Yenari, D. Cheng, R. M. Sapolsky and G. K. Steinberg, *J. Neurochem.*, 2003, **85**, 1026–1036.
24. Y. B. Ouyang and R. G. Giffard, *Cell Calcium*, 2004, **36**, 303–311.
25. G. Martinez, G. Musumeci, C. Loreto and M. L. Carnazza, *J. Mol. Histol.*, 2007, **38**, 295–302.
26. K. Niizuma, H. Endo, C. Nito, D. J. Myer and P. H. Chan, *Stroke*, 2009, **40**, 618–625.
27. L. Liu, X. Yu, X. Guo, Z. Tian, M. Su, Y. Long, C. Huang, F. Zhou, M. Liu, X. Wu and X. Wang, *Mol. Med. Rep.*, 2012, **5**, 753–760.
28. A. Cimmino, G. A. Calin, M. Fabbri, M. V. Iorio, M. Ferracin, M. Shimizu, S. E. Wojcik, R. I. Aqeilan, S. Zupo, M. Dono, L. Rassenti, H. Alder, S. Volinia, C. G. Liu, T. J. Kipps, M. Negrini and C. M. Croce, *Proc. Natl. Acad. Sci. U. S. A.*, 2005, **102**, 13944–13949.
29. R. Singh and N. Saini, *J. Cell Sci.*, 2012, **125**, 1568–1578.
30. V. Pileczki, R. Cojocneanu-Petric, M. Maralani, I. B. Neagoe and R. Sandulescu, *Clujul Med.*, 2016, **89**, 50–55.

31. W. Zhu, X. Shan, T. Wang, Y. Shu and P. Liu, *Int. J. Cancer*, 2010, **127**, 2520–2529.
32. W. Zhu, H. Xu, D. Zhu, H. Zhi, T. Wang, J. Wang, B. Jiang, Y. Shu and P. Liu, *Cancer Chemother. Pharmacol.*, 2012, **69**, 723–731.
33. Y. Tang, J. Zheng, Y. Sun, Z. Wu, Z. Liu and G. Huang, *Int. Heart J.*, 2009, **50**, 377–387.
34. X. Wang, P. Liu, H. Zhu, Y. Xu, C. Ma, X. Dai, L. Huang, Y. Liu, L. Zhang and C. Qin, *Brain Res. Bull.*, 2009, **80**, 268–273.
35. K. J. Yin, Z. Deng, M. Hamblin, Y. Xiang, H. Huang, J. Zhang, X. Jiang, Y. Wang and Y. E. Chen, *J. Neurosci.*, 2010, **30**, 6398–6408.
36. S. Yadav, A. Pandey, A. Shukla, S. S. Talwelkar, A. Kumar, A. B. Pant and D. Parmar, *J. Biol. Chem.*, 2011, **286**, 37347–37357.
37. K. J. Yin, Z. Deng, H. Huang, M. Hamblin, C. Xie, J. Zhang and Y. E. Chen, *Neurobiol. Dis.*, 2010, **38**, 17–26.
38. Y. Ouyang, Y. Lu, S. Yue, L. Xu, X. Xiong, R. White, X. Sun and R. Giffard, *Neurobiol. Dis.*, 2012, **45**, 555–563.
39. G. Chen, W. Zhu, D. Shi, L. Lv, C. Zhang, P. Liu and W. Hu, *Oncol. Rep.*, 2010, **23**, 997–1003.
40. J. Moon, L. Xu and R. Giffard, *J. Cereb. Blood Flow Metab.*, 2013, **33**, 1976–1982.
41. L. J. Xu, Y. B. Ouyang, X. Xiong, C. M. Stary and R. G. Giffard, *Exp. Neurol.*, 2015, **264**, 1–7.
42. L. B. Frankel and A. H. Lund, *Carcinogenesis*, 2012, **33**, 2018–2025.
43. J. Xiao, X. Zhu, B. He, Y. Zhang, B. Kang, Z. Wang and X. Ni, *J. Biomed. Sci.*, 2011, **18**, 35.
44. L. Zhang, X. Wang and P. Chen, *BMC Cancer*, 2013, **13**, 290.
45. X. Zhang, S. L. Cheng, K. Bian, L. Wang, X. Zhang, B. Yan, L. T. Jia, J. Zhao, N. Gammoh, A. G. Yang and R. Zhang, *Oncotarget*, 2015, **6**, 2277–2289.
46. S. Valastyan, F. Reinhardt, N. Benaich, D. Calogrias, A. M. Szasz, Z. C. Wang, J. E. Brock, A. L. Richardson and R. A. Weinberg, *Cell*, 2009, **137**, 1032–1046.
47. Y. Nan, L. Han, A. Zhang, G. Wang, Z. Jia, Y. Yang, X. Yue, P. Pu, Y. Zhong and C. Kang, *Brain Res.*, 2010, **1359**, 14–21.
48. I. Alim, L. Teves, R. Li, Y. Mori and M. Tymianski, *J. Neurosci.*, 2013, **33**, 17264–17277.
49. J. D. Bell, J. Ai, Y. Chen and A. J. Baker, *Brain*, 2007, **130**, 2528–2542.
50. J. D. Bell, E. Park, J. Ai and A. J. Baker, *Cell Death Differ.*, 2009, **16**, 1665–1680.
51. M. M. Harraz, S. M. Eacker, X. Wang, T. M. Dawson and V. L. Dawson, *Proc. Natl. Acad. Sci. U. S. A.*, 2012, **109**, 18962–18967.
52. J. Karr, V. Vagin, K. Chen, S. Ganesan, O. Olenkina, V. Gvozdev and D. E. Featherstone, *J. Cell Biol.*, 2009, **185**, 685–697.
53. R. Saba, P. H. Storchel, A. Aksoy-Aksel, F. Kepura, G. Lippi, T. D. Plant and G. M. Schratt, *Mol. Cell Biol.*, 2012, **32**, 619–632.

54. Y. Wang, Y. Zhang, J. Huang, X. Chen, X. Gu, Y. Wang, L. Zeng and G. Y. Yang, *BMC Neurol.*, 2014, **14**, 77.
55. P. Kaur, F. Liu, J. R. Tan, K. Y. Lim, S. Sepramaniam, D. S. Karolina, A. Armugam and K. Jeyaseelan, *Brain Sci.*, 2013, **3**, 360–395.
56. A. Majdi, J. Mahmoudi, S. Sadigh-Eteghad, M. Farhoudi and S. S. Shotorbani, *Neurol. Sci.*, 2016, **37**, 1765–1771.
57. M. D. Hill, R. H. Martin, D. Mikulis, J. H. Wong, F. L. Silver, K. G. Terbrugge, G. Milot, W. M. Clark, R. L. Macdonald, M. E. Kelly, M. Boulton, I. Fleetwood, C. McDougall, T. Gunnarsson, M. Chow, C. Lum, R. Dodd, J. Poublanc, T. Krings, A. M. Demchuk, M. Goyal, R. Anderson, J. Bishop, D. Garman, M. Tymianski and E. T. investigators, *Lancet Neurol.*, 2012, **11**, 942–950.
58. K. Jeyaseelan, K. Y. Lim and A. Armugam, *Stroke*, 2008, **39**, 959–966.
59. P. Follert, H. Cremer and C. Beclin, *Front. Mol. Neurosci.*, 2014, **7**, 5.
60. L. Zhao, C. Sun, L. Xiong, Y. Yang, Y. Gao, L. Wang, H. Zuo, X. Xu, J. Dong, H. Zhou and R. Peng, *J. Mol. Neurosci.*, 2014, **53**, 222–230.
61. A. S. Nudelman, D. P. DiRocco, T. J. Lambert, M. G. Garelick, J. Le, N. M. Nathanson and D. R. Storm, *Hippocampus*, 2010, **20**, 492–498.
62. B. W. Luikart, A. L. Bensen, E. K. Washburn, J. V. Perederiy, K. G. Su, Y. Li, S. G. Kernie, L. F. Parada and G. L. Westbrook, *PLoS One*, 2011, **6**, e19077.
63. H. Kawashima, T. Numakawa, E. Kumamaru, N. Adachi, H. Mizuno, M. Ninomiya, H. Kunugi and K. Hashido, *Neuroscience*, 2010, **165**, 1301–1311.
64. M. Tavaria, T. Gabriele, I. Kola and R. L. Anderson, *Cell Stress Chaperones*, 1996, **1**, 23–28.
65. R. G. Giffard, R. Q. Han, J. F. Emery, M. Duan and J. F. Pittet, *Anesthesiology*, 2008, **109**, 339–348.
66. R. G. Giffard, L. Xu, H. Zhao, W. Carrico, Y. Ouyang, Y. Qiao, R. Sapolsky, G. Steinberg, B. Hu and M. A. Yenari, *J. Exp. Biol.*, 2004, **207**, 3213–3220.
67. M. A. Yenari, J. Liu, Z. Zheng, Z. S. Vexler, J. E. Lee and R. G. Giffard, *Ann. N. Y. Acad. Sci.*, 2005, **1053**, 74–83.
68. M. Wang, R. Ye, E. Barron, P. Baumeister, C. Mao, S. Luo, Y. Fu, B. Luo, L. Dubeau, D. R. Hinton and A. S. Lee, *Cell Death Differ.*, 2010, **17**, 488–498.
69. J. Li, M. Ni, B. Lee, E. Barron, D. R. Hinton and A. S. Lee, *Cell Death Differ.*, 2008, **15**, 1460–1471.
70. Y. Oida, H. Izuta, A. Oyagi, M. Shimazawa, T. Kudo, K. Imaizumi and H. Hara, *Brain Res.*, 2008, **1208**, 217–224.
71. T. Kudo, S. Kanemoto, H. Hara, N. Morimoto, T. Morihara, R. Kimura, T. Tabira, K. Imaizumi and M. Takeda, *Cell Death Differ.*, 2008, **15**, 364–375.
72. Y. B. Ouyang, L. J. Xu, J. F. Emery, A. S. Lee and R. G. Giffard, *Mitochondrion*, 2011, **11**, 279–286.

73. Y. B. Ouyang, Y. Lu, S. Yue, L. J. Xu, X. X. Xiong, R. E. White, X. Sun and R. G. Giffard, *Neurobiol. Dis.*, 2012, **45**, 555–563.
74. G. Szabadkai, K. Bianchi, P. Varnai, D. De Stefani, M. R. Wieckowski, D. Cavagna, A. I. Nagy, T. Balla and R. Rizzuto, *J. Cell Biol.*, 2006, **175**, 901–911.
75. Y. B. Ouyang and R. G. Giffard, *Int. J. Cell Biol.*, 2012, **2012**, 1–8.
76. Z. X. Shan, Q. X. Lin, Y. H. Fu, C. Y. Deng, Z. L. Zhou, J. N. Zhu, X. Y. Liu, Y. Y. Zhang, Y. Li, S. G. Lin and X. Y. Yu, *Biochem. Biophys. Res. Commun.*, 2009, **381**, 597–601.
77. C. Xu, Y. Lu, Z. Pan, W. Chu, X. Luo, H. Lin, J. Xiao, H. Shan, Z. Wang and B. Yang, *J. Cell Sci.*, 2007, **120**, 3045–3052.
78. X. P. Ren, J. Wu, X. Wang, M. A. Sartor, J. Qian, K. Jones, P. Nicolaou, T. J. Pritchard and G. C. Fan, *Circulation*, 2009, **119**, 2357–2366.
79. C. Yin, F. N. Salloum and R. C. Kukreja, *Circ. Res.*, 2009, **104**, 572–575.
80. C. Eroglu and B. Barres, *Nature*, 2010, **468**, 223–231.
81. M. Yenari, T. Kauppinen and R. Swanson, *Neurotherapeutics*, 2010, **7**, 378–391.
82. J. Kim, N. Kim and M. Yenari, *CNS Neurosci. Ther.*, 2015, **21**, 309–319.
83. E. Benarroch, *Mayo Clin. Proc.*, 2005, **80**, 1326–1338.
84. A. Araque, R. Sanzgiri, V. Parpura and P. Haydon, *Can. J. Physiol. Pharmacol.*, 1999, **77**, 699–706.
85. Z. Liu and M. Chopp, *Prog. Neurobiol.*, 2015, 103–120, DOI: 10.1016/j.pneurobio.2015.09.008.
86. E. Blanco-Suarez, A. Caldwell and N. Allen, *J. Physiol.*, 2016, 1903–1916, DOI: 10.1113/JP270988,.
87. L. Clarke and B. Barres, *Nat. Rev. Neurosci.*, 2013, **14**, 311–321.
88. J. Zuchero and B. Barres, *Development*, 2015, **142**, 3805–3809.
89. A. Schousboe, N. Westergaard, U. Sonnewald, S. Petersen, A. Yu and L. Hertz, *Prog. Brain Res.*, 1992, **94**, 199–211.
90. E. Bushong, M. Martone, Y. Jones and M. Ellisman, *J. Neurosci.*, 2002, **22**, 183–192.
91. N. Rouach, J. Glowinski and C. Giaume, *J. Cell Biol.*, 2000, **149**, 1513–1526.
92. G. Cheung, O. Chever and N. Rouach, *Front. Cell. Neurosci.*, 2014, **8**, 348.
93. G. Arcuino, J. H. Lin, T. Takano, C. Liu, L. Jiang, Q. Gao, J. Kang and M. Nedergaard, *Proc. Natl. Acad. Sci. U. S. A.*, 2002, **99**, 9840–9845.
94. S. Y. Gordleeva, S. V. Stasenko, A. V. Semyanov, A. E. Dityatev and V. B. Kazantsev, *Front. Comput. Neurosci.*, 2012, **6**, 92.
95. V. Parpura and P. G. Haydon, *Proc. Natl. Acad. Sci. U. S. A.*, 2000, **97**, 8629–8634.
96. A. Perez-Alvarez and A. Araque, *Curr. Drug Targets*, 2013, **14**, 1220–1224.
97. T. Takano, N. Oberheim, M. Cotrina and M. Nedergaard, *Stroke*, 2009, **40**, S8–S12.
98. J. Shields, D. Kimbler, W. Radwan, N. Yanasak, S. Sukumari-Ramesh and K. Dhandapani, *Transl. Stroke Res.*, 2011, **2**, 633–642.

99. C. Rose and B. Ransom, *Glia*, 1997, **20**, 299–307.
100. J. Vella, C. Zammit, G. Di Giovanni, R. Muscat and M. Valentino, *Front. Cell. Neurosci.*, 2015, **9**, 108.
101. K. Hayakawa, E. Esposito, X. Wang, Y. Terasaki, Y. Liu, C. Xing, X. Ji and E. H. Lo, *Nature*, 2016, **535**, 551–555.
102. J. L. Zamanian, L. Xu, L. C. Foo, N. Nouri, L. Zhou, R. G. Giffard and B. A. Barres, *J. Neurosci.*, 2012, **32**, 6391–6410.
103. S. A. Liddelow, K. A. Guttenplan, L. E. Clarke, F. C. Bennett, C. J. Bohlen, L. Schirmer, M. L. Bennett, A. E. Munch, W. S. Chung, T. C. Peterson, D. K. Wilton, A. Frouin, B. A. Napier, N. Panicker, M. Kumar, M. S. Buckwalter, D. H. Rowitch, V. L. Dawson, T. M. Dawson, B. Stevens and B. A. Barres, *Nature*, 2017, **541**, 481–487.
104. A. Jovicic, R. Roshan, N. Moisoi, S. Pradervand, R. Moser, B. Pillai and R. Luthi-Carter, *J. Neurosci.*, 2013, **33**, 5127–5137.
105. Y. Ouyang, L. Xu, Y. Lu, X. Sun, S. Yue, X. Xiong and R. Giffard, *Glia*, 2013, **61**, 1784–1794.
106. E. Hutchison, E. Kawamoto, D. Taub, A. Lal, K. Abdelmohsen, Y. Zhang, W. R. Wood, E. Lehrmann, S. Camandola, K. Becker, M. Gorospe and M. Mattson, *Glia*, 2013, **61**, 1018–1028.
107. Y. Ouyang, L. Xu, S. Yue, S. Liu and R. Giffard, *Neurosci. Lett.*, 2014, **565**, 53–58.
108. Y. Ouyang, Y. Lu, S. Yue and R. Giffard, *Mitochondrion*, 2012, **12**, 213–219.
109. Y. Ouyang, C. Stary, R. White and R. Giffard, *Antioxid. Redox Signallig*, 2014, 187–202, DOI: 10.1089/ars.2013.5757.
110. Y. Ye, J. R. Perez-Polo, J. Qian and Y. Birnbaum, *Physiol. Genomics*, 2011, **43**, 534–542.
111. S. Khanna, C. Rink, R. Ghoorkhanian, S. Gnyawali, M. Heigel, D. Wijesinghe, C. Chalfant, Y. Chan, J. Banerjee, Y. Huang, S. Roy and C. Sen, *J. Cereb. Blood Flow Metab.*, 2013, **33**, 1197–1206.
112. C. Stary, X. Sun, Y. Ouyang, L. Li and R. Giffard, *Mitochondrion*, 2016, **30**, 248–254.
113. A. Karthikeyan, R. Patnala, S. P. Jadhav, L. Eng-Ang and S. T. Dheen, *Curr. Med. Chem.*, 2016, **23**, 3528–3546.
114. C. Benakis, L. Garcia-Bonilla, C. Iadecola and J. Anrather, *Front. Cell. Neurosci.*, 2014, **8**, 461.
115. M. K. McCoy and M. G. Tansey, *J. Neuroinflammation*, 2008, **5**, 45.
116. W. Li, B. Katz and S. Spinola, *Infect. Immun.*, 2012, **80**, 4426–4434.
117. C. Parisi, G. Napoli, P. Pelegrin and C. Volonté, *Mediators Inflammation*, 2016, 2989548.
118. L. Zhang, L. Dong, Y. Li, Z. Hong and W. Wei, *J. Neuroinflammation*, 2012, **9**, 211.
119. L. Zhang, Y. Li, X. Wu, Z. Hong and W. Wei, *J. Neurochem.*, 2015, **132**, 713–723.
120. P. Mitchell, R. Parkin, E. Kroh, B. Fritz, S. Wyman, E. Pogosova-Agadjanyan, A. Peterson, J. Noteboom, K. O'Briant, A. Allen, D. Lin,

N. Urban, C. Drescher, B. Knudsen, D. Stirewalt, R. Gentleman, R. Vessella, P. Nelson, D. Martin and M. Tewari, *Proc. Natl. Acad. Sci. U. S. A.*, 2008, **105**, 10513–10518.
121. R. Chandran, S. Mehta and R. Vemuganti, *Neurochem. Int.*, 2017, DOI: 10.1016/j.neuint.2017.01.015.
122. B. Martinez and P. Peplow, *Neural. Regen. Res.*, 2016, **11**(9), 1375–1378.
123. K. Tan, A. Armugam, S. Sepramaniam, K. Lim, K. Setyowati, C. Wang and K. Jeyaseelan, *PLoS One*, 2009, **4**, e7689.
124. L. Zeng, J. Liu, Y. Wang, L. Wang, S. Weng, Y. Tang, C. Zheng, Q. Cheng, S. Chen and G. Yang, *Front. Biosci.*, 2011, **3**, 1265–1272.
125. G. C. Jickling, B. P. Ander, X. Zhan, D. Noblett, B. Stamova and D. Liu, *PLoS One*, 2014, **9**, e99283.
126. A. Dharap, K. Bowen, R. Place, L. Li and R. Vemuganti, *J. Cereb. Blood Flow Metab.*, 2009, **29**, 675–687.
127. S. Lee, K. Chu, K. Jung, H. Yoon, D. Jeon, K. Kang, K. Park, E. Bae, M. Kim, S. Lee and J. Roh, *Stroke*, 2010, **41**, 1646–1651.
128. C. Stary, L. Xu, X. Sun, Y. Ouyang, R. White, J. Leong, J. Li, X. Xiong and R. Giffard, *Stroke*, 2015, **46**, 551–556.
129. da Z. Liu, G. Jickling, B. Ander, H. Hull, X. Zhan, C. Cox, N. Shroff, C. Dykstra-Aiello, B. Stamova and F. Sharp, *J. Cereb. Blood Flow Metab.*, 2016, **36**, 1374–1383.
130. Z. B. Yang, T. B. Li, Z. Zhang, K. D. Ren, Z. F. Zheng, J. Peng and X. J. Luo, *Intern. Med.*, 2016, **55**, 1279–1286.
131. S. Sepramaniam, J. R. Tan, K. S. Tan, D. A. DeSilva, S. Tavintharan, F. P. Woon, C. W. Wang, F. L. Yong, D. S. Karolina, P. Kaur, F. J. Liu, K. Y. Lim, A. Armugam and K. Jeyaseelan, *Int. J. Mol. Sci.*, 2014, **15**, 1418–1432.
132. W. He, S. Chen, X. Chen, S. Li and W. Chen, *J. Stroke Cerebrovasc. Dis.*, 2016, **25**, 1753–1759.
133. J. R. Tan, K. S. Tan, Y. X. Koo, F. L. Yong, C. W. Wang, A. Armugam and K. Jeyaseelan, *Int. J. Mol. Sci.*, 2013, **14**, 2072–2084.
134. S. H. Li, S. Y. Su and J. L. Liu, *Curr. Neurovasc. Res.*, 2015, **12**, 214–221.
135. C. S. Gan, C. W. Wang and K. S. Tan, *Genet. Mol. Res.*, 2012, **11**, 147–152.
136. L. Jia, F. Hao, W. Wang and Y. Qu, *Cell Biochem. Funct.*, 2015, **33**, 314–319.
137. A. Dharap and R. Vemuganti, *J. Neurochem.*, 2010, **113**, 1685–1691.
138. T. A. Lusardi, C. D. Farr, C. L. Faulkner, G. Pignataro, T. Yang, J. Lan, R. P. Simon and J. A. Saugstad, *J. Cereb. Blood Flow Metab.*, 2010, **30**, 744–756.
139. Z. Peng, J. Li, Y. Li, X. Yang, S. Feng, S. Han and J. Li, *J. Neurosci. Res.*, 2013, **91**, 1349–1362.
140. M. Sun, T. Yamashita, J. Shang, N. Liu, K. Deguchi, J. Feng and K. Abe, *Cell Transplant.*, 2015, **24**, 367–376.
141. J. H. Shin, Y. M. Park, D. H. Kim, G. J. Moon, O. Y. Bang, T. Ohn and H. H. Kim, *Biochim. Biophys. Acta*, 2014, **1839**, 826–836.

CHAPTER 8

Combined Drug–Diagnostic Test Co-development for Predicting and Preventing Brain Impairments

GALINA IZYKENOVA*[a] AND SHEAN PHELPS[b]

[a] GRACE Laboratories, LLC, Atlanta, Georgia, USA; [b] Georgia Institute of Technology, Atlanta, Georgia, USA
*Email: izykenova@yahoo.com

8.1 Introduction

In 2015, more than 25% of new breakthrough therapy designations granted by the US Food and Drug Administration (FDA) were for personalized medicine.[1] Despite cerebrovascular accident (CVA) being the third-leading cause of death in the US and traumatic brain injury (TBI) rates in the US and worldwide occurring at epidemic proportions, to date, none of these designations have included personalized therapies (Rx) aimed at directly tackling these disease entities. Whereas greater than 140 000 attributable deaths occur among the approximately 795 000 annually occurring strokes in the US alone,[2] when combined with an estimated global burden of around 6.2 million lives per year, there exists a glaring unfilled niche for so-called "companion diagnostics" (CDx) to such personalized therapies.[3] This gap should be addressed by FDA-approved, combination *in vitro* assays for the

Drug Discovery Series No. 62
Acute Brain Impairment: Scientific Discoveries and Translational Research
Edited by Philip V. Peplow, Svetlana A. Dambinova, Thomas A. Gennarelli and Bridget Martinez

Published by the Royal Society of Chemistry, www.rsc.org

assessment and treatment of cerebrovascular disorders (CVDs) and acute TBI. But there still remain a number of difficult challenges in the integration of such advances, particularly in the areas of scientific discovery, regulatory diagnostics, and policy. These have precluded the development and delivery of such "Rx–CDx" technologies for the intervention of stroke and TBI to the realm of clinical practice.

To add further complication to an already difficult scientific and regulatory environment, neurological impairments caused by CVDs or acute TBI exhibit heterogeneous and multilayer pathologies that are not readily amenable to the application of other emerging personalized medicine technologies (*e.g.*, genomics). While such technologies have revolutionized diagnosis and treatment in areas such as cancer,[4] comparable interventions in the diagnosis and treatment of CVD and TBI are more difficult, in that each poses special challenges based, in part, on the diverse manifestations and phenotypes of pathophysiology, cerebral structure, and hemodynamic variation, as well as due to the acute and chronic clinical presentations therein. Furthermore, clinical outcomes often depend partly or significantly on genetic or acquired individual predispositions.[4,5]

In this chapter, we discuss how CVDs and TBI-centric personalized therapeutics (*i.e.*, "Rx" or "targeted drugs" and/or therapies) and CDx may offer significant potential with regard to improving clinical outcomes by proffering patient treatments that are tailored toward their distinctive biological characteristics. We also attempt to address the application of personalized medicine in the face of challenges surrounding predicting and preventing the development of CVDs in persons with pre-existing conditions or TBI in persons with pre-existing conditions and/or increased risk factors.

8.2 Companion Diagnostic Precedent

8.2.1 Public Health Perspective

Recent health reports suggest that 90% of currently prescribed medications worldwide are effective only in 30–50% of cases,[6] while 25–62% of patients who suffer from oncologic and nervous system diseases do not respond to conventional therapy.[7]

The routine diagnosis of TBI and stroke is based on neurological examination including medical history, clinical assessment scales, and neuroimaging, often performed at a hospital emergency department (ED) and focused primarily on acute stabilization of the patient following a traumatic incident or significant neurological event. Differential diagnosis of stroke within the ED setting remains a problem as there are several subtypes of stroke as well as non-vascular disorders (stroke mimics) that may display clinical signs and symptoms that appear identical. Distinguishing these mimics from true CVAs is increasingly important for the proper application of stroke therapies that carry potential adverse effects. For prompt treatment decisions, it is also critical to distinguish between ischemic stroke, hemorrhagic stroke, and the transitory ischemic attack (TIA).

As soon as a firm diagnosis is established, aggressive interventions are implemented to preserve neurological function, to minimize risk of complications, and to reduce the likelihood of decompensation/dysfunction. However, necessary interventions are often put on hold and thereby are less efficient due to the delay in recognition of symptoms, the delay in proper treatment, or the late arrival of the critically ill patient to a neuro-critical intensive care unit (ICU) or interventional ICU suite where definitive diagnosis, intervention, and management of life-threatening conditions can be optimized.

The most successful emergent medication for acute stroke, tissue plasminogen activator (tPA), is approved by the FDA for intravenous administration within 3.0–4.5 hours of the onset of stroke symptoms. Even so, tPA is used only in 5–6% of qualifying patients due primarily to delays in prompt recognition of symptoms with subsequent early arrival to an ER, thereby not facilitating administration of thrombolysis within the recommended time window.[8]

At present, biopharmaceutical research companies are developing between 4 and 20 candidate therapeutic agents for TBI and stroke, respectively, out of 420 new medicines intended for use in neurological disorders.[9,10] Table 8.1 depicts recently FDA-approved agents aimed at preventing CVD. In clinical trials, only three medications have been designed for the use of CDx surrogate end-points. The FDA has not cleared a single drug or device by 2015 for TBI assessment or treatment, and in 2016, the FDA approved a single system the Immediate Post-Concussion Assessment and Cognitive Testing (aka: "ImPACT" and "ImPACT Pediatric", intended to assess cognitive functions immediately following a suspected concussion) for marketing.[11]

In large part, this slow progress has been bound to our inability to precisely stratify the multi-factorial conditions of TBI or stroke and thus discover breakthrough pharmacological agents relevant to certain phenotypes of brain injury, fulfill demand for devices, and allow accurate triage of patients into clinical trials utilizing the Rx/CDx approach.

8.2.2 Personalized or Precision Medicine

Personalized medicine is tailored medical care, allowing improved medical decisions, practices, interventions, and/or therapies shaped for the individual based on a patient's genetic, anatomical, and physiological characteristics, and taking into consideration the patient's predicted response or risk of disease.[12]

Precision diagnosis and treatment of brain impairments demands converting and standardizing a vast array of existing clinical, imaging, physiological, and different types of "omics" data into practicable information.[13,14] Multifaceted approaches include translational bioinformatics, emerging advanced neuroimaging, genomics, and protein biomarkers that would

Table 8.1 FDA-approved drugs for vascular diseases in 2015–2016.

Drug[a]	Composition	Mechanism of action	Intended use	CDx
Byvalson, Allergan	Fixed-dose combination of nebivolcl and valsartan 5 mg/80 mg tablet	β-adrenergic receptor blocking agent (β1 selective), and angiotensin II receptor blocker	For the treatment of hypertension	Systolic/diastolic blood pressure
Yosprala, Aralez Pharmaceuticals	Aspirin and omeprazole	Anti-platelet agent and proton pump inhibitor	For the prevention of cardiovascular and cerebrovascular events	—
Praluent, Sanofi Aventis	Alitocumab	PCSK9 (proprotein convertase subtilisin kexin type 9) inhibitor antibody	For the treatment of hypercholesterolemia or atherosclerotic cardiovascular disease	Low-density lipoprotein cholesterol
Prestalia, Symplmed Pharmaceuticals	Combination of perindopril and amlodipine	Angiotensin-converting enzyme inhibitor and dihydropyridine calcium channel blocker	For the treatment of hypertension	Systolic/diastolic blood pressure
Savaysa, Daiichi Sankyo	Edoxaban	Factor Xa inhibitor that works as an anti-coagulant	For the treatment of deep vein thrombosis, pulmonary embolism, and risk of stroke and embolism due to atrial fibrillation	—
Uptravi, Actelion Pharmaceuticals	Selexipag	Prostacyclin receptor agonist that exerts vasodilating effects	For the treatment of pulmonary arterial hypertension	—

[a]Cited from website CenterWatch: https://www.centerwatch.com/drug-information/fda-approved-drugs/therapeutic-area/1/cardiology-vascular-diseases.

guide the development of targeted therapy for the intensive care of brain impairments.[14,15]

Recent implementations in rapid cancer biomarkers and targeted therapy provide a template for a non-conventional approach to drug–test co-development for CVDs and TBI. As a tactic for assessing disease distinctiveness, specific *in vitro* assays or CDx might be used to detect and target pathological vascular "signatures", customizing the development of personalized therapies or Rx.

8.2.3 Drug-test Co-development Approach for Brain Impairments

Ongoing research will continue to identify biomarkers of brain damage, including those that can detect asymptomatic microvessel dysfunction that may assess stroke and TBI. The approach of combining several biomarkers to calculate a multi-marker score for detecting cerebral dysfunction has been explored for stroke outcome.[16] Multi-marker assays (Triage® Stroke Panel) were correlated with mortality at 120 days from stroke onset, while earlier assays showed no diagnostic validity in acute magnetic resonance imaging-proven ischemic stroke.[17]

Biomarkers such as glutamate, homocysteine,[18] and *N*-methyl-D-aspartate (NMDA) receptor peptide/antibodies (Abs)[19,20] have been found to be independently associated with prior multiple cerebrovascular events and can be measured simultaneously in blood.[21] A multiple panel of biomarkers for stroke that combines NMDA receptor markers with, for example, glial fibrillary acidic protein (GFAP)[22,23] and/or anticardiolipin Abs[24] could theoretically not only improve diagnostic certainty by evaluating the contributions of thrombotic, hemorrhagic, and neurotoxic events, but also offer clues as to the optimal pathways for both preventive and acute treatment.

Seven of the world's top experimental TBI centers combined their expertise in the Operation Brain Trauma Therapy (OBTT) Consortium to identify and bring forward new pharmacological agents for the assessment and treatment of TBI.[25] The consortium used a two-tier (Tier "A" and Tier "B") screening approach to evaluate potential new drugs. The OBTT Consortium utilizes three established TBI models[26–28] combined with the biomarkers GFAP and ubiquitin C-terminal hydrolase-L1 (UCH-L1) in Tier "A" to identify and recommend promising agents like levetiracetam[25] and human neural stem cell engraftment[29] for advancement into clinical translation for early post-TBI detection and management (Tier "B").

Frequently, the missing or limited link for prospective CDx therapy is a clinical study that proves a correlation between marker reductions secondary to drug therapy and event occurrence. Clinical studies must also be undertaken to show that the blood level of a selected marker is an appropriate cut-off for stratifying patients for treatment.

8.3 Regulatory Criteria for Drug-test Co-development

8.3.1 FDA Rx/CDx Concept

In April 2005, the FDA published a concept paper[30] followed by draft guidance[31] that described the steps needed to develop a drug in combination with a diagnostic test. Two pathways were outlined: (1) a biomarker is identified during clinical development of a new molecule; and (2) a drug and its CDx test are developed simultaneously. If the diagnostic marker is not identified until drug development begins, the drug and the marker need to be co-developed. Included in the paper are procedures for validating diagnostic tests, how to determine biomarker cut-off levels for drug therapy, statistical guidelines for selecting biomarker cut-off levels, and designs for prospective clinical trials that must be conducted to prove the biomarker's clinical utility. Results of clinical trials must not only validate the marker, but also directly influence its clinical utility, which is reflected in the indication and label on both the marker and on the product being co-developed.

The FDA's solution to the cut-off issue is to require the sponsor to conduct clinical testing in patients who are below and above the proposed cut-off for the biomarker. According to the FDA, it will often be necessary to validate the marker in patients for whom the test is negative in order to "improve the benefit/risk of the drug in the selected and unselected groups." Thus, a four-arm prospective clinical trial will be required in many instances, with two arms (placebo and active) devoted to patients who test negative for the marker and two arms (placebo and active) devoted to patients who test positive for the marker.

If a commercial version of a diagnostic test is developed prior to the onset of drug development and the relationship between biomarker and disease has been reasonably well explored, co-development of a drug/diagnostic is less complex. The success of the project very much depends on the strength of the biomarker and concrete hypotheses surmised from early research of molecular pathology, mechanisms of drug action, and preclinical data tested through prototype assays. In order to give an idea of the predictive potential of the prototype, assays should pass clinical trials and regulatory approval.

8.3.2 Intended Use and Indication of Use

As soon as a novel medication or device is entered into the product approval pipeline of the adjudicating regulatory body, questions of "intended use" and/or "product labeling" are raised. The application should bring together the applicable body of research data on safety and efficacy collected for the drug/device during the discovery stage with its actual use for the condition and its purpose for and use by physicians and consumers, while addressing any and all legal claims utilized by the manufacturer to promote the product.[32]

An issue that often generates confusion is: what is "intended use" *vs.* "indication of use"? *Intended use* for a drug or medical device means *exactly what your product is used for*, while *indication of use* is a statement concerning the *precise circumstances or medical condition for which the particular product should be used* (http://blog.greenlight.guru/intended-use-and-indications-of-use). For example, a medical device's *intended use* is the quantitative detection of a brain biomarker in certain human biological fluids (plasma, serum, cerebrospinal fluid, *etc.*). At the same time, the *indication of use* for this device could be differentiation between acute stroke and stroke mimics, suited for deployment in emergency rooms and in ambulance services.

As an example, S100B has been a leading biomarker in brain damage research over several decades, yet it could not pass regulatory requests as a test for brain injury since it has an absence of an "intended use."[33] Lately, however, the S100B test has been considered as a potential candidate in mild pediatric TBI to rule out unnecessary computer tomography imaging.[34]

In contrast, cyclosporine, which was cleared in 1995 by the FDA as an immunosuppressant with an indication to prevent organ rejection in organ transplant recipients, demonstrates neuroprotective activity after crossing the blood–brain barrier.[35] Since 2012, cyclosporine (sponsored by NeuroVive Pharmaceutical AB, Sweden) has been in Phase II clinical trials for the treatment of moderate and severe TBI.[36] However, this indication was not supported by the OBTT Consortium due to a "lack of behavioral protection in any model, indicators of toxicity, and a narrow therapeutic index."[37]

8.3.3 Trends in Point-of-care Testing and Rx

The volume of point-of-care (POC) testing is steadily growing, driven by changes in healthcare that are aimed at delivering less costly care closer to the patient.[38] There exists a challenge in providing more effective care for stroke and TBI where POC testing might play an important role, especially in the ER and neuro-critical care setting. Biosensors and lateral flow strips utilizing immobilized Abs to determine a range of parameters for cardiac markers are becoming more and more readily available and could be expanded for use in current and emerging stroke and TBI biomarkers as well (see Chapter 18).

The primary goal of neuroprotection aims to prevent salvageable neurons from dying. There are a number of neuroprotective treatments under investigation in clinical trials (*e.g.*, clinical hypothermia and ebselen).[39] Hypothermia remains a potentially viable strategy for neuroprotective therapy for ischemic stroke and is currently being investigated as a personalized approach in brain protection.[40]

As for development of neuro-critical therapy for brain impairments, ebselen, a glutathione peroxidase mimic, has been shown to provide protection in transient forebrain ischemia.[41] A Phase III trial exploring the efficacy of ebselen in patients with a cortical infarct is currently ongoing.[42] Magnesium produced significant protection against neurotoxicity in preclinical animal models of stroke and TBI.[43] However, the results of recent trials showed that magnesium is ineffective regarding clinical outcome after subarachnoid hemorrhage and TBI.[44,45]

Further concepts to improve translation include the investigation of mutual neuroprotectants and POC testing in multicenter preclinical studies with improved research models, as well as close alignment between clinical trial and preclinical methodologies. Future successful translation will require both new concepts for preclinical testing and innovative approaches based on mechanistic insights into the neurotoxicity cascade (see Chapter 9).

8.4 Advantages of Combined Brain Impairment Drug–Diagnostic Test use in the Clinical Setting

8.4.1 Optimization of Neuro-critical Care by Rx/CDx

The difference between critical care medicine and emergency medicine is that emergency medicine focuses on the acute stabilization of the patient as initial care following an accident or event, which leads the person to seek care in the hospital emergency room. Then, the critically ill patient with life-threatening diseases of the nervous system (the brain, spinal cord, and nerves) will be admitted to a neuro-critical care unit (a special facility available in Centers of Excellence for Stroke or TBI), also known as the neuro-ICU, where the diagnosis and management of these conditions requires sophisticated organ support and invasive monitoring.

In order to facilitate the selection of patients with acute ischemic stroke who can benefit from treatment with tPA, it is necessary to rapidly triage patients at high risk of suffering hemorrhagic transformation that causes high morbidity and mortality. It was demonstrated that a significant increase in semicarbazide-sensitive amine oxidase (SSAO) activity, presumably vascular adhesion protein 1 (VAP-1), in the plasma of patients who have suffered ischemic stroke and have been treated with intravenous tPA is significantly correlated with a higher probability of hemorrhagic transformation in these patients.[46]

Levels of NR2 peptide, an acute vascular biomarker for ischemic stroke, were significantly increased in patients with neurological adverse events (NAEs) compared to those without neurological deficits who underwent carotid endarterectomy and carotid artery stenting.[47] Clinical events were correlated with NR2 peptide levels in NAE patients before, during, and 24 hours after procedures with a sensitivity of 99%.[47]

Recently approved by the FDA, direct oral anticoagulants (DOACs; dabigatran, rivaroxaban, and others) are rapidly acting, target-specific inhibitors of free and bound activated serine proteases that are more robust than warfarin or heparin. They usually do not need laboratory testing for dose adjustment except before surgical or invasive procedures, on the occasion of hemorrhagic or thrombotic events, and when immediate reversal of anticoagulation is needed.[48,49] Routinely available prothrombin time (PT), activated thromboplastin time, and thrombin clotting time (TCT) tests should be primarily based on their best performances for respective DOACs.[48] Accordingly, the TCT is best suited for dabigatran and the PT is most responsive to the rivaroxaban direct factor Xa inhibitor.[50] More specialized assays such as the Hemoclot, a direct thrombin inhibition assay, showed good reproducibility and concordance with expected drug levels assessed by mass spectrometry and were effective for quantifying drug levels.

8.4.2 Rx/CDx for Preventing and Predicting Chronic CVDs

The Framingham Heart Study found that 26% of stroke survivors over 65 years of age end up in a nursing home. The annual direct and indirect cost of strokes in the US is approximately $50 billion.

The best cure for CVDs is preventing the degree of impairment following stroke and preventing CVDs themselves. Generally, practitioners are focused on research into identifying the genetic causes of CVDs, stroke, and stroke disability, identifying biological and behavioral CVD risk factors, and developing improved diagnostic (CDx), treatment (Rx), and rehabilitation techniques.

Lipoprotein-associated phospholipase A2 (Lp-PLA2) is a vascular-specific inflammatory enzyme that increases the risk of CVD events and stroke. After 12 weeks of lipid-lowering statins like pravastatin (40 mg daily) and atorvastatin (20 mg daily), Lp-PLA2 levels decreased by 22.1% and 47%, respectively.[51,52] Statins typically reduce Lp-PLA2 levels by 20–30% in patients with atherosclerosis and coronary artery disease.[53,54]

Serial measurements of blood levels of NR2 Abs in serum samples from persons with pre-existing conditions for stroke (diabetes mellitus, atherosclerosis, or valve replacement surgery) and persons who have suffered TIA/stroke preventively treated with antiplatelet drugs (aspirin or pentoxifylline) demonstrated that the assay was capable of assessing the risk of CVDs at 1–3 months before the event.[55] The levels of NR2 Abs were significantly reduced (below the cut-off of 2 $ng\,mL^{-1}$) in persons with no CVA consequences at 1 year of treatment follow-up. A drastic increase above the cut-off values of the biomarker (4.0–6.0 $ng\,mL^{-1}$) indicated an episode of a TIA that developed into an acute ischemic stroke in one patient.

Preoperative serum concentrations of NR2 Abs, but not S100B or C-reactive protein (CRP), were predictive of severe NAEs after cardiac surgery with cardiopulmonary bypass (CPB).[56] Patients with a positive NR2 Ab test ($\geq$2.0 $ng\,mL^{-1}$) preoperatively were nearly 18-times more likely to experience a postoperative neurological event than patients with a negative test (<2.0 $ng\,mL^{-1}$). The use of erythropoietin for neuroprotection before, during, and after CPB demonstrated an approximately 20% decrease in NR2 Abs.[57] Increases in serum levels of Abs to NMDA glutamate receptors reflected hypoxic–ischemic brain lesions in children with chronic post-traumatic headaches and dictated the need for metabolic therapy.[58]

8.5 Conclusion

While some of the best examples of personalized medicine are in the fields of oncology and anti-viral agents, opportunities are also present in other therapeutic areas, including stroke and TBI.

Triaging of patients based on results acquired from CDx utilizing specific biomarkers reflecting neurovascular unit strength (see Chapter 9) complementing and customizing CVD treatment would allow for a reduction in healthcare costs, aid in the prevention of CVDs in individuals with pre-existing conditions, improve outcome for patients in the acute stage of the disease, and prevent regression to chronic neurological disorders.

A gradual approach to obtaining deeper insights into the notable distinctions of neurovascular component functions (brain biomarkers) in the complex brain architecture (diffusion-weighted imaging, susceptibility-weighted imaging, and diffusion-weighted imaging radiography) should be considered when designing pharmacological agents. In that regard, the search for CDx using distinct biomarkers such as tissue-based evidence of early signs of cerebral blood flow insufficiency, microvasculature lesions, and edema formation would be invaluable. Targeting these distinctions as therapeutics for CVD patients in clinical practice may be fundamental to counteracting their disease course.

Currently, there are two major approaches to Rx/CDx that are unfolding in the neuro-critical care for stroke and TBI. The first deals with the search of diagnostic assays to already FDA-approved medications for other conditions, such as immunosuppressant cyclosporine for TBI treatment. The second involves designating existing acute biomarker assays and tests (FDA cleared or in translational research) to be used in acute-setting procedures, such as assessment of the heart failure biomarker brain natriuretic peptide (BNP) in patients with stroke.

A diagnostic marker or panel of biomarkers with the ability to distinguish ischemic *vs.* hemorrhagic stroke[23] and to predict the risk of CVDs in persons with pre-existing conditions and those who have suffered a TBI could support the development of preventive and emergent therapies and supply practitioners with real-time navigation to personalized treatment for reducing the risk of stroke.

References

1. Personalized Medicine Coalition 2015 Progress Report Online, http://www.personalizedmedicinecoalition.org/Userfiles/PMC-Corporate/file/2015_Progress_Report_PM_at_FDA.pdf (accessed on March 28, 2017).
2. CDC Centers for Disease Control and Prevention, National Center for Health Statistics. Online Database, http://wonder.cdc.gov/ucd-icd10.html. (accessed on March 28, 2017).
3. World Health Organization 2011 Report Online, http://www.who.int/nmh/publications/ncd_report_full_en.pdf (accessed on March 28, 2017).

4. J. D. Hinman, N. S. Rost, T. W. Leung, J. Montaner, K. W. Muir, S. Brown, J. F. Arenillas, E. Feldmann and D. S. Liebeskind, *J. Neurol. Neurosurg. Psychiatry*, 2017, **88**, 54.
5. Y. Wang, H. Liu, Y. Lin, G. Liu, H. Chu, P. Zhao, X. Yang, T. Zheng, M. Fan, X. Zhou, J. Meng and C. Sun, *Sci Rep.*, 2017, **7**, 40137, DOI: 10.1038/srep40137.
6. C. Womack, Online: http://www.pathsoc.org/files/meetings/edinburghprestues/ 1150%20Tues%20Fintry%20Womack.pdf (accessed on 30 January, 2017).
7. I. Akhmetov, R. Ramaswamy, I. Akhmetov and P. K. Thimmaraju, *J. Pers Med.*, 2015, **5**, 213.
8. A. S. Go, D. Mozaffarian, V. L. Roger, E. J. Benjamin, J. D. Berry, M. J. Blaha, S. Dai, E. S. Ford, C. S. Fox, S. Franco, H. J. Fullerton, C. Gillespie, S. M. Hailpern, J. A. Heit, V. J. Howard, M. D. Huffman, S. E. Judd, B. M. Kissela, S. J. Kittner, D. T. Lackland, J. H. Lichtman, L. D. Lisabeth, R. H. Mackey, D. J. Magid, G. M. Marcus, A. Marelli, D. B. Matchar, D. K. McGuire, E. R. Mohler 3rd, C. S. Moy, M. E. Mussolino, R. W. Neumar, G. Nichol, D. K. Pandey, N. P. Paynter, M. J. Reeves, P. D. Sorlie, J. Stein, A. Towfighi, T. N. Turan, S. S. Virani, N. D. Wong, D. Woo, M. B. Turner and American Heart Association Statistics Committee and Stroke Statistics Subcommittee, *Circulation*, 2014, **129**, e28S.
9. PhRMA Report on Disorders of the Brain, Spinal Cord and Nerves, Online, http://www.phrma.org/report/medicines-in-development-for-neurological-disorders-2015-report (accessed on 31 January, 2017).
10. R. Diaz-Arrastia, P. M. Kochanek, P. Bergold, K. Kenney, C. E. Marx, C. J. Grimes, L. T. Loh, L. T. Adam, D. Oskvig, K. C. Curley and W. Salzer, *J. Neurotrauma*, 2014, **31**, 135.
11. D. Kotz, FDA News Release, Online, http://www.fda.gov/NewsEvents/Newsroom/PressAnnouncements/ucm517526.htm (assessed on 28 March, 2017).
12. FDA USA, Online, https://www.fda.gov/downloads/ScienceResearch/SpecialTopics/PersonalizedMedicine/UCM372421.pdf (assessed on 28 March, 2017).
13. J. D. Tenenbaum, *Genomics, Proteomics Bioinf.*, 2016, **14**, 31.
14. D. K. Menon and A. Ercole, *Handb. Clin. Neurol.*, 2017, **140**, 239.
15. S. Makarenko, D. E. Griesdale, P. Gooderham and M. S. Sekhon, *J. Clin. Neurosci.*, 2016, **26**, 8.
16. F. Iemolo, E. Sanzaro, G. Duro, A. Giordano and M. Paciaroni, *Immun. Ageing*, 2016, **13**, 19.
17. C. Knauer, K. Knauer, S. Müller, A. C. Ludolph, D. Bengel, H. P. Müller and R. Huber, *BMC Neurol.*, 2012, **12**, 14.
18. N. Hasan, P. McColgan, P. Bentley, R. J. Edwards and P. Sharma, *Br. J. Clin. Pharmacol.*, 2012, **74**, 230.
19. S. A. Dambinova, K. Bettermann, T. Glynn, M. Tews, D. Olson, J. D. Weissman and R. L. Sowell, *PLoS One*, 2012, **7**, e42362, DOI: 10.1371/journal.pone.

20. J. D. Weissman, G. A. Khunteev, R. Heath and S. A. Dambinova, *J. Neurol. Sci.*, 2011, **300**, 97.
21. S. A. Dambinova, G. A. Khounteev and A. A. Skoromets, *Stroke*, 2002, **33**, 1181.
22. C. Foerch, M. Niessner, T. Back, M. Bauerle, G. M. De Marchis, A. Ferbert, H. Grehl, G. F. Hamann, A. Jacobs, A. Kastrup, S. Klimpe, F. Palm, G. Thomalla, H. Worthmann, M. Sitzer and BE FAST Study Group, *Clin. Chem.*, 2012, **58**, 237.
23. D. M. Stanca, I. C. Mărginean, O. Soriţău, C. Dragoş, M. Mărginean, D. F. Mureşanu, J. C. Vester and A. Rafila, *J. Cell. Mol. Med.*, 2015, **19**, 2253.
24. X. Bu, H. Peng, C. Zhong, T. Xu, Y. Peng, C. S. Chen, J. Wang, Z. Ju, Q. Li, D. Geng, Y. Sun, D. Zhang, J. Zhang, J. Chen, Y. Zhang and J. He, *Stroke*, 2016, **47**, 2742.
25. P. M. Kochanek, H. M. Bramlett, D. A. Shear, C. E. Dixon, S. Mondello, W. D. Dietrich, R. L. Hayes, K. K. Wang, S. M. Poloyac, P. E. Empey, J. T. Povlishock, A. Mountney, M. Browning, Y. Deng-Bryant, H. Q. Yan, T. C. Jackson, M. Catania, O. Glushakova, S. P. Richieri and F. C. Tortella, *J. Neurotrauma*, 2016, **33**, 606.
26. K. C. Van and B. G. Lyeth, *Methods Mol. Biol.*, 2016, **1462**, 231.
27. N. D. Osier and C. E. Dixon, *Front. Neurol.*, 2016, **7**, 134.
28. D. A. Shear, X. C. Lu, R. Pedersen, G. Wei, Z. Chen, A. Davis, C. Yao, J. Dave and F. C. Tortella, *J. Neurotrauma*, 2011, **28**, 2185.
29. M. S. Spurlock, A. I. Ahmed, K. N. Rivera, S. Yokobori, S. W. Lee, P. N. Sam, D. A. Shear, M. P. Hefferan, T. G. Hazel, K. K. Johe, S. Gajavelli, F. C. Tortella and R. M. Bullock, *J. Neurotrauma*, 2017, DOI: 10.1089/neu.2016.4602.
30. MMWR Weekly, Online, https://www.cdc.gov/mmwr/preview/mmwrhtml/mm6120a5.htm (accessed on 28 March, 2017).
31. FDA DRAFT GUIDANCE, Online, https://www.fda.gov/downloads/MedicalDevices/DeviceRegulationandGuidance/GuidanceDocuments/UCM510824.pdf (assessed on March 12, 2017).
32. Federal Register, Online: https://www.gpo.gov/fdsys/pkg/FR-2015-09-25/pdf/2015-24313.pdf (accessed on February 1, 2017).
33. K. Heidari, A. Vafaee, A. M. Rastekenari, M. Taghizadeh, E. G. Shad, R. Eley, M. Sinnott and S. Asadollahi, *Brain Inj.*, 2015, **11**, 1.
34. S. Manzano, I. B. Holzinger, C. J. Kellenberger, L. Lacroix, D. Klima-Lange, M. Hersberger, G. La Scala, S. Altermatt and G. Staubli, *Emerging Med. J.*, 2016, **33**, 42.
35. T. J. Kilbaugh, S. Bhandare, D. H. Lorom, M. Saraswati, C. L. Robertson and S. S. Margulies, *J. Neurotrauma*, 2011, **28**, 763.
36. S. Campbell, E. Elmér and M. Bronnegard, *Intern Neurotrauma* 2012, 31: http://www.internationalbrain.org/tbi-pharmaceuticals-the-long-odyssey-of-cyclosporine/ (accessed on March 12, 2017).
37. C. E. Dixon, H. M. Bramlett, W. D. Dietrich, D. A. Shear, H. Q. Yan, Y. Deng-Bryant, S. Mondello, K. K. Wang, R. L. Hayes, P. E. Empey, J. T. Povlishock, F. C. Tortella and P. M. Kochanek, *J. Neurotrauma*, 2016, **33**, 553.

38. A. St John and C. P. Price, *Clin. Biochem. Rev.*, 2014, **35**, 155.
39. J. Minnerup, B. A. Sutherland, A. M. Buchan and C. Kleinschnitz, *Int. J. Mol. Sci.*, 2012, **13**, 11753, DOI: 10.3390/ijms130911753.
40. A. Pastukhov, N. Krisanova, V. Maksymenko and T. Borisova, *EPMA J.*, 2016, **7**, 26.
41. H. Koizumi, H. Fujisawa, E. Suehiro, S. Shirao and M. Suzuki, *Neurol. Med. Chir.*, 2011, **51**, 337.
42. Stroke Trials Registry Home Page, Online, online: http://www.strokecenter.org/trials/clinicalstudies/ebselen-trial-phase-iii (accessed on 14 March, 2017).
43. V. E. O'Collins, M. R. Macleod, S. F. Cox, L. Van Raay, E. Aleksoska, G. A. Donnan and D. W. Howells, *J. Cereb. Blood Flow Metab.*, 2011, **31**, 962.
44. S. M. Dorhout Mees, A. Algra, W. P. Vandertop, F. van Kooten, H. A. Kuijsten, J. Boiten, R. J. van Oostenbrugge, R. Al-Shahi Salman, P. M. Lavados, G. J. Rinkel, W. M. van den Bergh and MASH-2 Study Group, *Lancet*, 2012, **380**, 44.
45. R. Gultekin, S. Huang, O. Clavisi, L. Pattuwage, T. C. König and R. Gruen, *Injury*, 2016, **47**, 516.
46. V. J. Montaner, M. Hernandez-Guillamon and M. Unzeta Lopez, *Eur Pat.*, 2 357 252 A1, 2011.
47. S. A. Dambinova, *Clin. Lab. Inter.*, 2008, **32**, 7.
48. D. M. Adcock and R. Gosselin, *Thromb. Res.*, 2015, **136**, 7.
49. H. Mani, *Int. J. Lab Hematol.*, 2014, **36**, 261.
50. R. Bonar, E. J. Favaloro, S. Mohammed, L. Pasalic, J. Sioufi and K. Marsden, *Pathology*, 2015, **47**, 355.
51. M. A. Albert, R. J. Glynn, R. L. Wolfert and P. M. Ridker, *Atherosclerosis*, 2005, **182**, 193.
52. S.-H. Lee, S.-M. Kang, S. Park, Y. Jang, N. Chung and D. Choi, *Clin. Cardiol.*, 2011, **34**, 108.
53. S. Racherla and R. Arora, *Am. J. Ther.*, 2012, **19**, 115.
54. J. C. Charniot, R. Khani-Bittar, J. P. Albertini, P. Giral, C. Cherfils, C. Cosson, E. Guillerm, P. Leprince, I. Gandjbakhch and D. Bonnefont-Rousselot, *Int. J. Cardiol.*, 2013, **168**, 132.
55. S. A. Dambinova, *US Pat.,* 20100210816, 2007.
56. P. M. Bokesch, G. A. Izykenova, J. B. Justice, K. A. Easley and S. A. Dambinova, *Stroke*, 2006, **37**, 1432.
57. M. Skitek and A. Jerin, *Lab Med.*, 2013, **44**, 56.
58. A. V. Goryunova, N. A. Bazarnaya, E. G. Sorokina, N. Y. Semenova, O. V. Globa, Zh. B. Semenova, V. G. Pinelis, L. M. Roshal' and O. I. Maslova, *Neurosci. Behav. Physiol.*, 2007, **37**, 761.

CHAPTER 9

Advances in Diagnostics and Treatment of Neurotoxicity after Sports-related Injuries

THOMAS A. GENNARELLI,[a,b] SVETLANA A. DAMBINOVA*[c] AND JOSEPH D. WEISSMAN[c]

[a] Medical College of Wisconsin, West Chester, PA, USA; [b] George Washington University, Washington, DC, USA; [c] DeKalb Medical, Decatur, GA, USA
*Email: dambinova@aol.com

9.1 Introduction

It has been estimated that about 80% of the annual neurological trauma cases in the United States result from concussions, which may also be defined as mild traumatic brain injuries (mTBIs).[1] Sports-related and other mTBIs with Glasgow Coma Scale (GCS) scores in the 13–15 range occur in 1.6–3.8 million young healthy individuals.[2] This GCS score range covers a broad severity of symptoms from subtle to more serious conditions, in some cases accompanied with temporary loss of consciousness.[3]

The assessment of mTBI is complicated, with many cases going unrecognized and often not reported. The impact of various forces on the head can cause primary acute (within 24 hours), sub-acute (up to 14 days), or secondary (beyond 2 weeks) impairments.[4] Symptomatically mTBIs may lead to executive brain dysfunctions, cognitive deficits (problems with

Drug Discovery Series No. 62
Acute Brain Impairment: Scientific Discoveries and Translational Research
Edited by Philip V. Peplow, Svetlana A. Dambinova, Thomas A. Gennarelli and Bridget Martinez

Published by the Royal Society of Chemistry, www.rsc.org

memory, intellect, concentration, and attention), mental status declines (confusion, amnesia, insomnia, or sleepiness), or behavioral changes (aggressiveness, irritability, or drowsiness).[5] The involvement of molecular or cellular events in concussion is revealed in effects at the neuronal and microvascular levels with characteristic patterns of subtle injury (cortical, subcortical, or brainstem areas). Recovery can be complete or partial depending on the extent and severity of injury, perhaps also modulated by genetic factors.[6]

Although there has been considerable recent focus on the axon as a locus in injury, it has become clear that other elements can also respond to the application of mechanical energy to them.[7] Clearly, when mechanical energy is applied to the brain, all of its components are subject to injury if the mechanical load is appropriate. Thus, not only can the axon be injured, but also the neuronal soma, dendrites, synapses, intraneuronal components, and extraneuronal cells, including microglia, oligodendroglia, astrocytes, and vascular elements.

Current therapeutic options mainly involve neuroprotective agents with a mechanism of action directed to the restoration of neuronal function, leaving still elusive the problem of cerebral blood flow (CBF) reperfusion or restoration. Indicators of blood flow restrictions due to physical impact could be a common denominator of the neurovascular unit (NVU) complex (microglia, astrocytes, neurons, pericytes, endothelial cells, and basal lamina).[8] The latter might be associated with biomarkers of the neurotoxic (immunoexcitotoxic) cascade that expands from the area of the original damage to nearby tissues. Consequently, the biomarker values might reflect alterations in velocity and vasoconstriction/vasodilation of capillaries and small- and medium-sized arteries and veins. Measuring the metabolic changes in the NVU as reflected in CBF in the arteriovenous system could enhance development of novel agents for the treatment of mTBI.

Consequently, there is a medical need for innovative chemical agents to expand the diagnostic and therapeutic options for concussion and mTBI through the protection of general NVU functions. The limited knowledge of the NVU role in the neurotoxic cascade is a main reason why there is no specific treatment for mTBI. In addition, there is still no established diagnostic biomarker blood test for expediting diagnosis and treatment. Drug–test co-development may be a key approach to refocusing drug discovery, targeting treatment options for sustainable NVU homeostasis (see Chapter 8 in this book).

The present review is focusing on the possible role of neurotoxicity/immunoexcitotoxicity biomarkers in directing emergent brain-targeted therapy after sports-related brain injury. Trends, pharmacological agents, and non-pharmacological approaches that could optimize NVU homeostasis and improve CBF are discussed.

9.2 Reversible and Irreversible Cerebrovascular Changes after mTBI

9.2.1 The NVU Concept

Minor physical impacts to the brain can cause alteration in capillary CBF and more severe changes in small- and medium-sized vessels. Reversible asymptomatic shifts in the microvessel matrix can be resolved spontaneously without complications, while impairment of small branching vessels in the cerebral vasculature could lead to irreversible changes in the deep brain vasculature accompanied by progressive neurological symptoms. At present, more attention is directed to NVU dysfunction as a reaction to capillary impairments.

The capillary network is a part of the angioarchitecture for all brain structures. It controls the blood supply delivering nutrients and oxygen to nerve cells. The interaction of micro-cerebral perfusion and nervous cell metabolism is known as the NVU and is concentrated in cortical and subcortical areas. NVU functioning in these territories is associated with a range of cognitive processes: attention and inhibitory control, working memory, and cognitive flexibility, as well as reasoning, problem solving, and planning.[9]

The NVU can be compromised by mTBI disruption of the blood–brain barrier (BBB) that produces increased permeability of the brain vasculature. This may result in reversible or irreversible changes depending on the severity of damage.[10]

Degradation of NVU function in the cortex after physical impact is reflected in insufficiency of glucose supply as an energy source, which reflects as transient symptom presentation.[9] Energy deficit could be transient or reversible and could lead to ion balance disturbances, shifts of pH to lower values, and increased release of glutamate, which plays a dual role in the nervous system as an energetic and signal transduction substance. A prolonged state of energy deficit might lead to formation of small-sized lesions that, depending on severity, could result in complete cortical infarction and worsen to symptomatic transient ischemic attack (TIA)/stroke.[11]

9.2.2 Alterations in Small- and Medium-sized Vessels

mTBI may affect arterial microvessel velocity and contractility, as well as causing significant dysfunction—vasospasm in small arteries of subcortical structures due to mechanical effects or to hypoxia.[12,13] Increased microvascular tone is a compensatory mechanism against hypoxic events in NVU, enhancing autoregulation and redistribution of blood flow.[13] If NVU functional disruption is not resolved spontaneously, it can induce white matter micro-hemorrhages with a high risk of extensive hemorrhagic stroke in subcortical areas.[14]

Moreover, sports-related injuries can be associated with more severe TBI with temporary unconsciousness[2] and involvement of transient reversible changes implying small- to medium-sized vessels in deep brain structures.[13] Acute inflammation might contribute to advancement of focal hyper-intensities (2–3 mm) defined by T2-weighted magnetic resonance images (MRIs). Additionally, certain inflammation biomarkers and antibodies (Abs) are detected within 12 hours.[15]

Genetic predisposition to CBF insufficiency, particularly in the posterior venous system, originates from subcortical medium-sized vessels that might be concomitant to more severe conditions.[16] Vertebral/basilar arterial circulation provides substantial blood flow through lateral, medial, and intracranial territories.[17] The carotid vessels can be a source of nasal bleeding caused by physical impact. Delayed consequences to NVU signaling after mTBI might be associated with chronic impairments to medium-sized arteries in cortical/subcortical areas[18–36] or medium-sized venous circulation in ventricular/brainstem NVU territories (see Table 9.1).[37,38]

Medium-sized artery dysfunction linked to compromised BBB, arterio-venous autoregulation, and neurovascular coupling in subcortical and brainstem areas is accompanied by various neurological symptoms including headaches, confusion, and short-term memory problems.[39]

After a forceful brain impact and neck area injury, the immediate formation of vasogenic or cytotoxic edema has been observed.[14] Depending on ischemic or hemorrhagic damage, impaired balance, contusion, temporary visual loss, and short-term unconsciousness have been observed, particularly in boxing (see Table 9.1).

9.3 BBB Dysfunction and Neuroinflammation after mTBI

9.3.1 Compromised BBB

The brain is protected by endothelial and astroglial cellular barriers containing glutamate-dependent metabolic pathways (enzymes, transporters, receptors, and mitochondrial energy activities). The tight junctions between the vascular endothelial cells that form the BBB have the relatively low vascular permeability that limits some macromolecules crossing into or out of the brain.[40]

Even a minor brain impact causes BBB compromise with the consequent release of abnormal catabolic protein fragments from the brain into the bloodstream. The immune system recognizes these fragments as "external antigens" and generates autoimmune reaction with Ab formation leading to the risk of brain autoimmune disorders.[41]

The pursuit in biomarkers research associated with BBB dysfunction resulted in the family of S100 proteins and their auto-Abs. S100B protein is an indicator of astrocytic damage after moderate-to-severe brain injury that reflects BBB disruption.[42] Multiple sub-concussive and/or concussive injuries may allow accumulation of S100B auto-Abs that could be associated with persistent white matter alterations.[43]

Table 9.1 Delayed consequences of glutamatergic dysfunction in NVU signaling.

Psycho-neurological deficit	NVU territory involved[a]	Abs to GluR subtype	Ref.
Autism spectrum disorders	Frontal cortex	NMDAR: NR1	18, 19
Intellectual disability: cognitive deficit, vascular dementia, Alzheimer's disease		NMDAR: NR2, mGluR, AMPAR	20, 21
Schizophrenia	Frontal/occipital	NMDAR: NR1-2	22
Depression	Cortical/subcortical	NMDAR: NR1-2A NR2A; AMPAR: GluR1	23
TIA/stroke	Cortical	NMDAR: NR2	24, 25
mTBI, contusion,	Subcortical	AMPAR: GluR1-2	26
vision impairment,	Cortical/subcortical	AMPAR, NMDAR	27, 28
ocular disorders	Occipital		
Epilepsy	Hippocampus, pons, corpus callosum	AMPAR: GluR2-3,	29, 30
Lacunar stroke	White matter, ventricle	AMPAR	31
Sleep apnea	Subcortical	AMPAR: GluR1-2	32
Drug and alcohol addictions, nicotine and opiate dependence	Cortical/subcortical limbic system	AMPAR: GluR1-3 NMDAR: NR1	33–35
Lupus erythematosus, neurovasculitis	Periventricular, white matter	NMDAR: NR2	36
Intracerebral micro-hemorrhage	Ventricular, basilar	AMPAR/kainite receptor	37
Disorders of unconsciousness	Brainstem, cervical	Kainite receptor	38

[a]Defined by MRI or magnetic resonance angiography (MRA).

9.3.2 Neuroinflammation

Neuroinflammation includes two major mechanisms: (1) immunocompetent reactions intrinsic to the central nervous system (CNS) that are associated with NVU repair; and (2) external reactions of the immune system to small vessel injury. mTBI triggers inflammatory changes within the nervous system that up-regulate cytokines, inflammatory genes, and immune responses within nervous tissues after acute and sub-acute impacts.[44] A neuroinflammatory unified concept relating glutamate release and activation of immune receptors to oxidative stress and potentially later cellular injury has been proposed and termed "immunoexcitotoxicity."[45] This process involves secondary irreversible injury to medium-sized vessels located in deep brain areas due to a physical impact. External autoimmune reactions result in delayed chronic neurological complications (see Table 9.1). It might result in the development of CNS vasculitis (inflammation of cerebral vessel walls in small- and medium-sized cerebral arteries) or lupus erythematosus, a vasculopathy (heterogeneous disease characterized by multisystem autoimmunity,[36] leading to an array of clinical presentations) with involvement of CNS.[46,47] Both disease conditions might cause diffuse neuronal/axonal loss, microemboli, blood vessel remodeling, parenchymal edema, global ischemic changes, micro- and macrohemorrhages, and resolved intracranial hemorrhages requiring an emergent treatment.[46,47]

9.3.3 Vasogenic and Cytotoxic Edema Formation

Cerebral edema is a dramatic alteration in the intra/extracellular ion and water balance, causing increased brain tissue volume and intracranial pressure as the major immediate consequences of neurotrauma.[48] Vasogenic (developing within 48 hours after TBI) and cytotoxic (developing beyond 48 hours after neurotrauma) edemas are the two broad categories of edema. Whether these changes are transitory or long lasting, regional vascular permeability depends on the time course, location (mostly subcortical and deep brain structures), and severity of injury. Global edema usually results in unconsciousness, coma, and even sudden death after severe neurotrauma. This is often accompanied by subarachnoid hemorrhage or delayed cerebral ischemia.[49,50] These life-threatening conditions may require emergent neurosurgical intervention.[51]

9.4 Focal and Traumatic Axonal Injury Correlates with Altered CBF

9.4.1 Traumatic Axonal Injury and Altered CBF

The frontal cortical and subcortical areas that are impacted by mTBI are controlled by arterioles—small arteries participating in internal anterior circulation. Posterior blood circulation is associated with occipital medium-sized vessels in the vertebrobasilar distribution regulating venous fluid drainage.[17]

Both blood circulating systems control the oxygenation/deoxygenation cycle and glucose metabolite turnover in focal and vertebrobasilar diffuse neuronal damage.[52] Axonal injuries in these cases reflect the disconnection between neurons in a determined territory controlled by changed blood flow.[53]

Axonal stretching and shearing are changes that may be reversible, occur within 24 hours of injury, and may cause mitochondrial swelling, alterations in cell permeability transition, and cytoskeletal abnormalities.[52] Microtubule disruption can interfere with axonal transport, potentially influencing synaptic neurotransmission or, in severe cases, prompting temporary or prolonged axonal disconnections within cortical, subcortical, and brainstem areas.[54,55] Early hemodynamic responses of brain structure layers are distinct and represent a complex implication for outcome. Hypoperfusion in deep sub-cortical structures results in poor tissue oxygenation and may lead to a more severe mTBI state.

9.4.2 Radiological Findings in mTBI

With advanced MRI, particularly susceptibility-weighted imaging and mapping technology, a better understanding and visualization of alterations in brain perfusion and oxygenation in impacted brain regions has occurred. Recently, increased CBF combined with induced venous oxygenation have been shown in persons with acute mTBI in contrast to regional hypoxia and cerebral hemodynamic changes in sub-acute and chronic stages.[13,53] It seems that increased blood flow to brain-impacted structures is not a long-lived response to mTBI that might be followed by CBF reduction correlating with delayed ischemic/hypoxic events within 1–2 weeks after mTBI.[14]

The additional information concerning microstructural rearrangements due to mTBI by advanced neuroimaging technologies should be helpful. Diffusion tensor imaging (DTI) may correlate with areas of traumatic axonal injury (TAI) with initially decreased fractional anisotropy (measuring linear water diffusion) in white matter tracts and myelin sheaths.[52] To date, only advanced 3 T DTI might reveal microscopic multiple lesions or TAI after mTBI with GCS scores of 13. Recent studies have reported changes in DTI images particularly in subcortical areas connected to cognitive impairment and amnesia.[56,57] However, some discrepancies between radiological findings and the clinical symptoms presentation were noted.[58–60]

This disagreement could be resolved by using cerebral vascular biomarkers that are navigating areas of interest in frontal cortical/subcortical or vertebrobasilar territories. It is evident that persons with brainstem injuries have a significantly higher risk for developing unconsciousness than those with forebrain impairment.[61,62] There is growing evidence that 3 T MRI (DTI and diffusion-weighted imaging [DWI] modalities) could be used in objective grading of mTBI severity in combination with NVU biomarkers sensitive to neurotoxicity/immunoexcitotoxicity.[63] It seems that the use of these biomarkers in addition to radiological findings might add to the assessment of brain damage severity. Primary NVU dysfunction is probably related to neuronal/glial synaptic disconnections in subcortical and/or brainstem areas and the likelihood of secondary cumulative neurological complications (see Table 9.1).

9.5 Biomarkers for Acute and Chronic Conditions after mTBI

9.5.1 Prospective Biomarkers of NVU

The autoregulation of CBF includes neuronal control by chemical signals implying excitatory amino acids (glutamate, aspartate, and their derivatives).[10] It was demonstrated that abundant concentrations of these amino acids (mainly derived from the extracellular space) in brain biological fluids lead to massive neuronal death in severe cases by overloading the glutamatergic system, named the neurotoxicity/immunoexcitotoxicity cascade.[45]

The glutamatergic system consists of a major family of approximately 25 proteins comprising channel-activated ionotropic receptors (fast GluR) or G-protein-regulated metabotropic receptors (slow GluR). Both types of receptor are strategically located either on the surfaces of cerebral vessels like fast GluR (cortical, subcortical, and brainstem areas)[64–66] or slow metabolic GluR regulating energy flow on mitochondrial levels of astrocytes and thrombocytes.[67,68]

In acute neurotrauma, GluR peptide fragments carrying immune active epitopes are capable of passing through the compromised BBB into biological fluids, accumulating in detectable amounts.[69] The protective effects of a compromised BBB impacting neurotoxicity are exacerbated further when accompanied by a delayed immunological response generating peripheral anti-CNS Abs.[70–72] These Abs (see Table 9.1) may be utilized as potential prognostic biomarkers of brain injury.[45]

Another essential BBB-protective mechanism against mTBI involves glial fibrillary acidic protein (GFAP), a monomeric intermediate filament expressed by astrocytes. The protein may serve better as a biomarker of hemorrhage in that elevated concentrations might be related to the number of boxing bouts.[73] Studies in adults with mTBI have also demonstrated that a breakdown product of GFAP (GFAP-BDP) was increased in patients with intracranial lesions on computed tomography (CT) and was able to be used to triage patients for neurosurgical intervention.[74–76] This compound demonstrated very good predictive ability (area under curve = 0.87), as well as significant discrimination of injury severity.[76] However, it was recently stated that the GFAP biomarker is not specific enough for detecting CT-negative mTBI.[77]

9.5.2 GluR Peptide and Ab Translational Research

In recent years, the role of GluR subtypes in CBF regulation in distinct brain territories has become clear. Indeed, NR2 subtypes of N-methyl-D-aspartate receptors (NMDARs) regulate predominantly gray matter (cortical) associated with arteriole/capillary functions,[64] the ionotropic α-amino-3-hydroxy-5-methyl-4-isoxazolepropionic acid receptor (AMPAR) subtypes most likely control small- to medium-sized artery tasks in white matter,[65] while kainate receptors (KARs) potentially affect vertebrobasilar venous circulation in the mid-brain and hind brain.[66,67]

Glutamate neuroreceptor biomarkers generally represent N-terminal peptide fragments of AMPARs, NMDARs, or KARs, with certain immune active epitopes that are severed from synaptic membrane surfaces by serine proteases.[69]

A feasibility study detecting the AMPAR peptide in a single blood draw taken on enrollment from all participants evaluated a preliminary cutoff (0.4 ng mL^{-1}) for the emergent diagnosis of mTBI before other diagnostic procedures.[26] The predictive value of the test was assessed as 91% (0.4 ng mL^{-1} cutoff) with a likelihood ratio of 11.9 for recognizing individuals with concussions. Additionally, longitudinal observation of the athlete with persistent symptoms demonstrated a number of white matter fiber tracts that were absent in frontal and parietal areas of the brain defined by 3 T DTI, indicating the severity of concussion and the need for treatment during the recovery.[37] The condition is accompanied by a steady increase in AMPAR peptide assays during the entire study.

In the same longitudinal study, AMPAR Ab values remained abnormally high in the blood of some athletes who had headaches and visual problems. It was suggested that this finding may reflect persistent changes in the subcortical areas of the brain.[37] The diagnostic potential of KAR Abs (sensitivity of 86–88% and specificity of 83–97% at 1.5 ng mL^{-1} cutoff) in the assessment of seizures defined by electroencephalogram as a consequence of sustained single or multiple mTBIs has been demonstrated for child and adult patients.[78–82]

The NR2 peptide assay indicated a capacity to recognize acute cerebrovascular accident with a sensitivity of 92–98% and a specificity of 72–97% at 0.5 and 1.0 ng mL^{-1} cutoffs.[83] A somewhat lower specificity (72% at 0.5 ng mL^{-1} cutoff) could be due to possible edema development in up to 14% of athletes with acute or sub-acute concussions in otherwise healthy young subjects.[37,84]

NR2 Abs were observed in 29 (34.5%) of 84 subjects with concussions; this was 2.5-times higher than that of the peptide values.[84] The latter might be a predictive factor for TIA with a sensitivity of 98% at a cutoff value of 2.0 ng mL^{-1}, with risk ratio of 33.[24,85] If NR2 Abs maintained an increased value beyond the cutoff (2.0 ng mL^{-1}) in athletes with sports-related concussions, then they should be directed to 3 T DWI/fluid-attenuated inversion recovery/DTI by the clinician to evaluate the presence of possible ischemic lesions.[38]

A pilot study of KAR peptide in active duty personnel ($n = 50$) with subacute mTBI showed a sensitivity of 90% and a specificity of 83% at the 1.0 ng mL^{-1} cutoff for brainstem impairments.[38] Preliminary assessment of the same peptide in patients with spinal cord injury has indicated the biomarker's diagnostic utility for assessing venous circulation in the cervical area of the spinal cord.[86]

9.6 Conventional Treatment Options for mTBI

The brain possesses protective forces that are able to heal and spontaneously restore up to 80% of impaired NVU functions after concussions, particularly in young athletes.[5] The reversible plasticity of NVU signaling could be evaluated by biomarkers associated with CBF assessing the degree of repair from vasogenic edema and inflammation revealed on neuroimaging scans.[52] There would be a main goal of rehabilitation after mTBI to stimulate NVU regulation and recovery of essential electrical signals in cortical/subcortical disconnections by pharmacological agents in severe cases. It would be plausible to use a non-pharmacological approach in concussion management to stabilize CBF and improve NVU functioning.

Conventional drug development has been a "one-size-fits-all" approach, with drugs developed based on antagonists of the glutamatergic system that are well established and characterized in translational research (see Table 9.2).[87–100] As long as a statistically significant benefit is seen in a target patient population, a drug can be approved for widespread use regardless of the probability that it will actually help a specific patient on an individual level.

Targeted therapy directed to curing NVU dysfunction due to mTBI is still in progress.[101] The extreme growth of sports-related injuries and the need to prevent neurological consequences demand personalized management of an otherwise healthy population.

9.6.1 Trends in Emergent Pharmacological Agents for mTBI

Efficient emergent treatments of acute mTBI should address the search for and selection of novel chemical agents that might have multiple effects on CBF circulation within 24 hours of the therapeutic window. Firstly, such treatments should be chemicals that easily penetrate the BBB (size <500 kDa) and rapidly reach the NVU in concentrations exhibiting neuroprotective and regenerative properties. Secondly, these chemicals should have lipophilic origin and bear positive polarity to pass the BBB. Thirdly, these agents should have rather short linear structures, not complex formations, and degrade to non-toxic and inactive simple molecules without side effects on executive brain functions. Then, these chemical residues must be easily captured and extracted by albumin through transport pathways.[102] These stipulations explain the difficulties in the area of mTBI drug discovery when pharmacological agents should satisfy major demands: secure the NVU velocity and protect higher brain functions.

Table 9.2 Selected compounds targeting neurotoxicity.

Medical condition	Drug	Pharmacological properties	Target	Therapeutic effects: Positive	Therapeutic effects: Negative	Ref.
In clinical use						
Alzheimer's disease Dementia Autism Migraine	Amantadine (memantine, rimantadine)	NMDAR antagonist D2 receptor agonist	Ca^{2+} channel pre-synaptic membranes	Decrease neuropathic pain, cognitive improvement	Dizziness, fainting, lightheadedness	87, 88
Depression and other psychiatric conditions	Dextromethorphan (DMX or "angel")	NMDAR/ quinidine antagonist, opioid receptor agonist	Inhibits receptors' enzyme degradation	Increases energy, decreases agitation	Sedation, euphoria, emotional disability	89, 90
Pain Sleep problems Depression	Ketamine	NMDAR and AMPAR non-competitive antagonist	Receptor binding site blocking, Ca^{2+} channel activation	Antidepressant, anesthetic	Hallucination, confusion, amnesia, addiction	91, 92
Schizophrenia Stroke TBI	Dizocilpine (MK801)	NMDAR and AMPAR antagonist	Ca^{2+} channel blocker	Anticonvulsant, anesthetic, sedation	Psychosis	93, 94
Epilepsy TBI	Aniracetam	AMPAR allosteric modulator	Inhibits AMPAR endocytosis	Antiepileptic, enhanced memory, attention	Insomnia, anxiety, pain, nausea, vertigo	95
Hypertension Cytotoxic edema	Cyclothiazide	Kynurenic acid antagonist	Induced long-term potentiation	Antihypertensive, diuretic	Dizziness, dehydration	96

Table 9.2 (*Continued*)

Medical condition	Drug	Pharmacological properties	Target	Therapeutic effects: Positive	Therapeutic effects: Negative	Ref.
Pre-clinical research (experimental drug)						
Anxiety Spatial memory deficit	Dihydrokainic acid	NMDAR/KAR antagonist	Selective inhibitor of Glu transporter and kynurenine oxy-gliosis	Reduce extracellular glutamate over-activation	N/A	97
Memory deficit Anxiety Disorders of unconsciousness	JM-6 G-111	NMDAR/KAR antagonist	Selective inhibitor Na^+/K^+ channels	Skeletal muscle relaxant, strong anticonvulsant	Sleep apnea	98
TBI Stroke	Oxaloacetate	mGluR antagonist, glutamate pump regulator	Transport diffusion, enzyme regulation	Decreases swelling, shuts down glutamate overload	N/A	99, 100

Numerous attempts in the development of drugs to cure neurotoxicity effects have sometimes yielded unwanted and unexpected side effects (see Table 9.2). These include cognitive problems, memory declines, mental disabilities, and other complications of executive brain functions.[87–97] A number of prior trials of various GluRs and their transporter antagonists and blockers failed due to safety issues and inefficacy.

Naturally derived inhibitors of GluR—small-sized peptides—are innovative chemical agents that could be used to promote the reduction of neurotoxicity/immune excitotoxicity in the NVU after mTBI. Therapeutic peptides have been designed by virtue of extended clinical studies of -electrical stimulations of subcortical and deep brain structures.[103] Pre-developmental and preclinical studies of the peptide for detecting origins, binding capabilities, safety, and treatment doses in experimental models of seizures, incomplete TBI, and transient cerebral ischemia were performed.[104] Metallopeptides have been designed with chelate properties that aid in removing and neutralizing Fe^{2+} release due to hemoglobin proteolysis when bleeding occurs.[105]

Intranasal administration of the metallopeptide Gafargin after acute mTBI decreases microbleeding and edema size, minimizes inflammation consequences, and reduces the acidity of molecules targeting the arteriovenous circulation.[7] Gafargin consists of Gly, Arg, and Gln[104] in complex with non-covalently bound Fe^{3+} that captures the free form of Fe^{2+} and deactivates its toxic effect, as shown by experimental research. Gafargin also affects metalloproteases and protects subcortical areas from microlesion development, decreasing the risk of abnormal spiking activity that frequently occurs following mTBI.[52]

At present, particular attention is being paid to the use of novel metallopeptides for nervous disorder therapies. Indeed, metallopeptides for the treatment of Alzheimer's disease (AD) demonstrated that Arg-2 and Lys-16 can inhibit cooper (Cu^{2+}) binding complexes found in large amounts in the β-amyloid plaques that have been attributed to AD.[106] Another metallopeptide that is specific to opioid receptors has been designed to improve drug delivery through the BBB into the brain to produce long-term pain relief with minimal addiction properties.[107] A metallopeptide agonist of melanocortin-3 and melanocortin-4 receptors binds ions of nickel (Ni^{2+}) or zinc (Zn^{2+}) in its complex and is capable of enhancing arteriovenous circulation in deep brain structures.[108] Refocusing attention on medications sustaining the NVU against the neurotoxic cascade might stimulate research into chemical agents that are capable of optimizing CBF and promoting nervous tissue functional recovery.

9.6.2 Non-pharmacological Treatment Options for Concussions

There are several approaches to the non-pharmacological management of healthy young athletes after sports-related injuries employing use of hyperbaric and normobaric oxygenation to improve oxygen delivery to the brain,[109] the addition of relevant nutrients,[110] and physical exercise tailored to the brain-impacted area.[111] In general, non-pharmacological programs like acupuncture[112] concentrate on NVU functions, improving CBF velocity, auto-pressure, and vasodilatation of the brain microvessel functions, facilitating oxygen and carbohydrate metabolic exchange.[113]

Nutrients (vitamins, minerals, and amino acids) and other natural agents like capsaicin,[114] eugenol,[115] and flavonoids[116] could optimize prolonged non-specific effects for the cerebral arteriovenous circulation. A dietary supplemental formula for concussion should include natural substances and specific agents targeting the recovery of energetic and osmotic balances and protecting the impacted capillary system from further functional decline.[112]

Non-pharmacological approaches to recovery after acute and sub-acute concussions have been offered by University of Pittsburgh Medical Center Sport Medicine.[117] After individual assessment of vestibular and ocular motor impairments (smooth pursuits, saccadic or rapid eye movements, near point of convergence, vestibular ocular reflex, and visual motion sensitivity), the medical expert identifies the dominant type of concussion (up to 90% accuracy).[118,119] Then, a tailored physical exercise program is offered according to the impaired brain structure. For example, vestibular therapy is an exercise-based program that particular addresses compensation in the brainstem and cerebellum, as well as visual and somatosensory sensations.[120] Several non-contact activities (swimming, walking, and running) have also been recommended for adjusting behavioral regulation (sleep, headache, and sensitivity to light and noise).

It seems that the addition of rapid blood testing could be useful for the objective assessment of concussion severity in order to single out groups with more serious conditions and to direct athletes for personalized exercising programs, reinforcing NVU functioning.

9.7 Conclusion

At present, a number of treatment demands in drug discovery for mTBI should be fulfilled, including corrections of acute capillary ionic imbalance and microbleeding consequences researched in conjunction with neuroimaging and key biomarkers of CBF in cortical, subcortical, and brainstem areas.

Current approaches optimizing mTBI treatment and avoiding limitations of drug usage in the CNS should take into account stabilization of hemodynamic responses in the acute state of mTBI. There are issues that should be solved in drug design that will employ the ability to pass the BBB and transform into a bioactive form within the targeted area as captured by specific navigators (Abs or gene vectors). Likewise, in order to prolong the effects of drugs, the agent carriers for lowering concentrations of bioactive molecules should be considered as bound to nanoparticles or inserted into liposomes. To evade toxic side effects of novel agents, metabolic digestion into simple inactive molecules has to be implemented.

Brain-derived biomarkers (GluR subunits) reflecting the neurotoxicity cascade in the NVU may aid in the stratification of mTBI management and diminish the risk of complications due to dysfunction of CBF circulation. In fact, the NR2 NMDAR subtypes involved in the regulation of cerebral arterial microvessel functions have shown sensitivity to energy depletion. Ionotropic GluR5-7 kainate receptors are susceptible to the carbon dioxide balance that potentially affects venous circulation. At the same time, oxygen-dependent ionotropic GluR1-3 AMPAR subtypes might indicate subcortical micro-bleeding and dendrite–axonal injury in white matter. The implementation of these key biomarkers might facilitate the drug discovery process.

Recent translational research of GluR metallopeptides that overcome most restrictions in drug design showed promising results in nervous cell proliferation. They might reduce neurotoxicity through reversible inhibition of glutamate receptors, stabilizing cerebral capillary functions. Other options in the management of more severe TBI could include stem cell therapies that rearrange and replace nervous cells in order to provide symptomatic relief (see Chapter 13 in this book).

The pharmacological agent research pathway and non-pharmacological approaches for the improvement of CBF after mTBI described herein should be further explored in order to protect executive brain functions and to secure the quality of life.

References

1. D. J. Harvey, J. Freeman, D. K. Broshek and J. T. Barth, in *Text book of traumatic brain injury*, ed. J. M. Silver, T. W. McAllister, S. C. Yudofsky, American Psychiatric Association, Arlington, 2nd edn, 2011, ch. 27, p. 427.
2. P. Leo and M. McCrea, in *Translational Research in Traumatic Brain Injury*, ed. D. Laskowitz, G. Grant, CRC Press/Taylor and Francis Group, Boca Raton, 2016, ch. 1, https://www.ncbi.nlm.nih.gov/books/NBK326730/ (last accessed March 2017).
3. T. A. Gennarelli, *Prog. Neurol. Surg.*, 2014, **28**, 243.
4. A. L. Petraglia, J. C. Maroon and J. E. Bailes, *Neurosurgery*, 2012, **70**, 1520.
5. J. C. Maroon, C. Mathyssek and J. Bost, *Prog. Neurol. Surg*, 2014, **28**, 1.
6. N. J. Porcher and T. J. Solecki, *J. Chiropr. Med.*, 2013, **12**, 260.
7. T. A. Gennarelli, presented in part at IRCOBI Conference, Lyon, September, 2015, http://www.ircobi.org/wordpress/downloads/irc15/pdf_files/02.pdf.
8. Y. Liu, B. Li, N. Feng, H. Pu, X. Zhang, H. Lu and H. Yin, *PLoS One*, 2016, DOI: 10.1371/ journal.pone.0156016.
9. C. O. Tan, W. P. Meehan, G. L. Iverson and J. A. Taylor, *Neurology*, 2014, **83**, 1665.
10. H. L. McConnell, C. N. Kersch, R. L. Woltjer and E. A. Neuwelt, *J. Biol. Chem.*, 2017, **292**, 762.

11. H. C. Abeysinghe, E. L. Phillips, H. Chin-Cheng, P. M. Beart and C. L. Roulston, *Int. J. Mol. Sci.*, 2016, **17**, 288.
12. D. R. Kramer, J. L. Winer, B. A. M. Rease, A. P. Amar and W. J. Mack, *Neurol. Res. Inter.*, 2013, DOI: 10.1155/2013/415813.
13. H. Doshi, N. Wiseman, J. Liu, W. Wang, R. D. Welch, B. J. O'Neil, C. Zuk, X. Wang, V. Mika, J. P. Szaflarski, E. M. Haacke and Z. Kou, *PLoS One*, 2015, **10**, e0127487.
14. L. Østergaard, T. S. Engedal, R. Aamand, R. Mikkelsen, N. K. Iversen, M. Anzabi, E. T. Næss-Schmidt, K. R. Drasbek, V. Bay, J. U. Blicher, A. Tietze, I. K. Mikkelsen, B. Hansen, S. N. Jespersen, N. Juul, J. C. Sørensen and M. Rasmussen, *J. Cereb. Blood Flow Metab.*, 2014, **34**, 1585.
15. J. Zhang, V. Puvenna and D. Janigro, in *Translational Research in Traumatic Brain Injury*, ed. D. Laskowitz, G. Grant, CRC Press/Taylor and Francis Group, Boca Raton, 2016, ch. 12, https://www.ncbi.nlm.-nih.gov/books/NBK326730/ (last accessed March 2017).
16. F. De Guio, A. Vignaud, S. Ropele, M. Duering, E. Duchesnay, H. Chabriat and E. Jouvent, *Stroke*, 2014, **45**, 2124.
17. J. V. Byrne, in *Tutorials in Endovascular Neurosurgery and Interventional Neuroradiology*, Springer, Berlin, 2012, ch. 2, p. 37.
18. Y. Hacohen, S. Wright, J. Gadian, A. Vincent, M. Lim, E. Wassmer and J. P. Lin, *Dev. Med. Child Neurol.*, 2016, **58**, 1092.
19. R. L. Blaylock and A. Strunecka, *Curr. Med. Chem.*, 2009, **16**, 157.
20. B. Di Marco, C. M. Bonaccorso, E. Aloisi, S. D'Antoni and M. V. Catania, *CNS Neurol. Disord. Drug. Targets*, 2016, **15**, 448.
21. H. Mehregan, H. Najmabadi and K. Kahrizi, *Arch. Iran Med.*, 2016, **19**, 363.
22. C. Hoffmann, S. Zong, M. Mané-Damas, P. Molenaar, M. Losen and P. Martinez-Martinez, *Antibodies*, 2016, **5**, DOI: 10.3390/antib5020009.
23. C. Levite, *J. Neural. Transm.*, 2014, **121**, 1029.
24. J. D. Weissman, G. A. Khunteev, R. Heath and S. A. Dambinova, *J. Neurol. Sci.*, 2011, **300**, 97.
25. D. M. Stanca, I. C Mărginean, O. Soriţău, C. Dragoş, M. Mărginean, D. F. Mureşanu, J. C. Vester and A. Rafila, *J. Cell. Mol. Med.*, 2015, **19**, 2253.
26. S. A. Dambinova, A. V. Shikuev, J. D. Weissman and J. D. Mullins, *Mil. Med.*, 2013, **178**, 285.
27. A. U. Brandt, T. Oberwahrenbrock, J. Mikolajczak, H. Zimmermann, H. Prüss, P. Friedemann and C. Finke, *Neurol. Neuroimmunol. Neuroinflamm.*, 2016, **3**, e198.
28. J. L. Cueva Vargas, I. K. Osswald, N. Unsain, M. R. Aurousseau, P. A. Barker, D. Bowie and A. Di Polo, *J. Neurosci.*, 2015, **35**, 12088.
29. A. Nibber, L. Clover, P. Pettingill, P. Waters, C. E. Elger, C. G. Bien, A. Vincent and B. Lang, *Eur. J. Paediatr. Neurol.*, 2016, **20**, 222.
30. H. Goldberg-Stern, Y. Ganor, R. Cohen, L. Pollak, V. I. Teichberg and M. Levite, *Psychoneuroendocrinology*, 2014, **40**, 221.

31. M. D. Valdés Hernández, X. Qiu, X. Wang, S. Wiseman, E. Sakka, L. C. Maconick, F. Doubal, C. L. Sudlow and J. M. Wardlaw, *Brain Behav.*, 2016, **7**, e00595.
32. D. L. Koo, J. Y. Kim, J.-S. Lim, H.-M. Kwon and H. Nam, *J. Clin. Sleep Med.*, 2017, **13**, 65.
33. F. Capone, W. Adriani, M. Shumilina, G. Izykenova, O. Granstrem, S. Dambinova and G. Laviola, *Psychopharmacology*, 2008, **197**, 535.
34. O. Granstrem, W. Adriani, M. Shumilina, G. Izykenova, S. Dambinova and G. Laviola, *Neurosci. Lett.*, 2006, **403**, 1.
35. S. A. Dambinova and G. A. Izykenova, *Ann. N. Y. Acad. Sci.*, 2002, **965**, 497.
36. E. H. Chang, B. T. Volpe, M. Mackay, C. Aranow, P. Watson, C. Kowal, J. Storbeck, P. Mattis, R. Berlin, H. Chen, S. Mader, T. S. Huerta, P. T. Huerta and B. Diamond, *E. Bio. Med.*, 2015, **2**, 755.
37. S. A. Dambinova, R. L. Sowell and J. C. Maroon, *J. Mol. Biomark Diagn.*, 2013, **S3**, DOI: 10.4172/2155-9929.S3-003.
38. S. A. Dambinova, J. D Mullins and T. Gennarelli, in *Neurotrauma: A Comprehensive Book on Traumatic Brain Injury and Spinal Cord Injury*, ed. K. K. Wang, Oxford University Press, New York, 2017, ch. XI, 4, in press.
39. L. Østergaard, T. S. Engedal, F. Moreton, M. B. Hansen, J. M. Wardlaw, T. Dalkara, H. S. Markus and K. W. Muir, *J. Cereb. Blood Flow. Metab.*, 2016, **36**, 302.
40. J. Laterra and G. W. Goldstein, in *Principles of Neural Science*, ed. E. R. Kandel, J. H. Schwartz and T. M. Jessell, McGraw-Hill, New York, 2000, 4th edn, p. 1288.
41. S. Amor and R. Huizinga, in *Neurological Disorders (Neurology-Laboratory and Clinical Research Developments)*, Novinka Books, Bel Air, 2011.
42. C. B. Jeter, G. W. Hergenroeder, M. J. Hylin, J. B. Redell, A. N. Moore and P. K Dash, *J. Neurotrauma*, 2013, **30**, 657.
43. J. J. Bazarian, T. Zhu, J. Zhong, D. Janigro, E. Rozen, A. Robert, H. Javien, K. Merchant-Borna, B. Abar and E. G. Blackman, *PLoS One*, 2014, **9**, e94734, DOI: 10.1371/journal.pone.0094734.
44. D. Lozano, G. S. Gonzales-Portillo, S. Acosta, I. de la Pena, N. Tajiri, Y. Kaneko and C. V. Borlongan, *Neuropsychiatr. Dis. Treat.*, 2015, **11**, 97.
45. R. L. Blaylock and J. C. Maroon, in *Biomarkers for TBI*, ed. S. A. Dambinova, R. L. Hayes, K. K. W. Wang, Royal Society of Chemistry, London, 2012, ch. 3, p. 45.
46. Y. Kaichi, S. Kakeda, J. Moriya, N. Ohnari, K. Saito, Y. Tanaka, F. Tatsugami, S. Date, K. Awai and Y. Korogi, *Am. J. Neuroradiol.*, 2014, **35**, 100.
47. D. S. Younger, *NeurolClin.*, 2016, **34**, 887.
48. P. H. Iffland II, G. A. Grant and D. Janigro, in *Vascular mechanisms in CNS trauma*, ed. E. H. Lo, J. Lok, M. Ning, M. Whalen, Springer Science, New York, 2014, p. 29.
49. A. Beaumont and T. Gennarelli, *Acta. Neurochir. Suppl.*, 2006, **96**, 30.

50. J. F. Burke, J. L. Stulc, L. E. Skolarus, E. D. Sears, D. B. Zahuranec and L. B. Morgenstern, *Neurology*, 2013, **81**, 33.
51. P. J. Hutchinson, A. G. Kolias, I. S. Timofeev, E. A. Corteen, M. Czosnyka, J. Timothy, I. Anderson, D. O. Bulters, A. Belli, C. A. Eynon, J. Wadley, A. D. Mendelow, P. M. Mitchell, M. H. Wilson, G. Critchley, J. Sahuquillo, A. Unterberg, F. Servadei, G. M. Teasdale, J. D. Pickard, D. K. Menon, G. D. Murray, P. J. Kirkpatrick and RESCUEi cp Trial Collaborators, *N. Engl. J. Med.*, 2016, **375**, 1119.
52. E. D. Bigler, in *Brain Neurotrauma: Molecular, Neuropsychological, and Rehabilitation Aspects*, ed. F. H. Kobeissy, CRC Press/Taylor & Francis, Boca Raton, 2015, ch. 31.
53. A. Pomschar, I. Koerte, S. Lee, R. P. Laubender, A. Straube, F. Heinen, B. Ertl-Wagner and N. Alperin, *PloS One*, 2013, **8**, e55447.
54. P. Dall'Acqua, S. Johannes, L. Mica, H. P. Simmen, R. Glaab, J. Fandino, M. Schwendinger, C. Meier, E. J. Ulbrich, A. Müller, L. Jäncke and J. Hänggi, *Front. Hum. Neurosci.*, 2016, **10**, DOI: 10.3389/fnhum.2016.00127.
55. L. Delano-Wood, K. J. Bangen, S. F. Sorg, A. L. Clark, D. M. Schiehser, N. Luc, M. W. Bondi, M. Werhane, R. T. Kim and E. D. Bigler, *Brain Imaging Behav.*, 2015, **9**, 500.
56. T. Wada, Y. Asano and J. Shinoda, *Am. J. Neuroradiol.*, 2012, **33**, 2117.
57. S. H. Jang, J. H. Yi and H. G. Kwon, *Brain Inj.*, 2016, **30**, 1271.
58. T. W. McAllister, J. C. Ford, S. Ji, J. G. Beckwith, L. A. Flashman, K. Paulsen and R. M. Greenwald, *Ann. Biomed. Eng.*, 2012, **40**, 127.
59. M. L. Lipton, N. Kim, M. E. Zimmerman, M. Kim, W. F. Stewart, C. A. Branch and R. B. Lipton, *Radiology*, 2013, **268**, 850.
60. E. M. Davenport, K. Apkarian, C. T. Whitlow, J. E. Urban, J. H. Jensen, E. Szuch, M. A. Espeland, Y. Jung, D. A. Rosenbaum, G. A. Gioia, A. K. Powers, J. D. Stitzel and J. A. Maldjian, *J. Neurotrauma*, 2016, **33**, 2133.
61. A. A. Potapov, N. E. Zarharova, V. N. Kornienko, I. N. Pronin, E. V. Alexandrova, O. S. Zaitsev, L. B. Likhterman, A. G. Gavrilov, G. V. Danilov, A. V. Oshorov, A. A. Sychev and A. A. Polupan, *Zh. Vopr. Neirokhir. Im. N. N. Burdenko*, 2014, **78**, 4.
62. E. V. Alexandrova, O. S. Zaitsev and A. A. Potapov, *Zh. Vopr. Neirokhir. Im. N. N. Burdenko*, 2014, **78**, 26.
63. S. A. Dambinova, J. C. Maroon, A. M. Sufrinko, J. D. Mullins, E. V. Alexandrova and A. A. Potapov, *Front. Neurol.*, 2016, **7**, 172.
64. C. D. Sharp, M. Fowler, T. H. Jackson 4th, J. Houghton, A. Warren, A. Nanda, I. Chandler, B. Cappell, A. Long, A. Minagar and J. S. Alexander, *BMC Neurosci.*, 2003, **4**, 28.
65. P. C. Christensen, Z. Samadi-Bahrami, V. Pavlov, P. K. Stys and G. R. Moore, *Neurosci. Lett.*, 2016, **630**, 1.
66. X. T. Jin and Y. Smith, *Adv. Exp. Med. Biol.*, 2011, **717**, 27.
67. A. L. Alhadeff, R. A. Holland, A. Nelson, H. J. Grill and B. C. De Jonghe, *J. Neurosci.*, 2015, **35**, 11094.

68. D. Gebremedhin, K. Yamaura, C. Zhang, J. Bylund, R. C. Koehler and D. R. Harder, *J. Neurosci.*, 2003, **23**, 1678.
69. S. A. Dambinova, in: *Biomarkers for TBI*, ed. S. A. Dambinova, R. L. Hayes, K. K. W. Wang, Royal Society of Chemistry, London, 2012, ch. 4, p. 66.
70. B. Engelhardt and R. M. Ransohoff, *Trends Immunol.*, 2005, **26**, 485.
71. A. Vincent, C. G. Bien, S. R. Irani and P. Waters, *Lancet Neurol.*, 2011, **10**, 759.
72. M. Raad, E. Nohra, N. Chams, M. Itani, F. Talih, S. Mondello and F. Kobeissy, *Neuroscience*, 2014, **28**, 16.
73. S. Neselius, H. Brisby, A. Theodorsson, K. Blennow, H. Zetterberg and J. Marcusson, *PLoS One*, 2012, **7**, e33606.
74. L. Papa, L. M. Lewis, J. L. Falk, Z. Zhang, S. Silvestri, P. Giordano, G. M. Brophy, J. A. Demery, N. K. Dixit, I. Ferguson, M. C. Liu, J. Mo, L. Akinyi, K. Schmid, S. Mondello, C. S. Robertson, F. C. Tortella, R. L. Hayes and K. K. Wang, *Ann. Emerg. Med.*, 2012, **59**, 471.
75. Z. Metting, N. Wilczak, L. A. Rodiger, J. M. Schaaf and J. van der Naalt, *Neurology*, 2012, **78**, 1428.
76. P. J. McMahon, D. M. Panczykowski, J. K. Yue, A. M. Puccio, T. Inoue, M. D. Sorani, H. F. Lingsma, A. I. Maas, A. B. Valadka, E. L. Yuh, P. Mukherjee, G. T. Manley, D. O. Okonkwo and TRACK-TBI Investigators, *J. Neurotrauma*, 2015, **32**, 527.
77. J. P. Posti, I. Hossain, R. S. K. Takala, H. Liedes, V. Newcombe, J. Outtrim, A. J. Katila, J. Frantzén, H. Ala-Seppälä, J. P. Coles, A. Kyllönen, H. R. Maanpää, J. Tallus, P. J. Hutchinson, M. van Gils, D. K. Menon and O. Tenovuo, *J. Neurotrauma*, 2017, DOI: 10.1089/neu.2016.4442.
78. S. A. Dambinova, G. A. Izykenova, S. V. Burov, E. V. Grigorenko and S. A. Gromov, *J. Neurol. Sci.*, 1997, **152**, 93.
79. S. A. Dambinova, O. K. Granstrem, A. Tourov, R. Salluzzo, F. Castello and G. A. Izykenova, *J. Neurochem.*, 1998, **71**, 2088.
80. A. V. Goryunova, N. A. Bazarnaya, E. G. Sorokina, N. Y. Semenova, O. V. Globa, Z. B. Semenova, V. G. Pinelis, L. M. Roshal and O. I. Maslova, *Neurosci. Behav. Physiol.*, 2007, **37**, 761.
81. J. D. Mullins, A. V. Shikuev, U. I. Danilenko and S. A. Dambinova, presented in part at Military Health System Research Symposium USAMRMC, Ft. Lauderdale, August 2012.
82. J. D. Mullins, presented in part at AMSUS 118th Annual Continuing Education Meeting, Seattle, November 2013.
83. S. A. Dambinova, K. Bettermann, T. Glynn, M. Tews, D. Olson, J. D. Weissman and R. L. Sowell, *PLoS One*, 2012, **7**, DOI: 10.1371/journal.pone.0042362.
84. S. A. Dambinova, S. Gill, L. St. Onge and R. L. Sowell, in *Biomarkers for TBI*, ed. S. A. Dambinova, R. L. Hayes, K. K. W. Wang, Royal Society of Chemistry, London, 2012, ch. 8, p. 134.
85. S. A. Dambinova, *Clin. Lab. Int.*, 2008, **32**, 7.
86. S. A. Dambinova and A. A. Skoromets, US Pat., 15/388, 2016.

87. K. Laver, S. Dyer, C. Whitehead, L. Clemson and M. Crotty, *Br. Med. J. Open*, 2016, **6**, e010767.
88. K. Spilovska, F. Zemek, J. Korabecny, E. Nepovimova, O. Soukup, M. Windisch and K. Kuca, *Curr. Med. Chem.*, 2016, **23**, 3245.
89. E. C. Lauterbach, *Psychopharmacol. Bull.*, 2016, **46**, 53.
90. C. P. Taylor, S. F. Traynelis, J. Siffert, L. E. Pope and R. R. Matsumoto, *Pharmacol. Ther.*, 2016, **164**, 170.
91. K. J. Bowers, K. B. McAllister, M. Ray and C. Heitz, *Acad. Emerg. Med.*, 2017, **8**, DOI: 10.1111/acem.13172.
92. G. J. Iacobucci, O. Visnjevac, L. Pourafkari and N. D. Nader, *Pain Physician*, 2017, **20**, E285.
93. E. Villegas, R. Estruch, G. Mengod and R. Cortés, *Addict. Biol.*, 2011, **16**, 163.
94. X. Y. Gao, J. O. Huang, Y. F. Hu, Y. Gu, S. Z. Zhu, K. B. Huang, J. Y. Chen and S. Y. Pan, *Sci. Rep.*, 2014, **4**, 7091.
95. A. G. Malykh and M. R. Sadaie, *Drugs*, 2010, **70**, 287.
96. J. Terhag, K. Gottschling and M. Hollmann, *Front. Mol. Neurosci.*, 2010, **3**, 117.
97. M. Gynther, A. Petsalo, S. H. Hansen, L. Bunch and D. S. Pickering, *Neurochem. Res.*, 2015, **40**, 542.
98. D. Zwilling, S. Y. Huang, K. V. Sathyasaikumar, F. M. Notarangelo, P. Guidetti, H. Q. Wu, J. Lee, J. Truong, Y. Andrews-Zwilling, E. W. Hsieh, J. Y. Louie, T. Wu, K. Scearce-Levie, C. Patrick, A. Adame, F. Giorgini, S. Moussaoui, G. Laue, A. Rassoulpour, G. Flik, Y. Huang, J. M. Muchowski, E. Masliah, R. Schwarcz and P. J. Muchowski, *Cell*, 2011, **145**, 863.
99. C. Rink, S. Gnyawali, R. Stewart, S. Teplitsky, H. Harris, S. Roy, C. K. Sen and S. Khanna, *FASEB J.*, 2017, DOI: 10.1096/fj.201601033R.
100. A. da Silva-Candal, A. Vieites-Prado, M. Gutiérrez-Fernández, R. I. Rey, B. Argibay, D. Mirelman, T. Sobrino, B. Rodríguez-Frutos, J. Castillo and F. Campos, *J. Cereb. Blood Flow Metab.*, 2015, **35**, 1206.
101. R. L. Vangilder, C. L. Rosen, T. L. Barr and J. D. Huber, *Pharmacol. Ther.*, 2011, **130**, 239.
102. S. Das, A. Carnicer-Lombarte, J. W. Fawcett and U. Bora, *Prog. Neurobiol.*, 2016, **14**, 1.
103. N. P. Bechtereva, *Magic of Brain and Labyrinths of Life*, Notabene, St. Petersburg, 2009.
104. U. I. Danilenko, G. A. Khunteev, A. Bagumyan and G. A. Izykenova, in: *Biomarkers for TBI*, ed. S. A. Dambinova, R. L. Hayes, K. K. W. Wang, Royal Society of Chemistry, London, 2012, ch. 5, p. 87.
105. Y. Zhang, J. F. Clark, G. Pyne-Geithman and J. Caruso, *Metallomics*, 2010, **2**, 628.
106. A. H. Hashim, PhD thesis, University South Florida, 2014.
107. S. Sharma, W. Yang and H.-Z. Cai, *US Pat.*, 2006, **11/291**, 389.
108. H.-Z. Cai, A. Shadiack, S. D. Sharma, Y. Q. Shi and W. Yang, WO Pat., 2002064091 A3, 2003.

109. P. G. Harch and V. McCullough, in *The Oxygen Revolution: Hyperbaric Oxygen Therapy: The Definitive Treatment of Traumatic Brain Injury (TBI) & Other Disorders*, Hatherleigh Press, New York, Third edn, 2016.
110. Committee on Nutrition, Trauma, and the Brain, Institute of Medicine, *Nutrition and Traumatic Brain Injury: Improving Acute and Subacute Health Outcomes in Military Personnel*, ed. J. Erdman, M. Oria and L. Pillsbury, National Academies Press, Washington, DC, 2011.
111. M. W. Collins, A. P. Kontos, D. O. Okonkwo, J. Almquist, J. Bailes, M. Barisa, J. Bazarian, O. J. Bloom, D. L. Brody, R. Cantu, J. Cardenas, J. Clugston, R. Cohen, R. Echemendia, R. J. Elbin, R. Ellenbogen, J. Fonseca, G. Gioia, K. Guskiewicz, R. Heyer, G. Hotz, G. L. Iverson, B. Jordan, G. Manley, J. Maroon, T. McAllister, M. McCrea, A. Mucha, E. Pieroth, K. Podell, M. Pombo, T. Shetty, A. Sills, G. Solomon, D. G. Thomas, T. C. Valovich McLeod, T. Yates and R. Zafonte, *Neurosurgery*, 2016, **79**, 912.
112. X. Zhou, G. Cui, H. H. Tseng, S. M. Lee, G. P. Leung, S. W. Chan, Y. W. Kwan and M. P. Hoi, *Evid Based Complement Alternat Med.*, 2016, 9627258.
113. M. Sircus, *Anti-Inflammatory Oxygen Therapy: Your Complete Guide to Understanding and Using Natural Oxygen Therapy*. Square One Publishers, Garden City Park, NY, 2015.
114. M. F. McCarty, J. J. Di Nicolantonio and J. H. O'Keefe, *Open Heart*, 2015, **2**, DOI: 10.1136/openhrt-2015-000262.
115. S. Fujisawa and Y. Murakami, in *Drug Discovery from Mother Nature. The series Advances in Experimental Medicine and Biology*, ed. S. C. Gupta, S. Prasad, B. B. Aggarwal, Springer Intern Publishing, Switzerland, 2016.
116. I. Jantan, W. Ahmad and S. N. Bukhari, *Front. Plant. Sci.*, 2015, **6**, 655.
117. E. Reynolds and M. W. Collins, *Prog. Neurol. Surg.*, 2014, **28**, 128.
118. B. A. Alsalaheen, S. L. Whitney, G. F. Marchetti, J. M. Furman, A. P. Kontos, M. W. Collins and P. J. Sparto, *Pediatr. Phys. Therapy*, 2014, **26**, 191.
119. A. Mucha, M. W. Collins, R. J. Elbin, J. M. Furman, C. Troutman-Enseki, R. M. DeWolf, G. Marchetti and A. P. Kontos, *Am. J. Sports Med.*, 2014, **42**, 2479.
120. M. W. Collins, A. P. Kontos, E. Reynolds, C. D. Murawski and F. H. Fu, *Knee Surg. Sports Traumatol. Arthroscopy*, 2014, **22**, 235.

CHAPTER 10

Functional Predictors for Prognosis and Recovery after Mild Traumatic Brain Injury

A. M. SUFRINKO,* N. SANDEL AND A. PUSKAR

University of Pittsburgh Medical Center Sports Concussion Program, Department of Orthopedic Surgery, 3200 S. Water St., Pittsburgh, PA 15203, USA
*Email: sufrinkoam@upmc.edu

10.1 Introduction

Although advanced brain imaging and metabolic biomarkers are gaining significant research interest, clinical evaluation of mild traumatic brain injury (mTBI) continues to rely primarily on functional assessment modalities. Functional assessment refers to the evaluation of skills and behaviors that are linked to the functions of particular cerebral areas, neural pathways, or cortical networks[1] and relate to human functioning in everyday life. Whereas acute evaluation and treatment of moderate and severe TBI hinges on neuroimaging findings, conventional imaging techniques (*e.g.*, computed tomography [CT] and magnetic resonance imaging [MRI]) typically do not reveal any structural abnormalities in patients with mTBI. It is not surprising that acute findings on imaging have significant bearing on recovery outcomes following moderate to severe TBI,[2] but functional assessment may be more beneficial in predicting recovery and prognosis following mTBI. Functional assessment requires clinical tools, such as neuropsychological testing, standardized self-report measures, and measures of vestibular and

Drug Discovery Series No. 62
Acute Brain Impairment: Scientific Discoveries and Translational Research
Edited by Philip V. Peplow, Svetlana A. Dambinova, Thomas A. Gennarelli and Bridget Martinez

Published by the Royal Society of Chemistry, www.rsc.org

oculomotor functioning. The purpose of this current chapter is to provide an overview of predictors for prognosis and recovery following mTBI. Particular injury characteristics and risk factors are associated with patterns of performance on assessment measures and other clinical outcomes, such as recovery duration following mTBI. Further, functional assessments are also used to monitor recovery and to make specific treatment recommendations.

10.2 Mild Traumatic Brain Injury

Approximately 80% of nonfatal TBIs are classified as mild,[3] with loss of consciousness (LOC) lasting less than 30 minutes, an initial Glasgow Coma Scale (GCS) score of 13–15, and post-traumatic amnesia (PTA) of less than 24 hours. However, there is no consensus on the definition of mTBI, and its terminology can be confusing.[4] mTBI is often considered to be synonymous to concussion, a term primarily used in a sports setting. Concussion is defined as a "complex pathophysiological process affecting the brain, induced by biomechanical forces (traumatically induced alteration in mental status with or without LOC)," as specified in the recent consensus statement on concussion in sports.[5] Several physical, cognitive, emotional, and sleep-related symptoms may be present following injury. Although a vast majority of concussions resolve within days to weeks,[6] it is often estimated that 10–20% of patients who sustain an mTBI will experience "post-concussion syndrome" or "post-concussion symptoms", a set of nonspecific symptoms that persist beyond the expected recovery period.[7] Several aspects of brain functioning can be compromised following mTBI and often correspond to subjective symptom reports or descriptions of day-to-day difficulties. For example, cognitive deficits on standardized neurocognitive testing often predict problems in completing certain work-related tasks,[8] while vestibular dysfunction often manifests as difficulty balancing or navigating busy environments.[9] Similarly, oculomotor impairments can make basic reading challenging.[10]

10.3 Standardized Evaluation Tools

The National Institute of Neurological Disorders and Stroke (NINDS) recommends that the evaluation of TBI includes assessment of neuropsychological impairment and neurobehavioral function.[11] Tools utilized to assess mTBI must be capable of repeated, serial administration over short time intervals in order to track recovery and guide treatment.[12,13] A standardized evaluation of mTBI in the acute phase of the injury can yield valuable prognostic information and assist in the delineation of deficits.[12,14–16] They should be tailored to the individual based on other personal factors, including demographic characteristics, environmental circumstances, and pre-morbid factors such as age, gender, injury circumstances, etc.

10.3.1 Neuropsychological Evaluation

Neuropsychological evaluation provides information about brain–behavior relationships. The etiology of brain injury (*i.e.*, diffuse/localized/anoxic), mechanism (*i.e.*, blast or blunt trauma), form of impact (*i.e.*, open *versus* closed), and brain anatomy affected (*e.g.*, frontal lobe, hippocampus, or temporal lobe) do not always directly translate to limitations observed after TBI.[17,18] Standardized neuropsychological tests assist in evaluating cognition and behavior in order to determine functional deficits, which can be connected to brain damage in a particular neural structure or pathway. Empirical studies indicate that cognitive impairment on neuropsychological testing is a strong predictive factor of daily functioning following mTBI.[19,20] Neuropsychological testing can be administered *via* paper-and-pencil or computerized formats, and typically includes a combination of cognitive tests and measures of behavioral/adaptive functioning measures.[21]

10.3.1.1 Neurocognitive Testing

Traditional, comprehensive paper-and-pencil neurocognitive testing is more commonly utilized for evaluating functional limitations after moderate/severe TBI, in which deficits may be persistent.[22] Neurocognitive test batteries for TBI often include measures of attention, learning, memory, language, visual–spatial and fine motor skills, executive functioning, intelligence/intellect and performance validity.[11] Neuropsychological impairments are believed to be related to underlying disruption in the connectivity of white matter tracts, such as connections in the corpus callosum.[23,24] The pattern of impairment on cognitive testing can elucidate certain environmental demands or tasks that may not be feasible for an individual after TBI. For instance, a child with deficits in attention on neuropsychological testing may have trouble sustaining focus while in the classroom,[25,26] or an adult with slowed processing speed may not be as productive in the work environment.[27]

Computerized neurocognitive testing (CNT) is a widely utilized measure of cognitive outcomes after mTBI.[5,28] Most computerized tools for evaluating mTBI are brief (20–30 minutes), randomize test stimuli to eliminate practice effects from repeated administration, and provide immediate scoring of test performance, all of which improve the feasibility of use for mTBI.[29–31] Studies have demonstrated that assessment using CNT within the first 1–2 weeks of a sport-related mTBI improves the ability to prognosticate whether an athlete will demonstrate a simple or complicated recovery from the injury.[15,32] For instance, athletes who demonstrate slowed reaction times on CNT within 2 weeks of sustaining an mTBI are expected to have a longer recovery time.[33] The particular pattern of cognitive deficits on CNT can also provide information regarding particular subtypes of mTBI and inform treatment.[12,16]

10.3.1.2 Self-report Measures

Symptom questionnaires assess physical, sleep, cognitive, and mood symptoms that may occur post-injury. A high symptom burden in the acute phase of injury following mTBI has consistently been shown to be a predictor of prolonged recovery,[15,33,34] even compared to other acute makers of injury, such as PTA. The presence of specific symptoms may also help identify post-injury impairments and guide treatment recommendations. For instance, a report of dizziness immediately after injury is predictive of prolonged recovery time[35] and linked to vestibular dysfunction[36–39] following sport-related mTBI. Similarly, identifying secondary affective or mood-related changes following concussion is important due to the increased risk of a complicated recovery when psychiatric sequelae occur following mTBI.[40] Informant symptom reports completed by parents or caretakers may also be helpful when insight or judgment is impaired due to brain injury.[11,21,41]

10.3.2 Vestibular/Balance

Central sensory/motor systems that are commonly affected following mTBI include vestibular/balance dysfunction and oculomotor function.[33,42] Neural control of balance/postural stability and vision stabilization during movement is maintained by the vestibular system. Greater than 50% of individuals with a sport-related mTBI report dizziness or imbalance after injury.[33,43] Dysfunction of the vestibular system after an mTBI can result in symptoms of dizziness, imbalance, vertigo, and visual disturbance.[36,38,44,45] Anatomical correlates of vestibular dysfunction include both peripheral (inner-ear labyrinth) and central (neural network) vestibular components. When vestibular dysfunction is observed following mTBI, it is typically related to a central vestibular disturbance. The neuroanatomy of the central vestibular pathways involves four vestibular nuclei (superior vestibular nucleus [SVN], medial vestibular nucleus [MVN], lateral vestibular nucleus [LVN], and descending vestibular nucleus [DVN]) that have multiple projections to other regions of the nervous system including the vestibular–ocular/optokinetic, vestibulospinal, vestibulocerebellar, and vestibular–autonomic systems.[46] The vestibular–ocular and vestibular–spinal tracts may be implicated after mTBI.[36,38] Individuals who demonstrate impairment on vestibular–ocular screening tools[38] at their initial visit following an mTBI take longer to recover from the injury[47] and may require formal treatment for rehabilitation.[36,45] There are several limitations associated with motion sensitivity due to vestibular dysfunction after mTBI, including an inability to tolerate driving, balance impairment, lightheadedness/feeling overwhelmed in complex environments, intolerance of dynamic physical exercise, and dizziness with head movement.[16,36,38,45]

10.3.2.1 Oculomotor

Recent studies have demonstrated that post-traumatic oculomotor abnormalities are present in about 24–42% of mTBIs when evaluated within 1 month of injury.[48,49] Nearly half of these individuals require formal rehabilitation to restore oculomotor function, and it may take 4–6 months to recover in order to complete ocular therapy.[48] Neuroanatomical correlates of oculomotor dysfunction after mTBI are likely associated with axonal stretching/neurometabolic changes in areas of the brain devoted to ocular function, including the brainstem, cerebellum, and cerebral cortex.[50,51] Due to the complexities of the visual system, specific brain regions implicated in oculomotor dysfunction after mTBI are far from fully understood. Post-traumatic oculomotor abnormalities can result in increased mTBI symptoms when engaging in tasks with a visual demand. Difficulties that are expected with oculomotor dysfunction include blurred/double vision or trouble focusing when reading, looking at screens (*e.g.*, televisions or computers), and adjusting between near and distance vision. Individuals with oculomotor impairment may have trouble navigating school and/or certain work environments in which reading and computer work are essential for performance (Table 10.1).

10.4 Pre-injury Risk Factors/Conditions

Several individual and demographic characteristics are associated with risk of sustaining an mTBI and with outcomes following injury. Gender influences both risk of injury[52] and recovery outcomes;[53] males are twice as likely as females to sustain an mTBI,[54] which may be related to engagement in riskier activities.[52,54] However, when comparing same sport athletes, females have been shown to have a higher incidence of injury.[52] Differences in neck strength and anatomic musculature may account for this discrepancy.[55] Females tend to take longer to recover from injury, with an increased likelihood of reporting continuing symptoms and impairments at greater than one month post-injury.[53,56,57] Females also demonstrate a lengthier recovery with greater impairments on neurocognitive testing post-injury.[53] Females are more likely to have a history of migraine, another risk factor linked to protracted recovery.[16]

The relationship between age and head injury outcomes is inconsistent. In the sport-related mTBI literature, age commonly refers to comparing youth and adolescent athletes (*i.e.*, less than 17 years of age) to older, adult-aged athletes (*i.e.*, college and professional athletes aged 18 years and older) when investigating risk and recovery. There is a general consensus that younger athletes are at an increased risk for mTBI and demonstrate longer recovery times compared to older, adult athletes.[58,59] Level of play and associated risk may account for this age difference, as younger athletes are more likely to be involved in no-check or non-contact leagues[60] and are less likely to be exposed to risk of injury. Controlling for injury exposure, younger athletes have been associated with increased injury risk, higher initial symptoms, and a lengthier recovery course.[59,61,62] Risk of fall is a common mechanism of injury in older individuals,[54,63] work place injuries are frequent in middle age,[63] and risk of motor vehicle accident carries across the lifespan.[54,63] However, there has been limited comprehensive and controlled study of age-related differences for mTBI outside of athletes.[64] Factors of conditioning

Table 10.1 mTBI impairments, assessment modalities, and related neuroanatomical correlates.

Impairment type	Example deficits	Examples of functional assessment measures	Related neuroanatomical correlates
Neurocognitive	• Poor memory • Poor attention • Mentally slow • Trouble multitasking	• National Hockey League battery (*e.g.*, Hopkins Verbal Learning Test, Trails, Symbol Digit Modality Test) • Immediate Post-concussion Assessment and Cognitive Testing (ImPACT)	Alterations in white matter connectivity[23,24]
Vestibular	• Dizziness • Lightheadedness • Nausea • Impairment in gaze stability • Imbalance	• Vestibular/Ocular-motor Screening (VOMS) • Balance Error Scoring System (BESS) • Sensory Organization Test (SOT) • Clinical Test of Dynamic Visual Acuity	Central vestibular nuclei (*i.e.*, SVN, MVN, LVN, and DVN) and pathways (vestibular–ocular, vestibulospinal, vestibulocerebellar, vestibule–autonomic systems)[36,38]
Oculomotor	• Convergence insufficiency • Accommodative insufficiency • Saccadic intrusions	• VOMS • King–Devick Test	Brainstem, cerebellum, and cerebral cortex[50,51]
Affective/ behavioral changes	• Anxiety • Panic attacks • Depressed mood • Irritability • Emotional lability • Avoidance behavior	• Beck Depression Inventory-II (BDI-II) • Generalized Anxiety Questionnaire-7 item (GAD-7) • Patient Health Questionnaire (PHQ-9) • Minnesota Multiphasic Personality Inventory-2—Restructured Form (MMPI-2-RF) • Strengths and Difficulties Questionnaire (SDQ)	Autonomic dysregulation (*e.g.*, alterations in heart rate variability and blood pressure) and metabolic alterations in the frontotemporal cortex, particularly the ventromedial and dorsolateral prefrontal cortex, and limbic circuitry[89–94]

and muscular resilience, environmental and social demands, stress exposure, and type of injury exposure are a limited subset of the many potential confounding factors when comparing across age groups. As a result, evaluating and treating mTBI requires comprehensive consideration of all dispositional factors and mechanisms of injury exposure.

There are several other pre-injury risk factors that can influence both the likelihood of sustaining a head injury and the course of injury recovery. Having sustained a previous head injury has been associated with an increased potential for protracted recovery and impaired neurocognitive performance, as well as an increased risk for subsequent injury.[57,65–67] Additionally, a history of premorbid headache or premorbid migraine headache has been associated with an increased likelihood of injury and poorer neurocognitive outcomes post-injury.[57,68] Further research is warranted to determine other pre-injury risk factors that may increase vulnerability to sustaining and/or experiencing a complicated recovery from mTBI.

10.4.1 Post-injury Risk Factors for Poor Recovery

Setting the expectation for a full recovery is an important and valid treatment approach that serves to promote a positive attitude that is receptive to treatment.[16] Recognition of potential issues early in the treatment process and taking steps to mitigate the influence of these factors can help to improve treatment outcomes and patient recovery.

Pre-existing or post-injury psychiatric concerns such as anxiety, depression, or post-traumatic stress disorder (PTSD) have also been associated with protracted recovery, increased symptom severity, and maintained clinical deficits.[69] Further, a highly negative emotional view of symptoms and the injury, or catastrophizing, has been associated with lengthier and more protracted recovery from mTBI.[70] Additionally, the role of litigation and potential secondary gain is very prevalent in individuals with a protracted recovery.[71] Whether due to the stress of a contentious litigation situation exacerbating symptoms and reducing the ability to access treatment or due to the deliberate maintenance of symptoms to achieve a more favorable outcome, litigation is a factor that clinicians must consider when faced with individuals experiencing a longer or more complex recovery.

10.4.2 Rehabilitation Factors for Recovery

A comprehensive, yet targeted, treatment approach is necessary to achieve optimal rehabilitation and recovery. Evidence-based treatment for mTBI is limited and clinical management relies on consensus-based guidelines, which recommend cognitive and physical rest to prevent re-injury and to reduce post-concussion symptoms.[72] However, research has been rapidly accumulating to support early physical activity among sport-related mTBI patients, as exercise may actually increase immune function, mitochondrial volume, and coronary artery size and decrease inflammation and blood coagulation.[73] Exercise has a positive effect on cognition, sleep, and mood, all of which are affected after mTBI.[69] Similarly, there is a growing body of evidence supporting vestibular therapy as a treatment modality for mTBI.[74] Cognitive rehabilitation has been shown to be most effective in patients with moderate and severe TBI.[75] However, it is unclear which patients may recover faster with therapies, and the ideal timeframe post-injury to start more active rehabilitation is unknown (Table 10.2).[76]

10.5 Summary and Future Directions

10.5.1 Assessment Modalities and Multidisciplinary Team Approach

TBI is a heterogeneous injury that requires a multifaceted and comprehensive evaluation approach that lends to an individualized injury treatment plan.[16] Many studies have demonstrated that a multimodal or multifaceted test battery is more sensitive than any standalone measure for diagnosing mTBI.[77–79] Different types of deficits recover at different rates.[80] Due to the heterogeneous nature of mTBI, assessment and management should be implemented in the context of a multidisciplinary team, including physicians, psychologists, and physical and occupational therapists. While an initial clinical assessment can help predict recovery and outcomes, pre-injury risk factors and conditions also greatly influence recovery.

10.5.2 Future Directions and Emerging mTBI Assessment Tools

With assessment for mTBI rapidly evolving, many of the currently used tools remain understudied, despite being widely used in clinical practice. Further research is needed to provide normative data and information on the validity, reliability, and practice effects of many newly proposed measures. Researchers have also started compiling normative data on healthy athletes for new assessment strategies, such as a tandem gait test,[81] and are finally improving the quality of normative data available for widely used tools, such as the Sport Concussion Assessment Tool – 3rd edition (SCAT-3).[82,83]

New neurobehavioral tests that can detect mTBI are under development. Many of these emerging technologies have been used in a research setting, but have not yet widely impacted clinical practice. For example, OptoGait motion analysis technology, described as an innovative system for movement analysis and the assessment of patients with normal or pathological conditions, has been tested in healthy collegiate athletes.[84] There has also been the creation of a virtual reality balance assessment[85] and increased use of force plate technology,[86] for example. While these tools may be more sophisticated and potentially more sensitive to identifying mTBI, the practicality of implementing such tools in actual clinical practice is unclear. There is also some interest in advancing oculomotor testing following mTBI

Table 10.2 Factors associated with the risk of complicated recovery following an mTBI.

Individual characteristics	Pre-injury conditions	Signs/injury factors	Testing predictors	Post-injury factors
• Sex[16,52,53,58] • Age[58–61,63,64] • Level of fitness, neck strength[55]	• Migraine[16,57,68] • Prior injury[65]	• Remaining in play during sporting event[95,96] • Unaware of injury/ blind-sided[97] • Motor vehicle accident[56,98]	• Elevated cognitive–migraine–fatigue symptoms • Increased symptom levels[16,56,98] • Greater levels of impairment on neurocognitive testing[13,16,56,98]	• Lack of rehabilitation[99] • Lack of support[100] • Poor adjustment[69,100] • Litigation[71] • Motivation[56] • PTSD[69,101,102] • Catastrophizing[70]

due to the rapidly growing body of literature supporting impairments in saccades and vergence following injury.[87] Eye-Tracking Rapid Attention Computation (EYE-TRAC) examines dynamic visuomotor synchronization between gaze and the target during predictive visual tracking. An initial study showed mTBI patients demonstrated dynamic visuomotor scores that were worse than 95% of normal subjects, and concussed patients improved toward the normal range across time.[88]

10.6 Conclusion

Clinical evaluation of mTBI relies primarily on functional assessment modalities. Performance on assessment measures in the acute stage of recovery from mTBI can be useful in prognosis and in predicting recovery, although several pre-injury risk factors have a significant bearing on recovery following mTBI as well. A comprehensive, multimodal assessment battery implemented within the context of a multidisciplinary treatment team is the goal for assessing and directing treatment for mTBI. Current research is focused on developing neurobehavioral tools that can be used to both diagnose and manage treatment following mTBI.

References

1. S. A. Dambinova, R. L. Sowell and J. C. Maroon, *J. Mol. Biomarkers Diagn.*, 2013, **2013**, 3.
2. D. I. Katz and M. P. Alexander, *Archiv. Neurol.*, 1994, **51**, 661.
3. J. F. Kraus and P. Nourjah, *J. Trauma Acute Care Surg.*, 1988, **28**, 1637.
4. M. Wintermark, P. C. Sanelli, Y. Anzai, A. J. Tsiouris, C. T. Whitlow and A. H. I. Institute, *J. Am. Coll. Radiol.*, 2015, **12**, e1.
5. P. McCrory, W. H. Meeuwisse, M. Aubry, B. Cantu, R. J. Echemendia, L. Engebretsen, K. Johnston, J. S. Kutcher and M. Raftery, *Br. J. Sports Med.*, 2013, **47**, 250.
6. L. C. Henry, R. Elbin, M. W. Collins, G. Marchetti and A. P. Kontos, *Neurosurgery*, 2015, **78**, 232–241.
7. L. M. Ryan and D. L. Warden, *Int. Rev. Psychiatry*, 2003, **15**, 310.
8. M. Wäljas, G. L. Iverson, R. T. Lange, S. Liimatainen, K. M. Hartikainen, P. Dastidar, S. Soimakallio and J. Öhman, *J. Head Trauma Rehabil.*, 2014, **29**, 443.
9. G. Zhou and J. R. Brodsky, *Otolaryngol.-Head Neck Surg.*, 2015, 0194599815576720.
10. C. L. Master, M. Scheiman, M. Gallaway, A. Goodman, R. L. Robinson, S. R. Master and M. F. Grady, *Clin. Pediatr.*, 2015, 0009922815594367.
11. National Institute of Neurological Disorders and Stroke, https://www.commondataelements.ninds.nih.gov/TBI.aspx#tab=Data_Standards, accessed February 16, 2017.

12. L. C. Henry, R. J. Elbin, M. W. Collins, G. Marchetti and A. P. Kontos, *Neurosurgery*, 2016, **78**, 232.
13. T. Covassin, R. J. Elbin and Y. Nakayama, *Phys. Sportsmed.*, 2010, **38**, 87.
14. A. P. Kontos, R. Elbin, B. Lau, S. Simensky, B. Freund, J. French and M. W. Collins, *Am. J. Sports Med.*, 2013, **41**, 1497.
15. B. C. Lau, M. W. Collins and M. R. Lovell, *Neurosurgery*, 2012, **70**, 371.
16. M. Collins, A. Kontos, E. Reynolds, C. Murawski and F. Fu, *Knee Surg. Sports Traumatol. Arthrosc.*, 2014, **22**, 235.
17. D. A. Cahn-Weiner, S. T. Farias, L. Julian, D. J. Harvey, J. H. Kramer, B. R. Reed, D. Mungas, M. Wetzel and H. Chui, *J. Int. Neuropsychol. Soc.*, 2007, **13**, 747.
18. P. K. Dash, J. Zhao, G. Hergenroeder and A. N. Moore, *Neurotherapeutics*, 2010, **7**, 100.
19. Y. Rassovsky, P. Satz, M. S. Alfano, R. K. Light, K. Zaucha, D. L. McArthur and D. Hovda, *J. Clin. Exp. Neuropsychol.*, 2006, **28**, 567.
20. Y. Rassovsky, P. Satz, M. S. Alfano, R. K. Light, K. Zaucha, D. L. McArthur and D. Hovda, *J. Clin. Exp. Neuropsychol.*, 2006, **28**, 581.
21. M. D. Lezak, *Neuropsychological Assessment*, Oxford University Press, USA, 2004.
22. L. Ruttan, K. Martin, A. Liu, B. Colella and R. E. Green, *Archiv. Phys. Med. Rehabil.*, 2008, **89**, S69.
23. I. D. Croall, C. J. Cowie, J. He, A. Peel, J. Wood, B. S. Aribisala, P. Mitchell, A. D. Mendelow, F. E. Smith and D. Millar, *Neurology*, 2014, **83**, 494.
24. J. J. Bazarian, T. Zhu, B. Blyth, A. Borrino and J. Zhong, *Magn. Reson. Imaging*, 2012, **30**, 171.
25. G. A. Gioia, J. C. Schneider, C. G. Vaughan and P. K. Isquith, *Br. J. Sports Med.*, 2009, **43**(Suppl 1), i13.
26. M. D. Sady, C. G. Vaughan and G. A. Gioia, *Phys. Med. Rehabil. Clin. North Am.*, 2011, **22**, 701.
27. J. Ponsford, K. Draper and M. Schönberger, *J. Int. Neuropsychol. Soc.*, 2008, **14**, 233.
28. K. A. Kinnaman, R. C. Mannix, R. D. Comstock and W. P. Meehan III, *Pediatr. Emerg. Care*, 2014, **30**, 458.
29. M. Collins, M. R. Lovell, G. L. Iverson, T. Ide and J. Maroon, *Neurosurgery*, 2006, **58**, 275.
30. C. T. Gualtieri and L. G. Johnson, *Arch. Clin. Neuropsychol.*, 2006, **21**, 623.
31. W. R. Cole, J. P. Arrieux, K. Schwab, B. J. Ivins, F. M. Qashu and S. C. Lewis, *Arch. Clin. Neuropsychol.*, 2013, **28**, 732.
32. B. C. Lau, M. W. Collins and M. R. Lovell, *Am. J. Sports Med.*, 2011, **39**, 1209.
33. B. Lau, M. R. Lovell, M. W. Collins and J. Pardini, *Clin. J. Sport Med.*, 2009, **19**, 216.
34. W. P. Meehan III, R. C. Mannix, A. Stracciolini, R. Elbin and M. W. Collins, *J. Pediatr.*, 2013, **163**, 721.

35. B. C. Lau, A. P. Kontos, M. W. Collins, A. Mucha and M. R. Lovell, *Am. J. Sports Med.*, 2011, **39**, 2311.
36. B. A. Alsalaheen, A. Mucha, L. O. Morris, S. L. Whitney, J. M. Furman, C. E. Camiolo-Reddy, M. W. Collins, M. R. Lovell and P. J. Sparto, *J. Neurol. Phys. Therapy*, 2010, **34**, 87.
37. S. P. Broglio, M. W. Collins, R. M. Williams, A. Mucha and A. P. Kontos, *Clin. Sports Med.*, 2015, **34**, 213.
38. A. Mucha, M. W. Collins, R. J. Elbin, J. M. Furman, C. Troutman-Enseki, R. M. DeWolf, G. Marchetti and A. P. Kontos, *Am. J. Sports Med.*, 2014, **42**, 2479.
39. M. J. Ellis, D. Cordingley, S. Vis, K. Reimer, J. Leiter and K. Russell, *J. Neurosurg. Pediatr.*, 2015, **16**, 248.
40. J. Ponsford, P. Cameron, M. Fitzgerald, M. Grant, A. Mikocka-Walus and M. Schonberger, *Neuropsychology*, 2012, **26**, 304.
41. E. A. Wilde, G. G. Whiteneck, J. Bogner, T. Bushnik, D. X. Cifu, S. Dikmen, L. French, J. T. Giacino, T. Hart and J. F. Malec, *Archiv. Phys. Med. Rehabil.*, 2010, **91**, 1650.
42. C. L. Master, M. Scheiman, M. Gallaway, A. Goodman, R. L. Robinson, S. R. Master and M. F. Grady, *Clin. Pediatr.*, 2016, **55**, 260.
43. A. P. Kontos, R. J. Elbin, P. Schatz, T. Covassin, L. Henry, J. Pardini and M. W. Collins, *Am. J. Sports Med.*, 2012, **40**, 2375.
44. M. E. Hoffer, K. R. Gottshall, R. Moore, B. J. Balough and D. Wester, *Otol. Neurotol.*, 2004, **25**, 135.
45. B. A. Alsalaheen, S. L. Whitney, A. Mucha, L. O. Morris, J. M. Furman and P. J. Sparto, *Physiother. Res. Int.*, 2013, **18**, 100.
46. J. M. Goldberg, *The Vestibular System: A Sixth Sense*, OUP USA, 2012.
47. A. J. Anzalone, D. Blueitt, T. Case, T. McGuffin, K. Pollard, J. C. Garrison, M. T. Jones, R. Pavur, S. Turner and J. M. Oliver, *Am. J. Sports Med.*, 2016, **45**, 36.
48. E. P. Storey, S. R. Master, J. E. Lockyer, O. E. Podolak, M. F. Grady and C. L. Master, *Optom. Vision Sci.*, 2017, **94**, 96.
49. K. L. Pearce, A. Sufrinko, B. C. Lau, L. Henry, M. W. Collins and A. P. Kontos, *Am. J. Sports Med.*, 2015, **43**, 3055.
50. K. J. Ciuffreda, N. Kapoor, D. Rutner, I. B. Suchoff, M. E. Han and S. Craig, *Optom.-J. Am. Optometr. Assoc.*, 2007, **78**, 155.
51. P. D. Gamlin and K. Yoon, *Nature*, 2000, **407**, 1003.
52. S. Abrahams, S. Mc Fie, J. Patricios, M. Posthumus and A. V. September, *Br. J. Sports Med.*, 2014, **48**, 91.
53. T. Covassin, R. Elbin, A. Bleecker, A. Lipchik and A. P. Kontos, *Am. J. Sports Med.*, 2013, **41**, 2890.
54. J. D. Cassidy, L. Carroll, P. Peloso, J. Borg, H. Von Holst, L. Holm, J. Kraus and V. Coronado, *J. Rehabil. Med.*, 2004, **36**, 28.
55. J. T. Eckner, Y. K. Oh, M. S. Joshi, J. K. Richardson and J. A. Ashton-Miller, *Am. J. Sports Med.*, 2014, **42**, 566.
56. J. J. Bazarian, T. Wong, M. Harris, N. Leahey, S. Mookerjee and M. Dombovy, *Brain Inj.*, 1999, **13**, 173.

57. C. C. Reddy and M. W. Collins, *Curr. Sports Med. Rep.*, 2009, **8**, 10.
58. T. Covassin, R. J. Elbin III, E. Larson and A. P. Kontos, *Clin. J. Sport Med.*, 2012, **22**, 98.
59. M. Field, M. W. Collins, M. R. Lovell and J. Maroon, *J. Pediatr.*, 2003, **142**, 546.
60. C. A. Emery and W. H. Meeuwisse, *Am. J. Sports Med.*, 2006, **34**, 1960.
61. K. M. Guskiewicz, N. L. Weaver, D. A. Padua and W. E. Garrett, *Am. J. Sports Med.*, 2000, **28**, 643.
62. E. J. Pellman, M. R. Lovell, D. C. Viano and I. R. Casson, *Neurosurgery*, 2006, **58**, 263.
63. S. R. Laker, *PM&R*, 2011, **3**, S354.
64. J. Bruns and W. A. Hauser, *Epilepsia*, 2003, **44**, 2.
65. B. E. Gavett, R. A. Stern and A. C. McKee, *Clin. Sports Med.*, 2011, **30**, 179.
66. K. M. Guskiewicz, S. W. Marshall, J. Bailes, M. McCrea, R. C. Cantu, C. Randolph and B. D. Jordan, *Neurosurgery*, 2005, **57**, 719.
67. K. M. Guskiewicz, M. McCrea, S. W. Marshall, R. C. Cantu, C. Randolph, W. Barr, J. A. Onate and J. P. Kelly, *J. Am. Med. Assoc.*, 2003, **290**, 2549.
68. J. P. Mihalik, J. E. Stump, M. W. Collins, M. R. Lovell, M. Field and J. C. Maroon, *J. Neurosurg.*, 2005, **102**, 850.
69. D. K. Broshek, A. P. De Marco and J. R. Freeman, *Brain Inj.*, 2015, **29**, 228.
70. G. Chaput, S. P. Lajoie, L. M. Naismith and G. Lavigne, *Pain Res. Manag.*, 2016, **2016**, 1.
71. J. E. Karr, C. N. Areshenkoff and M. A. Garcia-Barrera, *Neuropsychology*, 2014, **28**, 321.
72. C. C. Giza, J. S. Kutcher, S. Ashwal, J. Barth, T. S. Getchius, G. A. Gioia, G. S. Gronseth, K. Guskiewicz, S. Mandel and G. Manley, *Neurology*, 2013, **80**, 2250.
73. S. P. Chrisman and F. P. Rivara, *J. Am. Med. Assoc.*, 2016, **316**, 2491.
74. K. J. Schneider, W. H. Meeuwisse, A. Nettel-Aguirre, L. Boyd, K. M. Barlow and C. A. Emery, *Br. J. Sports Med.*, 2013, **47**, e1.
75. H. Hallock, D. Collins, A. Lampit, K. Deol, J. Fleming and M. Valenzuela, *Front. Human Neurosci.*, 2016, **10**, 537.
76. M. W. Collins, A. P. Kontos, D. O. Okonkwo, J. Almquist, J. Bailes, M. Barisa, J. Bazarian, O. J. Bloom, D. L. Brody and R. Cantu, *Neurosurgery*, 2016, **79**, 912.
77. S. P. Broglio, S. N. Macciocchi and M. S. Ferrara, *Neurosurgery*, 2007, **60**, 1050.
78. J. E. Resch, C. N. Brown, J. Schmidt, S. N. Macciocchi, D. Blueitt, C. M. Cullum and M. S. Ferrara, *Br. Med. J. Open Sport Exercise Med.*, 2016, **2**, e000012.
79. J. K. Register-Mihalik, K. M. Guskiewicz, J. P. Mihalik, J. D. Schmidt, Z. Y. Kerr and M. A. McCrea, *J. Head Trauma Rehabil.*, 2013, **28**, 274.
80. L. C. Henry, R. Elbin, M. W. Collins, G. Marchetti and A. P. Kontos, *Neurosurgery*, 2016, **78**, 232.

81. J. R. Oldham, M. S. DiFabio, T. W. Kaminski, R. M. DeWolf and T. A. Buckley, *Sports Health A Multidiscip. Approach*, 2016, 1941738116680999.
82. T. R. Snedden, M. A. Brooks, S. Hetzel and T. McGuine, *Clin. J. Sport Med.*, 2017, DOI: 10.1097/JSM.0000000000000389.
83. T. Hänninen, M. Tuominen, J. Parkkari, M. Vartiainen, J. Öhman, G. L. Iverson and T. M. Luoto, *J. Sci. Med. Sport*, 2016, **19**, 636.
84. M. A. Engelson, R. Bruns, C. J. Nightingale, K. M. Bardwell, C. A. Mason, S. Tu, L. Nelson and S. A. Butterfield, *J. Chiropractic Med.*, 2017, DOI: 10.1016/j.jcm.2016.12.001.
85. E. F. Teel, M. R. Gay, P. A. Arnett and S. M. Slobounov, *Clin. J. Sport Med.*, 2016, **26**, 162.
86. C. Doherty, L. Zhao, J. Ryan, Y. Komaba, A. Inomata and B. Caulfield, *Clin. Biomechanics*, 2017, **42**, 79.
87. J. McDevitt, K. Appiah-Kubi, R. Tierney and W. Wright, *Int. J. Sports Med.*, 2016, **37**, 738.
88. J. Maruta, J. Tong, S. W. Lee, Z. Iqbal, A. Schonberger and J. Ghajar, SPIE Defense, Security, and Sensing, 2012.
89. K. A. Kirk, M. Shoykhet, J. H. Jeong, E. C. Tyler-Kabara, M. J. Henderson, M. J. Bell and E. L. Fink, *Dev. Med. Child Neurol.*, 2012, **54**, 759.
90. J. Leddy, A. Hinds, D. Sirica and B. Willer, *PM&R*, 2016, **8**, S91.
91. G. A. Davis, G. L. Iverson, K. M. Guskiewicz, A. Ptito and K. M. Johnston, *Br. J. Sports Med.*, 2009, **43**(Suppl 1), i36.
92. C. W. Sung, H. C. Lee, Y. H. Chiang, W. T. Chiu, S. F. Chu, J. C. Ou, S. H. Tsai, K. H. Liao, C. M. Lin and J. W. Lin, *Psychophysiology*, 2015, 349.
93. I. J. Baguley, R. E. Heriseanu, I. D. Cameron, M. T. Nott and S. Slewa-Younan, *Neurocrit. Care*, 2008, **8**, 293.
94. J.-K. Chen, K. M. Johnston, M. Petrides and A. Ptito, *Archiv. General Psychiatry*, 2008, **65**, 81.
95. M. W. Collins, G. L. Iverson, M. R. Lovell, D. B. McKeag, J. Norwig and J. Maroon, *Clin. J. Sport Med.*, 2003, **13**, 222.
96. M. W. Collins, M. R. Lovell, G. L. Iverson, R. C. Cantu, J. C. Maroon and M. Field, *Neurosurgery*, 2002, **51**, 1175.
97. S. L. Zuckerman, D. J. Totten, K. E. Rubel, A. W. Kuhn, A. M. Yengo-Kahn and G. S. Solomon, *Neurosurgery*, 2016, **63**, 102.
98. A. Seiger, E. Goldwater and E. Deibert, *J. Head Trauma Rehabil.*, 2015, **30**, E52.
99. M. W. Collins, A. P. Kontos, E. Reynolds, C. D. Murawski and F. H. Fu, *Knee Surg. Sports Traumatol. Arthrosc.*, 2014, **22**, 235.
100. C.-C. Chiang, S.-E. Guo, K.-C. Huang, B.-O. Lee and J.-Y. Fan, *Qual. Life Res.*, 2016, **25**, 2009.
101. H. G. Belanger, Z. Proctor-Weber, T. Kretzmer, M. Kim, L. M. French and R. D. Vanderploeg, *Clin. Neuropsychol.*, 2011, **25**, 702.
102. C. A. Luethcke, C. J. Bryan, C. E. Morrow and W. C. Isler, *J. Int. Neuropsychol. Soc.*, 2011, **17**, 36.

CHAPTER 11

Predictors for Prognosis and Recovery from Unconsciousness Due to Brain Trauma

A. A. POTAPOV,[a] E. V. ALEXANDROVA,*[a] S. A. DAMBINOVA,[b] N. E. ZAKHAROVA[a] AND G. V. DANILOV[a]

[a] N.N. Burdenko National Scientific and Practical Centre for Neurosurgery, 4th Tverskaya-Yamskaya 16, 125047, Moscow, Russia; [b] DeKalb Medical Centre, Atlanta, GA, USA
*Email: ealexandrova.nsi@gmail.com

11.1 Introduction

Severe traumatic brain injury (sTBI) is characterized by a comatose condition that may be followed by the development of post-comatose unconsciousness of varying duration. Evaluation of the possibility and limits of consciousness recovery is the key to assessing the therapeutic and rehabilitation potential and to predicting long-term outcomes of injury. However, the existing clinical criteria for diagnosing unconscious states are not always appropriate for an accurate interpretation of the patient's state of consciousness. Recent studies have demonstrated that about 40% of vegetative state (VS) patients actually exhibit signs of a higher level of consciousness—a minimally conscious state (MCS).[1]

Recently, the ethical aspect of consciousness recovery after sTBI has also been debated. In particular, the issue is whether to interrupt or prolong the life of these patients. A recent survey of physicians and related specialists in

Drug Discovery Series No. 62
Acute Brain Impairment: Scientific Discoveries and Translational Research
Edited by Philip V. Peplow, Svetlana A. Dambinova, Thomas A. Gennarelli and Bridget Martinez

Published by the Royal Society of Chemistry, www.rsc.org

Europe demonstrated that 66% of them supported euthanasia of patients in a chronic VS, and only 28% of them favored euthanizing patients in the MCS.[2]

In this chapter, we introduce an integrative approach that includes the clinical, neuroimaging and biochemical basics of comatose states, as well as the factors that affect the prognosis of consciousness recovery after severe brain injury.

11.2 Coma and Post-comatose Unconsciousness

11.2.1 Consciousness and Unconsciousness

Consciousness is the superior integrative brain function that includes a quantitative component—the level of consciousness—and a qualitative component—the content of consciousness (awareness, in particular orientation).[3] Unconscious and conscious processes in the brain are interrelated and strictly hierarchically organized into a single system. Visual manifestations by observers are not sufficient to conclude whether behavior is conscious or unconscious. The major problem in assessing patients in an unconscious state is to objectively evaluate the degree of consciousness impairment and to monitor consciousness recovery.

Brain systems that maintain the level of consciousness and provide the content of consciousness are different. The level of consciousness is controlled by the reticular formation of the brainstem, hypothalamus, midline portions of the thalamus, basal parts of the forebrain (ascending activating system), association areas of the frontal and parietal cortices, cingulate cortex, and precuneus (medial occipital cortex). Injury to these structures leads to depression of consciousness (up to coma), disturbance of the sleep–wake cycle, and reduced concentration, which also severely disrupts memorizing current information.[4] The content of consciousness is provided by the multi-level sensorimotor cortical–subcortical systems, medial portions of the temporal lobes and diencephalic region involved in the information storage processes, and the limbic system, controlling emotions and motivation. Injury to these structures may lead to the development of different disorders, such as agnosia, aphasia, apraxia, *etc.*[4]

11.2.2 Post-traumatic Coma and Scales

Coma is the strongest impairment of the level of consciousness. The most common and simplest definition, given by Plum and Posner,[5] states that coma is unresponsiveness of the patient (*i.e.* the patient does not open the eyes either spontaneously or in response to any external stimuli).[5] A more complete definition is that coma is an unconscious state that lasts from several hours to several weeks (usually less than 4–6 weeks) and is characterized by lack of eye opening in response to external stimuli, lack of signs of conscious reactions, lack of sleep–wake cycle, and impaired regulation of vital body functions.

The absence of eye opening is a main sign differentiating coma from post-comatose disorders of consciousness. Coma can be caused both by direct damage to the main brain structures responsible for maintaining the level of consciousness and by dysfunction of the ascending activating system due to brainstem compression associated with dislocation processes caused by edema, intracranial hematomas, cerebrospinal fluid circulation disorders, *etc.*[6]

There are several scales available to score the depth of coma. The Glasgow Coma Scale (GCS) is the most widely used and reproducible scale.[7] This is the shortest scale that evaluates only three components (the sum of the three components or a total score of 3–15): eye opening (arousal), motor reactions, and verbal responses. According to the scale, moderate coma is scored 6–7, deep coma is scored 4–5, and deepest coma is scored 3. However, the scale has significant limitations in patients with orbital trauma or in the case of artificial lung ventilation. For this reason, other scales, along with the traditional GCS, are now used for more detailed evaluation. In particular, the Full Outline of UnResponsiveness (FOUR) scale, which includes four parameters (eye opening and tracking [E], motor responses [M], brainstem reflexes [B], and respiration [R]), is quite simple and reproducible.[8]

11.2.3 Post-comatose Unconsciousness

The rate of recovery from coma may be tentatively classified as rapid (a GCS score increase of 4 or more within 24 hours), moderate (a GCS score increase of 4 or more within 3 days), slow (a GCS score increase of 4 or more within 1 week), or prolonged recovery (no changes within weeks).[9]

In short-term coma (duration of several minutes to several hours), recovery to consciousness occurs in stages similar to awakening with a successive alteration: coma, sopor, stupor, clear consciousness. In this case, the shorter the coma duration, the faster is the rate of recovery from coma to clear consciousness.

Recovery from prolonged coma (from several days to 2–3 weeks) is a gradual, multistep process that includes seven consecutive stages: 1—eye opening; 2—gaze fixation and visual pursuit of others; 3—recognition of persons emotionally important for the patient; 4—understanding of directly addressed speech; 5—own speech activity; 6—amnestic confusion; 7—formally clear consciousness.

The VS (previously called persistent VS [PVS]) term was first introduced by Jennett and Plum in 1972[5] to describe wakefulness without awareness of self and the environment. The VS is characterized by the presence of a sleep–wake cycle, eye opening, and orientation reflexes, but a lack of conscious, purposeful actions.[10,11] Therefore, the VS is a unique state of brain function that lacks any content of consciousness but preserves the level of wakefulness (*i.e.* there is a dissociation between the two components of consciousness). To emphasize this dissociation, some authors suggest replacing the term VS with "unresponsive wakefulness syndrome".[12]

A MCS is the next stage of consciousness recovery. The term and diagnostic criteria were first proposed by Giacino *et al.* in 2002.[13] Currently, two MCS variants, MCS-minus and MCS-plus, are distinguished. MCS-minus is diagnosed if the patient recovers from the VS (*i.e.* there is consistent visual fixation and visual pursuit, and emotional reactions may be present, but signs of speech understanding [execution of instructions] are still missing).[1,13] MCS-plus is characterized by the ability to follow simple verbal instructions and the inability to speak. Recovery from this state occurs when the patient is capable of producing speech, responding with gestures, or using two or more external objects in accordance with their functional purposes.[13]

11.3 Structural and Functional Basis of Unconsciousness

11.3.1 Neuroimaging and Novel Magnetic Resonance Imaging-based Classification of sTBI

sTBI is a relevant model for studying structural and functional disintegration of the brain.[14,15] Compared to computed tomography (CT), advanced magnetic resonance imaging (MRI) modalities (diffusion-weighted imaging [DWI], diffusion tensor imaging [DTI], susceptibility-weighted angiography [SWAN], and T2-fluid attenuated inversion recovery [T2-FLAIR]) enable the detection of microstructural hemorrhagic and non-hemorrhagic brain lesions in severely unconscious patients. In TBI, the accuracy of MRI is essential for understanding the relationship between damage to certain deep brain structures and the severity of trauma.

The introduction of MRI into clinical practice has demonstrated that brainstem lesions are more common in brain injury and are associated with a high risk of a poor outcome; therefore, early visualization of these lesions has a prognostic value. It is also known that damage to the brainstem is not always accompanied by a comatose state and does not necessarily lead to unfavorable outcome.[16]

MRI is more sensitive for the detection of parenchymal non-hemorrhagic lesions and micro-hemorrhagic lesions, as well as being appropriate for imaging of the posterior fossa structures, including the brainstem, in acute, subacute, and chronic phases of TBI.

The classification of severe head injury based on MRI was initially done by Firsching.[17] In this study, all supratentorial lesions were grouped into one stratum. However, in this work, the prognostic value of different brainstem and hemispheric damage levels was evaluated without considering involvement of the deep hemispheric structures, such as the basal ganglia,

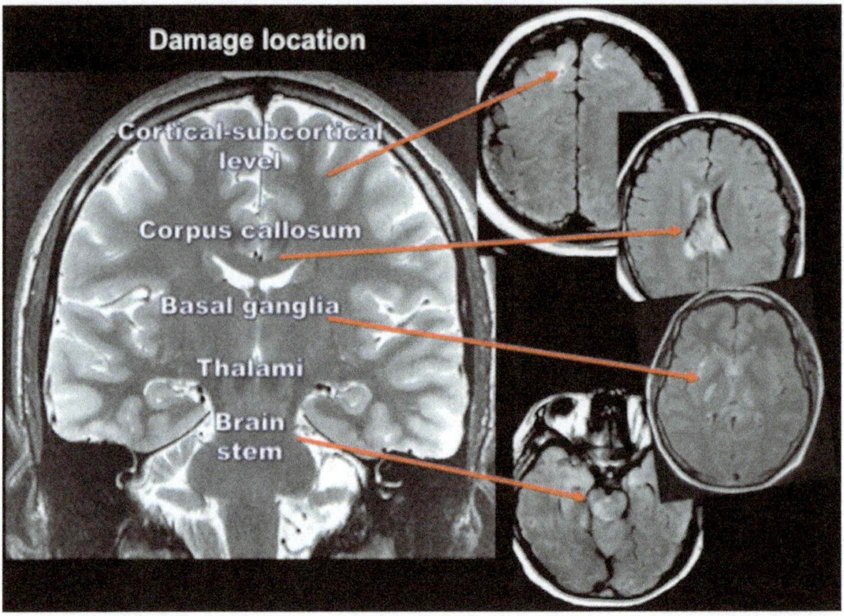

Figure 11.1 The stratification of focal, multifocal, and diffuse traumatic brain lesions by level and localization.

thalamus, and medulla oblongata, in patients with severe, moderate, and mild TBI (Figure 11.1). The classification of these different focal, multifocal, and diffuse brain injuries using opportunities of neuroimaging techniques is one of the major issues.

To address these issues, we analyzed clinical and MRI findings of 212 patients with acute TBI. The relationship between the level and localization of brain lesions and the severity and outcome of TBI was studied using advanced MRI.[18–20] Injury to the corpus callosum, basal ganglia, and brainstem was found to be significantly more frequent in comatose patients, while cortical and subcortical contusions prevailed in non-comatose patients. Involvement of the corpus callosum, basal ganglia, thalamus, and brainstem in the affected area (extension of lesions to the caudal direction) was associated with a higher risk of an unfavorable outcome (severe disability or death).

To establish the relationship between the depth of brain damage, severity of injury, and its outcome, we proposed an 8-score scale based on the location of lesions verified by MRI (Table 11.1). Table 11.1 shows that the extension of lesions to the caudal direction is associated with a higher risk of an unfavorable outcome: severe disability, VS, or death.

Seventeen patients without brain lesions were scored 1; one of the patients was in coma, but all patients had favorable outcomes. A third of 39 patients with cortical–subcortical lesions alone developed coma, and a small fraction of the patients (15%) had an unfavorable outcome.

Table 11.1 An MRI-based scale for traumatic brain lesions (a proposed classification of brain lesion level and localization).

Score	Lesion level/localization	Total	GCS ≤8	Unfavorable outcome
	Total	212	66%	48%
1	No damage	17	6%	0%
2	Cortical–subcortical	39	36%	15%
3	Corpus callosum ± 2	28	64%	25%
4	Basal ganglia, internal capsule, thalamus ± 2–3	28	64%	46%
5	Unilateral brainstem ± 2–4	38	84%	55%
6	Bilateral midbrain ± 2–4	34	91%	88%
7	Bilateral pons ± 2–6	26	96%	92%
8	Bilateral medulla oblongata ± 2–7	2	50%	50%

Injury to the corpus callosum significantly worsened the condition of patients: 64% of these patients developed coma. Injury to the basal ganglia resulted in an unfavorable outcome in almost every other patient.

A higher score on the proposed scale means that injury affects deeper brain structures, which does not exclude involvement of more superficial structures.

Injury to the brainstem structures resulted in a significant increase in the rate of coma and unfavorable outcome. Therefore, most patients with injury to the bilateral pons and midbrain had a poor outcome. However, brainstem injury did not always lead to coma. Two patients with medulla oblongata lesions were scored 8.

Strong and significant correlations between the proposed MRI-based scale and the GCS ($R = -0.65$, $p < 0.001$)/Glasgow outcome scale ($R = -0.65$, $p < 0.01$) were found, indicating a prognostic value of the proposed scale and the possibility of using it for classification of the level and localization of brain lesions.

Our findings demonstrate that injury caused by a motor vehicle accident significantly more often affects the corpus callosum as well as subcortical and brainstem structures compared to other injuries that usually affect superficial cortical structures. Patients in traumatic coma have a specific neuroanatomical pattern that is quite different from that in non-comatose patients. This paradigm supports a hypothesis proposed by Ommaya and Gennarelli more than 40 years ago.[83]

11.3.2 Cerebral Blood Flow Mapping

As neuroimaging techniques have been developed, cerebral blood flow has been studied using radioactive and stable xenon-enhanced CT, single-photon emission CT, perfusion and arterial spin labeling (ASL) MRI, positron emission tomography (PET), transcranial Doppler ultrasonography, and dynamic contrast-enhanced CT perfusion (CT perfusion or perfusion CT).[21,22]

Perfusion CT quantifies the volume of blood passing through unit brain mass per unit time: as a contrast bolus moves through the vasculature, images of the same brain level are acquired. Quantitative blood flow parameters are calculated from the time-varying density of a contrast tracer in the area of interest.[8] The traditional parameters determined by this method reflect the cerebral blood flow (CBF; mL 100 g^{-1} min^{-1}) in a particular region, regional cerebral blood volume in 100 g of brain substance (mL 100 g^{-1}), and mean transit time spent by a contrast tracer to pass through the area of interest (seconds).

Excessive fluctuations of blood flow (a decrease or increase in certain parameters) in the acute period of TBI lead to cerebral ischemia or hyperemia, respectively. Diffuse and focal lesions of the cerebral hemispheres may lead to a decrease in cerebral blood flow. According to Wintermark and coauthors,[23] CT perfusion demonstrated a significant decrease in all blood flow parameters in hemispheric foci of contusion. Similar observations were described by Potapov and coauthors,[24] who analyzed changes in blood flow in a group of 40 patients with sTBI, with allowance for a dominant type of diffuse or focal damage. Attempts to grade blood flow to determine the threshold of ischemic disorders in stroke and TBI were undertaken in several studies using PET.[25]

Despite a large variety of modern neuroimaging techniques, blood flow in TBI has been mainly studied in hemispheric structures. Danilov, Zakharova, Potapov *et al.* in 2014, 2016[26–28] comprehensively investigated the relationships between structural, clinical, and monitored parameters (intracranial, arterial, and cerebral perfusion pressure and the state of autoregulation) and blood flow parameters in the brainstem with allowance for the effect of surgical treatment on all CT perfusion parameters. The study demonstrated a high variability of brainstem blood flow parameters (from 4 to 76 mL 100 g^{-1} min^{-1}), with the mean values being lower compared to those for the hemispheres (26–28 mL 100 g^{-1} min^{-1}). The widest range of values was found in patients with sTBI (a GCS score of 3–4). This variability was associated with two main causes: structural damage to the brainstem and disorders of the cerebral blood flow regulation system. Brainstem blood flow reduction in primary brainstem injury was limited by the hemorrhagic lesion area (Figure 11.2). In contrast, severe secondary brainstem injury (compression) leads to autoregulation failure with extremely high brainstem blood flow (up to 76 mL 100 g^{-1} min^{-1}; Figure 11.3). This autoregulation breakdown was significantly related to baseline intracranial pressure greater than 15 mm Hg. The authors demonstrated that impairment of the brainstem circulation may occur earlier than signs of an axial dislocation developing, and indications for decompression surgery were defined.[29]

A CT perfusion examination may be recommended for TBI patients in whom the severity of the condition is not consistent with findings of routine brain CT, and MRI cannot be used to identify hypoperfusion and hyperperfusion areas in the hemispheres and brainstem. In this case, blood

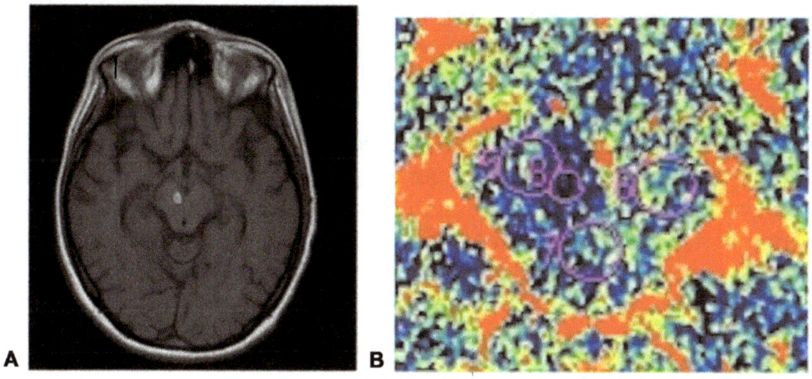

Figure 11.2 T1-MRI (A) and CT perfusion map (B) for a patient with a primary brainstem lesion. CBF inside the lesion was reduced to 4 mL 100 g^{-1} min^{-1}.

flow evaluation and its interpretation together with the data on the condition of the regulatory mechanisms of cerebral hemodynamics may have significant diagnostic and prognostic value.

11.3.3 Neurotransmitter Aspects of Unconsciousness

Currently, there are several major neurotransmitter theories explaining the development of different states of impaired consciousness. One of these is a mesocircuit hypothesis.[30] According to this hypothesis, functional abnormalities of the interactions between large neuronal connections of the forebrain develop first due to an overall reduction in excitatory neurotransmission.[30] Connections of the forebrain include a complex system of interactions among the frontal cortex, thalamus, and striatopallidal system and are most affected in multifocal brain injury. This hypothesis is based on dysfunction of the dorsal pathway where the thalamus is the key structure. The central thalamus is known to be highly innervated by cholinergic, serotonergic, and noradrenergic afferents of the brainstem ascending activating system,[31] as well as by descending projections from the frontal cortex that enables the executive functions underlying purposeful behavior. These ascending and descending influences on the central thalamus are believed to modulate the level of wakefulness depending on overall activity, different cognitive tasks, stress, and other processes typical of wakefulness.[31,32]

Striatal medium spiny neurons (MSNs) containing glutamate are the second key structures. They send inhibitory (GABAergic) projections to the globus pallidus internus that tonically inhibit the central thalamus in the absence of these inputs.[33] The MSN activity, in turn, is regulated by excitatory (glutamate) thalamostriatal, frontostriatal, and hippocampal projections.[34] The MSN activity is also regulated by the dopaminergic system.[35] Therefore, inhibition of the MSN activity caused by reduced dopaminergic or glutamatergic modulation of these neurons or their direct damage may play a key role in dysfunction of the forebrain neuronal systems and may lead to the development of impaired consciousness syndromes.

Another (dopamine) neurotransmitter hypothesis is based on frequent direct or indirect traumatic injury to the ascending (nigrostriatal and mesocorticolimbic) and descending (from the prefrontal cortex to the striatum and thalamus; from the anterior cingulate gyrus to the striatum) pathways of the dopaminergic system, which subsequently leads to the development of pronounced cognitive disorders.[36] The interaction between the dorsolateral prefrontal cortex and the striatum is shown to play an important role in the regulation of executive functions and working memory. This interaction, in turn, is modulated by the dopaminergic system.[36]

The third hypothesis is based on the disruption of the interaction between the noradrenergic and dopaminergic systems of the brain, which are known to jointly control mental activity and motor and endocrine functions. First, note that the systems are characterized by a similar topographic distribution and organization. For example, the major nuclei of both systems are situated in the brainstem from where long axons, innervating almost all parts of the central nervous system (CNS), originate. Individual nuclei (parabrachial and paranigral) contain both dopaminergic and noradrenergic neurons. It is important to note that the dopaminergic and noradrenergic systems are closely interacting as agonists or antagonists at virtually all CNS levels. This interaction between the systems is probably important for maintaining a consciousness level and holistic mental activity. The importance of the interaction between these systems for maintaining a consciousness level is evidenced by the different dynamics of norepinephrine and dopamine plasma levels in patients with severe TBI (sTBI) in unconscious states and after recovery of consciousness.[37,38]

11.4 Prognostic Value of Neuroimaging for Coma and Unconsciousness

11.4.1 MRI Features and Predictors in Unconsciousness

One of the leading causes of disability and mortality of patients with brain trauma is diffuse axonal injury (DAI). The anatomical structures most affected by DAI include the subcortical white matter, corpus callosum, dorsolateral midbrain, and subcortical structures.[39–42]

MRI in T2-FLAIR and T2* gradient echo modes enables a more accurate diagnosis and prognosis of TBI outcomes, depending on the level of injury to the hemispheric and brainstem structures.[17,43] DWI was shown to be capable of detecting lesions that were undetectable even with T2, T2*, and T2-FLAIR MRI.[44]

The implementation of DTI opened up new opportunities in quantitative and qualitative assessment of injuries to the conduction paths of the brain, the acquisition of their three-dimensional images, and thus the *in vivo* visualization of the severity of DAI in clinical conditions.[22,45,46]

DTI evaluates the diffusion parameters of the investigated medium as well as the direction of water diffusion (anisotropy), and thus provides information on the degree of integration of the white matter tracts.[47,48] Diffusional anisotropy is heterogeneous in different regions of the white matter and reflects the differences in myelination, diameter, and direction of fibers.[49] Pathological processes that alter the white matter microstructure (*e.g.* rupture, disorganization, and disconnection of fibers) and are associated with the destruction of myelin, the retraction of neurons, and an increase or decrease in the extracellular space affect the diffusion and anisotropy parameters.[50,51]

The most frequently used quantitative parameters for the evaluation of DWI and DTI are the apparent diffusion coefficient and fractional anisotropy (FA). DTI studies of TBI patients demonstrated that FA decreased in the first week after injury, while no changes in the white matter were detected by CT or routine MRI.[54] A decrease in FA in DAI was also found in the anterior and posterior corpus callosum (Figure 11.3), the posterior limb of the inner capsule, and the outer capsule, as well as in studies of overall parameters of the entire white matter.[53,54]

According to our studies, extensive changes in the structure of the corpus callosum and corticospinal tracts (CSTs) occur within the first 2–17 days after severe diffuse brain injury accompanied by the development of coma and, later, disability of varying degrees. This confirms the results of earlier neuromorphological studies of DAI.[55] The most sensitive indicators of conduction tract injury in DAI in the early post-injury period were FA parameters. A significant decrease in these parameters, in comparison to

A

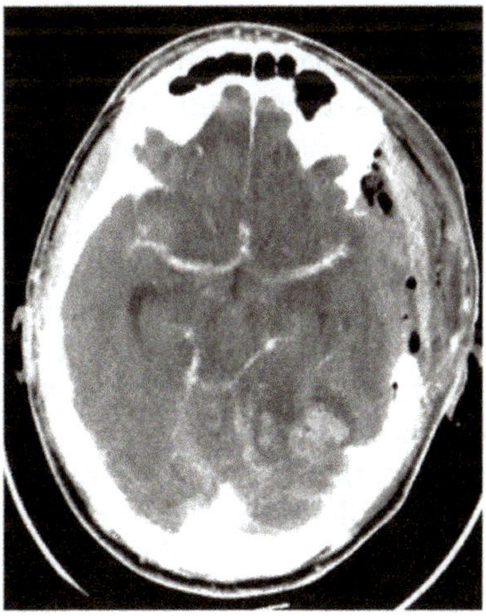

B

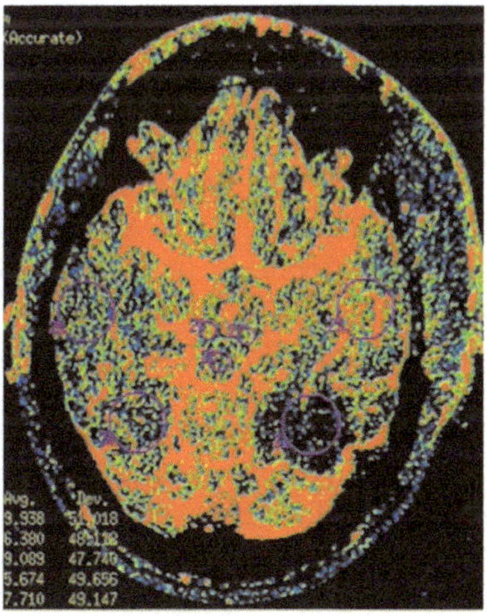

Figure 11.3 CT scan (A) and CT perfusion map (B) for a patient with secondary brainstem damage. CBF was dysregulated and increased to 76 mL 100 g^{-1} min^{-1}.

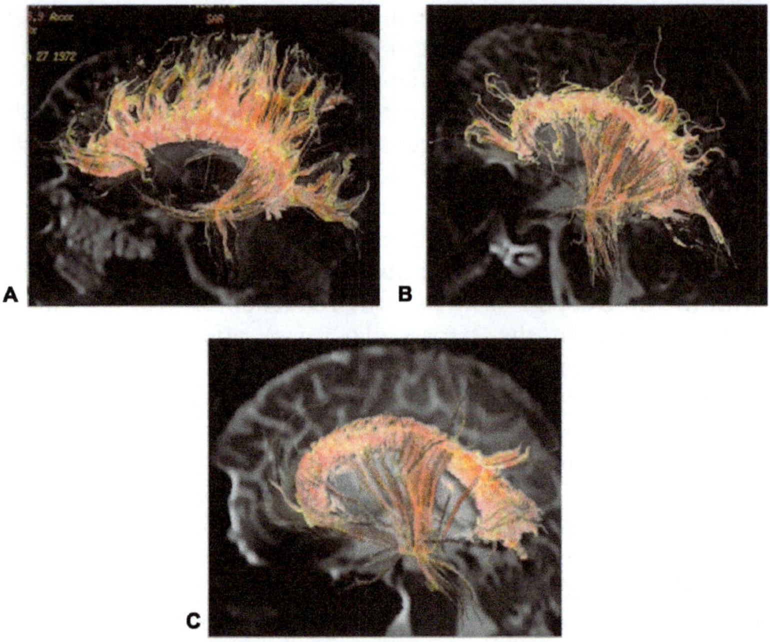

Figure 11.4 Dynamics of corpus callosum fibers degradation (by DT tractography) for a severely injured patient: 5 days (A), 25 days (B), and 19 months (C) after TBI.

normal ones, was revealed in both the corpus callosum and the CSTs, at different levels, in the entire group of patients with severe DAI. Similar data were obtained by Huisman *et al.*[52] in 20 patients in the first 7 days after TBI of varying severity in the splenium and posterior limb of the inner capsule (Figure 11.4).

In our study, the reliable correlation between DAI outcomes and FA parameters in the corpus callosum and along CSTs revealed on day 10 – 17 after injury indicates the high prognostic value of diffusional anisotropy. It may be supposed that primary injury to the conduction tracts (in CST and corpus callosum structures) in DAI leads to axonal degeneration that causes a more significant decrease in anisotropy 2–3 weeks after injury. These results confirm the data of van der Knaap[56] that primary brain injury, such as severe DAI, acts as a trigger for degenerative changes in the axons and myelin sheaths of the white matter in the brain, which lead to complete destruction and atrophy of the structures in 2–3 months after injury.

Long-term studies of the "split brain" in patients who underwent corpus callosotomy for severe epilepsy revealed the role of the corpus callosum in brain functioning.[57] These studies demonstrated that a split of the corpus callosum blocks interhemispheric transmission of sensory, motor, gnostic, and other information. These findings clarified the interhemispheric difference and interhemispheric interaction mechanisms. Studies of patients with local damage to the corpus callosum revealed that the posterior corpus callosum was predominantly involved in the transmission of sensory (visual, auditory, and somatosensory) information, while the anterior corpus callosum was involved in the transmission of cognitive information.[58,59]

It seems that classical models of congenital agenesis of the corpus callosum and a "split brain" may explain, to a certain extent, the differences in the rate and quality of consciousness recovery in patients with traumatic injury to the corpus callosum and its subsequent atrophy who were examined in our studies. However, it is quite obvious that diffuse axonal brain injury is a fundamentally different model of a multidimensional split of the brain caused by injury to both commissural (interhemispheric) and associative (intrahemispheric) as well as projection (cortical–subcortical and subcortical–brainstem) connections. Patients with MRI signs of injury and subsequent development of atrophy of the corpus callosum and brainstem may recover from the unconscious state either through the transitory VS to the MCS or through a prolonged period of disintegrated consciousness with persistent impairments of mental (orientation, memory, speech, *etc.*) functions. At the same time, the corpus callosum either remained intact or partially degenerated over time in patients who relatively quickly and completely recovered mental activity in a few weeks to months after DAI.

Our studies showed that injuries to the corpus callosum and CSTs with development of their marked atrophy in severe DAI occurred in patients with partial recovery of consciousness and mental and motor functions (MCS and disintegrated consciousness, tetraparesis, and hemiparesis) or in patients in a persistent VS.[22]

Further research using DTI at different times after injury will help determine qualitative and quantitative changes in the white matter tracts and their clinical correlations, which will provide a better understanding of the pathological basis of traumatic brain disease. The new tractography technologies such as the constrained-spherical deconvolution-high angular resolution diffusion imaging (CSD-HARDI) algorithm (Figure 11.5) provide a promising perspective for better evaluation of crossing brain fibers.

11.4.2 Prognostic Value of Non-structural Neuroimaging Data (functional MRI)

Functional MRI (fMRI) studies have demonstrated that patients in VSs have metabolism impairments in the structures involved in the frontoparietal interactions that include the polymodal association cortex (prefrontal cortex, Broca area, temporoparietal and posterior parietal cortices, and precuneus).[60] These interactions form the basis for conscious perception of information. In this case, the appearance of atypical cortical activity on fMRI scans may serve as a surrogate marker for a favorable prognosis,[61] and recovery of the patient from a VS is associated with functional recovery of these interactions.[6]

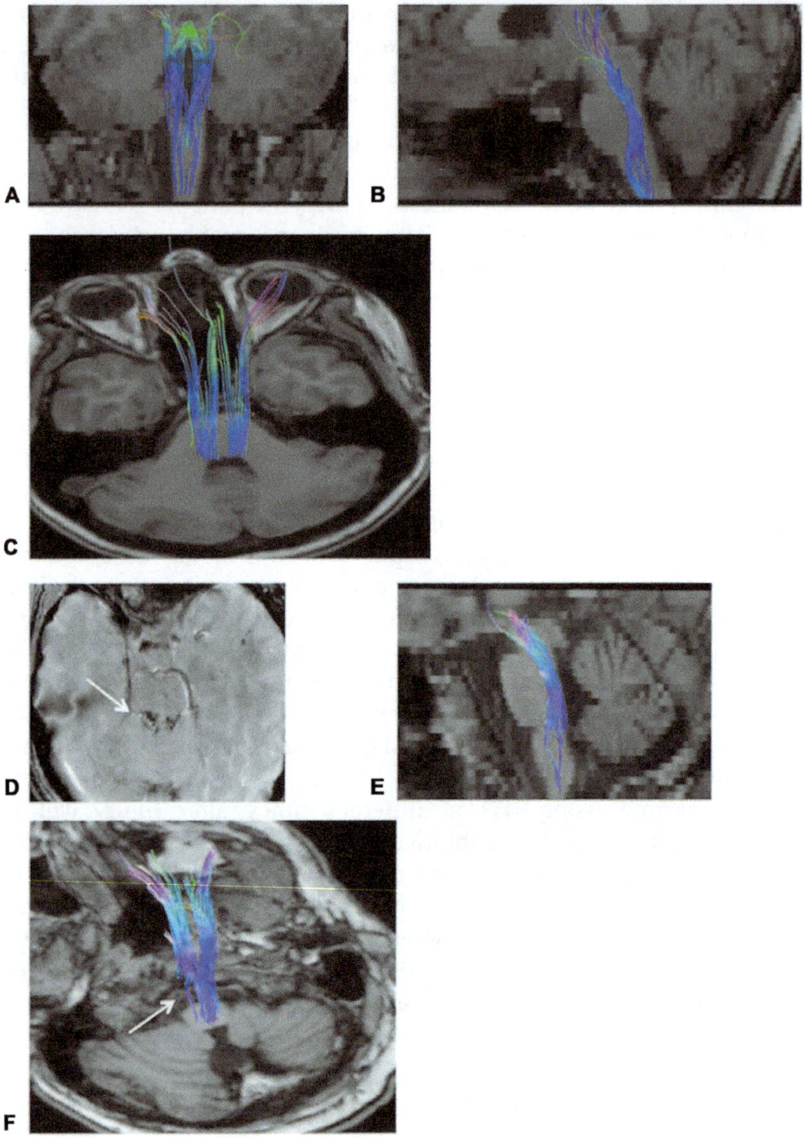

Figure 11.5 CSD-HARDI algorithm for reticular formation fiber tracking (from cholinergic nuclei area) in a normal brain (A, B, C) and in a patient with severe TBI (D, E, F). The arrow shows damage of the brainstem.

Changes in fMRI in response to significant external (pain or acoustic) stimulation or presentation of a task may precede clinical signs of a patient's transition from a VS to an MCS.[62] For example, pain stimuli are perceived by patients in both VS and MCS, but activation patterns on fMRI scans are different. In VS patients, only the brainstem, thalamus, and primary somatosensory cortex are activated, whereas in MCS patients, all structures involved in the perception of pain in a healthy subject (thalamus, primary and secondary somatosensory cortices, anterior cingulate cortex, and frontoparietal interconnections) are activated.[62] Acoustic stimuli are perceived similarly: activation in VS patients is limited to the primary auditory cortex, while additional activation of association areas occurs in MCS patients.[6,62] In this case, significant acoustic stimuli (*e.g.* a crying baby or calling by name) cause more pronounced activation of these cortical areas.[6,62]

The role of resting-state fMRI in assessing the preservation of cognitive neural networks in unconscious patients is under investigation.[63] This technique is based on assessing the activity of certain brain structures (default mode network),[64] including the precuneus, posterior parietal cortex, and medial prefrontal cortex. These structures are more active in a quiet wakefulness state and are inactive when solving cognitive tasks and concentrating attention. Further investigation of the activity of these structures in unconscious patients may be useful for differentiating the diagnosis between VS and MCS and predicting recovery of consciousness.

11.4.3 Neurometabolic Imaging and Prognostic Value of Unconsciousness

Magnetic resonance spectroscopy (MRS) is a noninvasive technique that provides information on neurochemical changes in the brain substance. Usually, MRS is used in studies of consciousness disorders to measure peaks of choline (Cho), creatine (Cr), *N*-acetylaspartate (NAA), and lactate (La). Cho, which is a cell membrane component, serves as a metabolic biomarker for membrane synthesis and destruction, which may indirectly reflect glial proliferation and the development of inflammatory processes. Its level is slightly higher in the white matter than in the gray matter.

Cr is found in both glial cells and neurons and is a biomarker for the aerobic metabolic pathway. Its values are mostly stable, therefore Cr is often used as a reference compound, particularly in calculations of NAA/Cr and Cho/Cr ratios. NAA occurs in both the gray and white matter and serves as a marker for neuronal density. In healthy subjects, there is an increase in the concentration gradient of NAA in the gray matter from ventral to dorsal and from the cerebral hemispheres to the spinal cord.[65] La is a marker for anaerobic glycolysis, which is almost not detected in healthy brain tissue and increases in trauma, hypoxia, or ischemia.

MRS in the acute period of severe TBI (in the first 24 hours) detects primary structural lesions and reversible changes (vasogenic and cytotoxic edema) in the brain. MRS in the early subacute period (1–13 days) enables evaluation of the volume and localization of structural brain lesions caused by intracranial hypertension, brain dislocation, and discirculatory disorders,[52,66] which is essential for predicting the outcomes of possible cognitive and behavioral disorders. The NAA/Cr ratio was shown to be preferable to the NAA/Cho ratio for evaluating the prognosis of TBI, despite cases of adverse outcomes where MRI and MRS findings were normal.[67] There is evidence of a correlation between a TBI outcome and NAA/Cr values in the gray and white matter of the parieto-occipital[68] and frontal lobes,[69] corpus callosum,[70] thalamus,[71] and pons.[72] Furthermore, MRS of the pons in the acute period of TBI enables the distinguishing of patients with a favorable prognosis from those with severe disability, VS, or death.[72]

To assess brain function in coma patients, MRS uses axial sections of the basal ganglia, thalamus, insular cortex, and periventricular white matter to visualize the structures involved in the regulation of consciousness.

11.5 Neurochemical Basis of Coma and Prognostic Biomarkers of Outcome

At present, more than 50 different neurotransmitters and modulators of synaptic transmission are known. The major neurotransmitters enabling mental and motor functions of the brain include glutamate, γ-aminobutyric acid (GABA), acetylcholine, catecholamines (dopamine, noradrenaline, and adrenaline), indolamines (serotonin and histamine), and glycine.

Mediators are synthesized in different brain structures. Glutamatergic and GABA neurons account for 25–30% and 10–20% of synapses of the entire brain, respectively, and are distributed diffusely (in the cortex, subcortical structures, brainstem, and cerebellum). Neurons of the cholinergic (5–10% of synapses), noradrenergic (1–2% of synapses), dopaminergic (1–2% of synapses), adrenergic (1–2% of synapses), serotonergic (1–2% of synapses), and histaminergic (1–2% of synapses) systems are grouped into separate nuclei located in the brainstem and hypothalamic and subcortical structures of the brain.[73]

These circumstances underlie more frequent damage to glutamatergic and GABA systems in any brain trauma. The cholinergic system, whose neurons are located in the neostriatum and basal parts of the forebrain, and also the dopaminergic system located in the midbrain and diencephalic region, are less affected. Injuries occur even less often in the serotonergic, noradrenergic, adrenergic, and (partially) cholinergic systems, the nuclei of which are situated more deeply, in the pons and medulla oblongata.[74]

A prolonged abnormal sleep state (coma) and post-coma unconsciousness occur due to disturbance of the interactions among different neurotransmitter systems in the brain. Each pathological condition in a particular patient is probably characterized by a specific pattern of imbalance among the major neurotransmitter systems involved in regulation of the sleep–wakefulness rhythm and mental functions.

Functional abnormalities of the interactions among large neuronal connections of the forebrain develop first due to reduced excitatory neurotransmission[75] and neurotoxicity development (see Chapter 9 in this book).

There is an intensive search for objective markers capable of predicting the outcome of coma, in particular awakening from unconsciousness.

Combining a post-resuscitation summated GCS score and acute-phase levels of interleukin (IL)-8 significantly improved the accuracy of a post-traumatic unfavorable outcome prediction (sensitivity of 100% and specificity of 96%) after severe TBI in children.[76] Higher levels of the microtubule-associated protein 2 (MAP-2) were released into the bloodstream of chronic TBI patients.[77]

Serum levels of an astroglial marker, the S100β protein, were significantly higher in brain-dead (BD) patients, particularly in an isolated TBI BD group, compared to those in surviving patients after sTBI and healthy volunteers.[78] The diagnostic properties of S100B reached a positive predictive value of 100%, as well as sensitivity and negative predictive values of 60% and 86.7%, respectively, which suggests using S100β serum levels in BD diagnosis as a potential confirmatory test.

Several studies indicated a role of genetic factors in determining the outcome after sTBI.[79–81] A recent explorative study identified a single-nucleotide polymorphism through altered mRNA expressions in cerebrospinal liquid of comatose patients that might be related to TBI consequences.[82]

In our pilot study, the level of antibodies (Abs) to the kainate receptor (KAR) in serum samples from patients in post-traumatic coma ($n = 23$) on

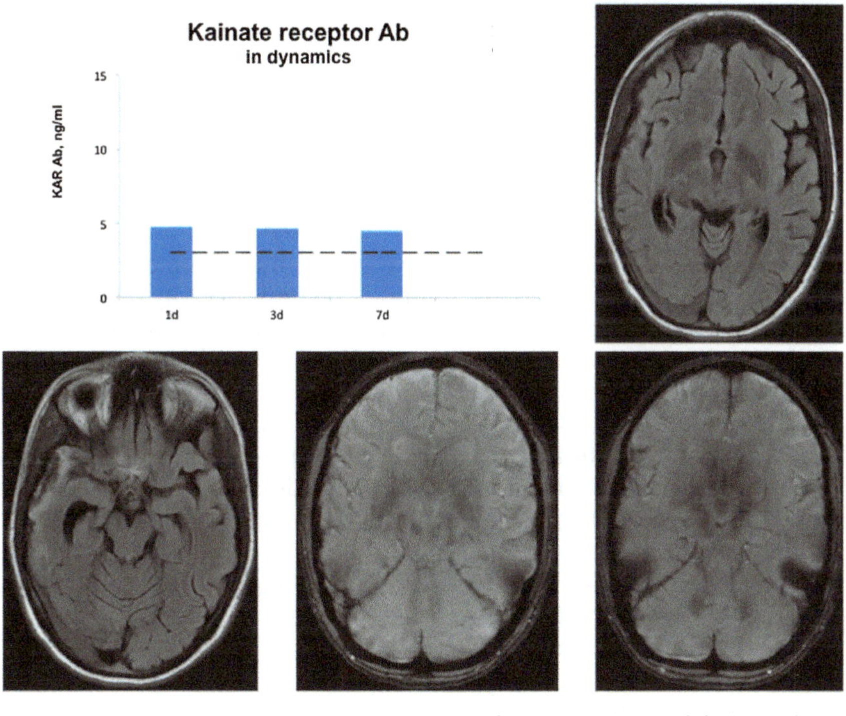

Figure 11.6 3T MRI of a patient with severe TBI (19 year old female) (GCS = 4). MRI shows diffuse axonal injury without brainstem damage. The diagram depicts the levels of antibodies to kainate receptor in acute period of severe TBI.

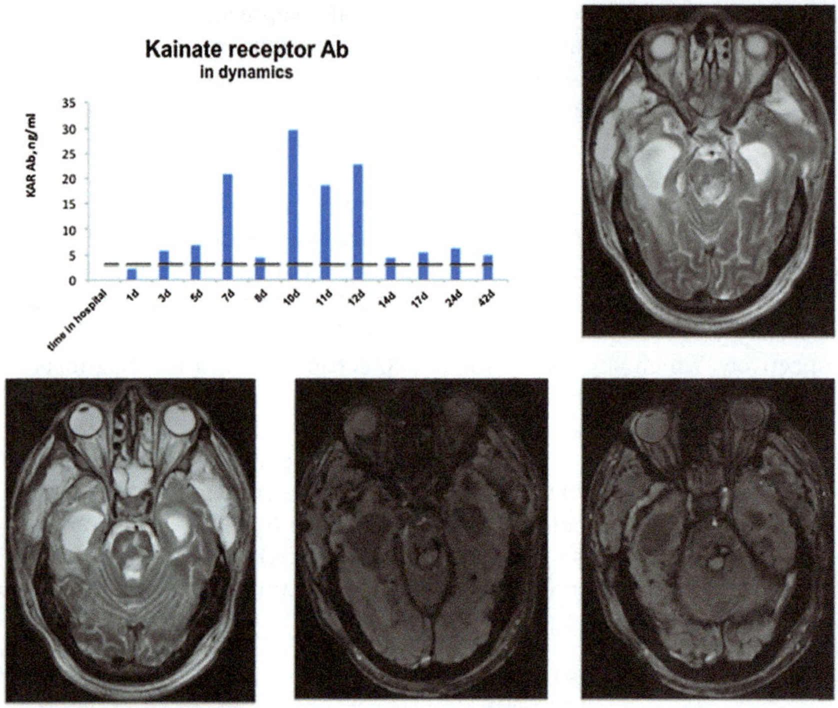

Figure 11.7 MRI of patient with severe TBI (39 year old male) (GCS = 3). MRI shows multiple hemispheric and severe brainstem damage. The diagram depicts the levels of antibodies to kainate receptor in acute period of severe TBI.

day 3–5 after sTBI was significantly increased ($p = 0.04$) in patients with brainstem injury compared to that in patients with other subcortical injuries. Evaluation of brainstem injury yielded a sensitivity of 73% and a specificity of 75% at a cut-off of 1.5 ng mL^{-1}. The relatively low performance characteristics may be due to the small power of the study, which should be expanded to add patients with different levels of unconsciousness.

Preliminary data on KAR Ab levels in patients in post-traumatic coma at different times indicate the opportunity of awakening from coma as the biomarker level reduces (Figures 11.6–11.8). In addition, dynamical changes in the KAR Ab level may be associated with the outcome of coma (moderate disability, VS, or death).

Combined application of clinical, radiological, and potential biomarker of coma may provide an integrative vision of unconsciousness and predict a positive or negative outcome.

11.6 Conclusion

Severe TBI is characterized by coma in the acute period, which is usually caused by brainstem dysfunction due to its direct traumatic injury or

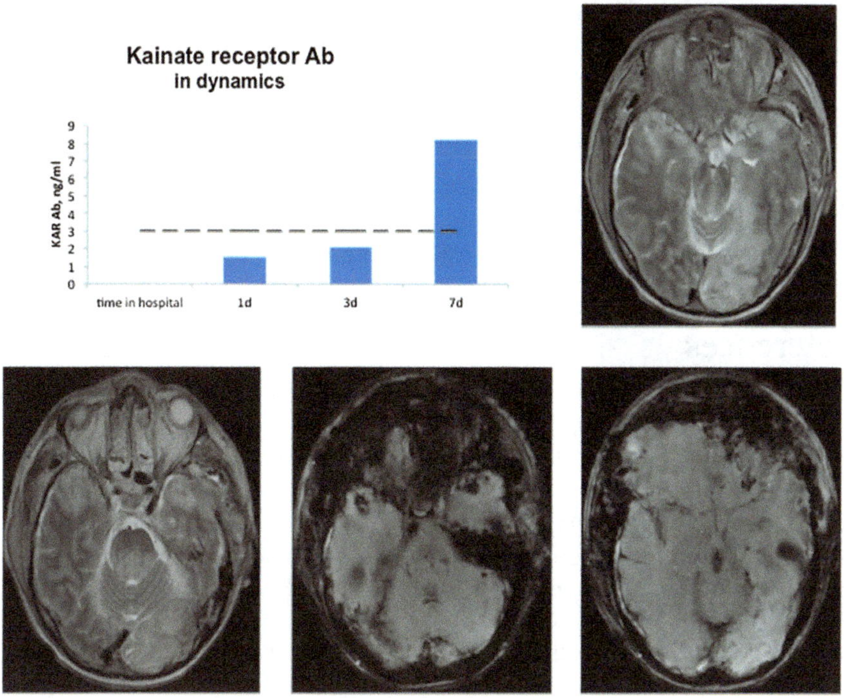

Figure 11.8 3T MRI of patient with severe TBI (44 year old male) (GCS = 3). MRI shows diffuse axonal injury with severe brain herniation and edema in ortical-subcortical and the brainstem areas.

concomitant secondary (non-traumatic) causes, such as edema, dislocation, intracranial hypertension, *etc*. Recovery from prolonged coma in about 35% of cases is accompanied by a transition to an unconscious state (VS or akinetic mutism), the duration of which affects the degree of further recovery of mental functions and working capability and depends on the presence and severity of injury to the deep brain structures (brainstem, basal ganglia, and thalamus).

The prognosis of a severe TBI outcome and recovery from an unconscious state includes a neurological evaluation using CT/MRI and is aimed at establishing the type, cause, and degree of consciousness depression. If the patient is in an unstable condition, cannot be transported outside the intensive care unit, and has contraindications (metal objects in the body), the most informative diagnostic technique—brain MRI—often cannot be used. In addition, the usual MRI modes do not evaluate biochemical processes in the brain.

Recovery from coma in the acute period of sTBI is difficult to predict using only conventional techniques. Determining the prognosis in coma is not only of medical but also of great economic importance because of the need to redistribute material resources and to limit intensive care measures in severe cases. As of now, rapid assessment of specific parameters of the brain state over time still remains an unresolved problem. In this regard, the identification of specific brain biomarkers that reflect the course of traumatic disease and can be measured in peripheral blood is an important task in monitoring the brain condition in unconscious states.

Acknowledgements

We acknowledge the support from RFBR grant no. 16-29-08304.

References

1. C. Schnakers and S. Laureys, *Coma and Disorders of Consciousness*, Springer, 2012, p. 169.
2. A. Demertzi, A. Soddu and S. Laureys, *Curr. Opin. Neurobiol.*, 2013, **23**, 239.
3. A. Zeman, *Brain*, 2001, **7**, 1263.
4. H. Blumenfeld, Sunderland (MA): Sinauer Associates; 2010.
5. B. Jerome Posner, B. Clifford, N. Schiff and F. Plum, Oxford University Press, 2007.
6. S. Laureys and G. Tononi, *The Neurology of Consciousness*, Elsevier Ltd, 2009, p. 423.
7. J. A. Opara, I. Małecka and M. Szczygiel, *J. Med. Life*, 2014, **7**, 124.
8. E. F. Wijdicks, *Rev. Neurol. Dis.*, 2006, **3**, 109.
9. E. A. Freeman, *The Catastrophe of Coma*, David Bateman Ltd., Australia, 1987.
10. K. Andrews, G. Beaumont, F. Danze, M. D. Dimancescu, E. A. Freeman, A. Fugl-Meyer, Z. Groswasser, B. Jennett and J. Kelly, *et al.*, International working party on the imanagement of the vegetative state: summary report, *Brain Injury.*, 1996, **10**, 797.
11. K. V. Wild, F. Gerstenbrand, G. Dolce and H. Binder, *et al.*, Guidelines for quality management of apallic syndrome/vegetative state, *Eur. J. Trauma Emerg. Surg.*, 2007, **33**, 268.
12. S. Laureys, G. G. Celesia and F. Cohadon, *et al.*, European Task Force on Disorders of Consciousness, *BMC Med.*, 2010, **8**, 68.
13. J. T. Giacino, S. Ashwal and N. Childs, *et al.*, *Neurology*, 2002, **58**, 349.
14. N. E. Zakharova, A. A. Potapov and V. N. Kornienko, *et al.*, *Zh. Vopr. Neirokhir. im. N. N. Burdenko*, 2010, **2**, 3.
15. A. A. Potapov, V. V. Krylov and A. G. Gavrilov, *et al.*, *Zh. Vopr. Neirokhir. im. N. N. Burdenko*, 2015, **79**, 100.
16. G. Danilov, N. Zakharova and E. Alexandrova, *et al.*, *MEDIMOND S R L PP*, 2015, 211.
17. A. A. Potapov, N. E. Zakharova and V. N. Kornienko, *et al.*, *Zh. Vopr. Neĭrokhir. im. N. N. Burdenko*, 2014, **78**, 4.
18. N. Zakharova, A. Potapov and V. Kornienko, *et al.*, EANS 2014 – 15th European Congress of Neurosurgery, Proceedings book, 2015, 211.

19. N. E. Zakharova, A. A. Potapov, V. N. Kornienko, I. N. Pronin, E. V. Alexandrova, G. V. Danilov, A. G. Gavrilov, O. S. Zaytsev, A. D. Kravchuk and A. A. Sychev, *RFBR Inf. Bull.*, 2016, **90**, 12.
20. A. A. Potapov, V. V. Krylov and A. G. Gavrilov, *et al.*, *Zh. Vopr. Neirokhir. im. N. N. Burdenko*, 2016, **80**, 98.
21. L. Axel, *Invest. Radiol.*, 1983, **18**, 94.
22. N. Zakharova, V. Kornienko, A. Potapov and I. Pronin, *Neuroimaging of Traumatic Brain Injury,* Springer, 2014. p. 159.
23. M. Wintermark, G. van Melle and P. Schnyder, *et al.*, *Radiology*, 2004, **232**, 211.
24. A. A. Potapov, N. E. Zakharova and I. N. Pronin, *et al.*, *Zh. Vopr. Neirokhir. im. N. N. Burdenko*, 2011, **75**, 3.
25. H. Ito, I. Kanno and H. Fukuda, *Ann. Nucl. Med.*, 2005, **19**, 65.
26. N. Zakharova, A. Potapov and V. Kornienko, *et al.*, *Neuroradiology*, 2012, **54**, 136.
27. G. Danilov, N. Zakharova and A. Potapov, *et al.*, *MEDIMOND S R L*, 2015, 225.
28. G. Danilov, N. Zakharova and A. Potapov, *GEOTAR-Media*, 2016, 176.
29. G. V. Danilov, N. E. Zakharova, A. A. Potapov, V. N. Kornienko, I. N. Pronin, A. G. Gavrilov, E. V. Alexandrova, A. V. Oshorov, A. A. Sychev and A. A. Polupan, *RFBR Inf. Bull.*, 2016, **90**, 33.
30. N. D. Schiff, *Trends Neurosci.*, 2010, **33**, 1.
31. N. D. Schiff, *Ann. N. Y. Acad. Sci.*, 2008, **1129**, 105.
32. Y. Nagai, H. D. Critchley and E. Featherstone, *et al.*, *Neuroimage*, 2004, **21**, 1232.
33. S. Grillner, J. Hellgren and A. Munard, *et al.*, *Trends Neurosci.*, 2005, **28**, 364.
34. Y. Smith, D. Raju and B. Nanda, *et al.*, *Brain Res. Bull.*, 2009, **78**, 60.
35. C. M. Pennartz and S. T. Kitai, *J. Neurosci.*, 1991, **11**, 2838.
36. J. W. Bales, A. K. Wagner and A. E. Kline *et al.*, *Neurosci. Biobehav. Rev.*, 2009, **33**, 981.
37. E. V. Aleksandrova, O. S. Zaytsev and V. D. Tenedieva, *et al.*, *Zh. Nevrol. i Psychiatry*, 2011, **3**, 58.
38. E. V. Alexandrova, O. S. Zaytsev, A. A. Potapov and N. N. Burdenko, *Zh. Vopr. Neirokhir. im. N. N. Burdenko*, 2014, **1**, 24.
39. A. Potapov and E. Gaytur, Biomechanics and basic links of pathogenesis of TBI, in *Clinical Guidelines on Traumatic Brain Injury*, ed. A. Konovalov, L. Likhterman and A. Potapov, 1998, Antidor, Moscow, pp. 152–165.
40. T. Gennarelli, J. Adams and D. Graham, Diffuse axonal injury – a new conceptual approach to an old problem, in *Mechanism of Secondary Brain Damage*, ed. A. Baethman, *et al.*, Plenum, New York, 1986, pp. 15–28.
41. J. Povlishock and J. Stone, Traumatic axonal injury, in *Head Trauma: Basic, Preclinical and Clinical Directions*, ed. L. Miller and R. Hayes, Wiley-Liss, New York, 2001, pp. 281–302.
42. P. Reilly, R. Bullock, Hodder Arnold, London, 2005.
43. R. Mannion, J. Cross and P. Bradley, *et al.*, *J. Neurotrauma*, 2007, **24**, 128.

44. T. Huisman, A. Sorensen and K. Hergan, *et al.*, *J. Comput. Assist. Tomogr.*, 2003, **27**, 5.
45. R. Benson, S. Meda and S. Vasudevan, *et al.*, *J. Neurotrauma*, 2007, **24**, 446.
46. J. Kim, B. Avants and S. Patel, *et al.*, *Neuroimage*, 2008, **39**, 1014.
47. I. Pronin, L. Fadeeva and N. Zakharova, *et al.*, *Ann. Clin. Exp. Neurol.*, 2008, **2**, 32.
48. N. Papadakis, K. Martin and M. Mustafa, *et al.*, *Magn. Reson. Med.*, 2002, **38**, 394.
49. C. Pierpaoli, P. Jezzard and P. Basser, *et al.*, *Radiology*, 1996, **201**, 637.
50. E. Wilde, Z. Chu and E. Bigler, *et al.*, *J. Neurotrauma*, 2006, **23**, 1412.
51. F. Grinberg, E. Farrher and J. Kaffanke, *Neuroimage*, 2011, **57**, 1087.
52. T. A. Huisman, L. H. Schwamm and P. W. Schaefer, *et al.*, *Am. J. Neuroradiol.*, 2004, **25**, 370.
53. M. Inglese, S. Makani and G. Johnson, *et al.*, *J. Neurosurg.*, 2005, **103**, 298.
54. R. Benson, S. Meda and S. Vasudevan, *et al.*, *J Neurotrauma*, 2007, **24**, 446.
55. S. Kasumova, Pathologic anatomy of traumatic brain injury, in *Clinical Guidelines on Traumatic Brain Injury*, ed. A. Konovalov, L. Likhterman and A. Potapov, 1998, vol. 1, Antidor, Moscow, p. 169–229.
56. M. van der Knaap, Wallerian degeneration and myelin loss secondary to neuronal and axonal degeneration, in *Magnetic Resonance of Myelination and Myelin Disorders*, ed. M. van der Knaap and J. Valk, Springer, Heidelberg, 3rd edn, 2005, p. 832–839.
57. M. Gazzaniga, *Nat. Rev. Neurosci.*, 2005, **6**, 653.
58. S. Buklina, *Zh. Neurol. Psikhiatr. im. SSKorsakova*, 2004, **5**, 8.
59. M. Fabri, *et al.*, *J. Cogn. Neurosci.*, 2001, **13**, 1071.
60. S. Laureys, M. Boly and P. Maquet, *J. Clin. Invest.*, 2006, **116**, 1823.
61. X. Di and R. Bullock, *J. Neurosurg.*, 1996, **85**, 655.
62. M. Boly, M. E. Faymonville and P. Peigneux, *et al.*, *Neuropsychol. Rehabil.*, 2005, **15**, 283.
63. M. Boly, L. Tshibanda and A. Vanhaudenhuyse, *et al.*, *Hum. Brain Mapp.*, 2009, **30**, 2393.
64. M. E. Raichle, A. M. MacLeod and A. Z. Snyder, *et al.*, *Proc. Natl. Acad. Sci. U. S. A.*, 2001, **98**, 676.
65. P. J. Pouwels and J. Frahm, *Magn. Reson. Med.*, 1998, **39**, 53.
66. S. Marino, E. Zei and M. Battaglini, *et al.*, *J. Neurol. Neurosurg. Psychiatry*, 1998, **78**, 501.
67. N. Weiss, D. Galanaud and A. Carpentier, *et al.*, *J Neurol*, 2008, **255**, 217.
68. S. D. Friedman, W. M. Brooks and R. E. Jung, *et al.*, *Neurology*, 1999, **52**, 1384.
69. M. R. Garnett, A. M. Blamire and R. G. Corkill, *et al.*, *Brain*, 2000, **123**, 2046.
70. G. Sinson, L. J. Bagley and K. M. Cecil, *et al.*, *Am. J. Neuroradiol.*, 2001, **22**, 143.

71. M. Uzan, S. Albayram and S. G. Dashti, *et al.*, *J. Neurol. Neurosurg. Psychiatry*, 2003, **74**, 33.
72. A. Carpentier, D. Galanaud and L. Puybasset, *et al.*, *J. Neurotrauma*, 2006, **23**, 674.
73. O. B. Halbach and R. Dermietzel, 2006 WILEY-VCH Verlag GmbH & Co. KGaA, Weinheim, Germany, 386P.
74. E. V. Alexandrova, V. D. Tenedieva and A. A. Potapov. Geotar-Media, Moscow, 2015, p. 394.
75. S. A. Dambinova, J. C. Maroon, A. M. Sufrinko, J. D. Mullins, E. V. Alexandrova and A. A. Potapov, *Front. Neurol.*, 2016, **5**, 172.
76. T. Y. Lo, P. A. Jones and R. A. Minns, *J. Neurotrauma*, 2010, **27**, 2139.
77. S. Mondello *et al.*, *Brain Inj.*, 2012, **26**, 1629.
78. Egea-Guerrero *et al.*, *Transplant. Proc.*, 2013, **45**, 3569.
79. T. W. McAllister, *Handb. Clin. Neurol.*, 2015, **128**, 723.
80. E. V. Aleksandrova, M. M. Iusupova and V. D. Tenedieva, *et al.*, *Zh. Vopr. Neirokhir. im. N. N. Burdenko*, 2014, **78**, 53.
81. E. Dardiotis, K. Paterakis and G. Tsivgoulis, *et al.*, *J. Neurotrauma*, 2014, **31**, 1920.
82. V. Di Pietro, M. Ragusa and D. J. Davies, *et al.*, *J. Neurotrauma*, 2017, **34**, 1948.
83. A. K. Ommaya and T. A. Gennarelli, Cerebral Concussion and Traumatic Unconsciousness. Correlation of Experimental and Clinical Observations on Blunt Head Injuries, *Brain*, 1974, **97**, 633.

CHAPTER 12

Neurotoxicity in Spinal Cord Impairments

GRIGORY V. PONOMAREV,*[a] SVETLANA A. DAMBINOVA[b] AND ALEXANDER A. SKOROMETS[a]

[a] I.P. Pavlov First State Medical University, St. Petersburg, Russia; [b] DeKalb Medical, Decatur, GA, USA
*Email: grigoryponomarev@yandex.ru

12.1 Introduction

Over the past decade, substantial progress has been made in experimental research into the molecular basis of hemodynamic and metabolic changes after spinal cord (SC) impairments. However, in human studies, early diagnosis and timely treatment remain unsatisfactory and require the search for novel functional, specific structural and molecular biomarkers.[1] This may be particularly valuable in the management of patients with SC impairments, since the current assessments and therapies of SC impairments are limited, lack prognostic utility, and are outdated.

SC ischemia or infarction is a rare but devastating disorder caused by a wide range of pathological conditions including neurotrauma and vascular insufficiency in glucose and oxygen supply to neural cells, resulting in myelopathy.[2] Early diagnosis of the myelopathy is difficult due to the often asymptomatic nature of benign disease, leaving it under-recognized and under-registered.[3] Besides, the diagnosis of this type of infarction is challenging due to the non-specific neurologic presentation and the complexities of current noninvasive imaging modalities.[4]

Drug Discovery Series No. 62
Acute Brain Impairment: Scientific Discoveries and Translational Research
Edited by Philip V. Peplow, Svetlana A. Dambinova, Thomas A. Gennarelli and Bridget Martinez

Published by the Royal Society of Chemistry, www.rsc.org

Functional and radiological assessments are performed when severe symptoms are presented that vary from paraplegia to minor weakness. About 70% of complaints of severe back pain are typically at the level of the lesion and are defined by the affected vascular territory.[5–7] Magnetic resonance imaging (MRI) reveals lesion areas, necrosis, and inflammation. In most cases, delayed diagnosis leads to the poor prognosis of SC infarction[8] with a risk to life and a mortality rate of up to 20–25%.[9]

Specific biomarkers for SC ischemia combined with coherent guidelines could aid the early diagnosis of disease, target treatment development, and reduce mortality. The present chapter is devoted to the role of the neurotoxicity cascade in the hemodynamics of SC impairment. Tendencies in early diagnosis and treatment of acute and chronic conditions are highlighted.

12.2 Pathological Features of SC Impairments

12.2.1 Hemodynamic Aspects of SC Injury

The SC is supplied by a number of radicular arteries (Figure 12.1) at the cervical, thoracic, and lumbar levels forming the anterior spinal artery (ASA) and the two posterior spinal arteries (PSAs).[10,11] The ASA supplies the anterior two-thirds of necessary blood flow of the medulla (anterior columns, anterior horns and roots) and the PSAs cover the posterior third of the blood supply to the SC (posterior columns, posterior roots and dorsal horns; Figure 12.2).[12] Branches from ASA and PSA surface vessels form a network that provides the dorsal median oxygenated blood supply (external portion) to the SC, including the large part of the posterior horns.[13]

There are several peculiarities of the SC blood supply: (i) the ASA forms the most vulnerable point where upward and downward blood flows come into contact (Figure 12.1); (ii) it could be a sole radicular artery supplying the SC at any level;[14] and (iii) SC blood flows in the cervical and lumbar white matter are 22–26 mL 100 g^{-1} min^{-1} and 15–24 mL 100 g^{-1} min^{-1}, respectively, which are less than cerebral blood flow (CBF).[15] These features determine the susceptibility of the SC to damage, especially due to ischemia, by a wide range of pathologic states.

The venous system contains intrinsic and exterior blood drainage (Figure 12.3) where the interior veins present as sulcal and radial veins, while the external veins are composed of the anterior and posterior spinal veins.[15] The venous drainage of the structures in the spinal canal is analogous to that of the brain venous circulation.

About 70–80% of SC ischemia cases are consequences of embolic/thrombotic events due to vascular risk factors, after aortic surgery, spontaneous vertebral artery dissection, systemic hypotension or arteriovenous malformation[16] and blood flow insufficiencies (Table 12.1). Plagues and/or emboli that cause artery occlusion are often the origins of SC infarction. Mild traumatic brain injury (TBI)[17] and SC injury (SCI)[18,19] might activate irreversible anoxic/hypoxic changes and lead to delayed spinal infarction.

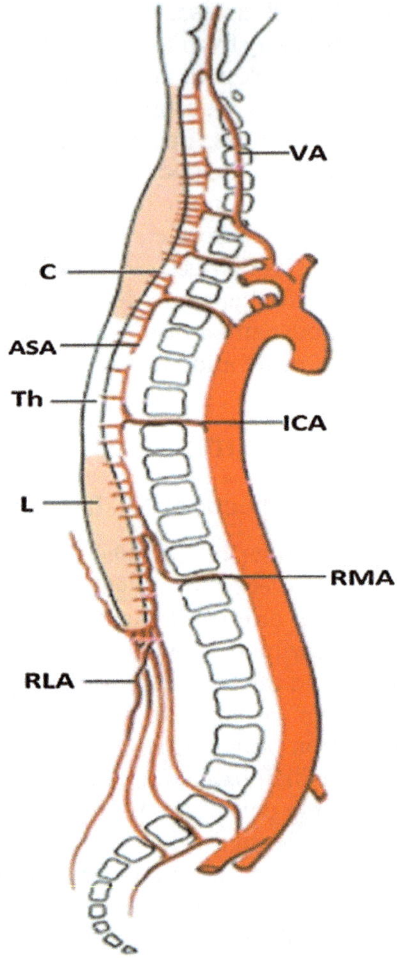

Figure 12.1 Scheme of SC arterial supply. ASA: anterior spinal artery; C: cervical area of spinal cord; ICA: intercostal artery; L: lumbar area of spinal cord; RLA: radicular lumbosacral arteries; RMA: big radiculo-medullary artery (artery of Adamkiewicz); Th: thoracic area of spinal cord; VA: vertebral arteries.

SC pathology (myelopathies) can be classified as acute, sub-acute and chronic. Acute spinal infarction (hours to days) is caused by restricted arterial and/or venous blood supply concomitant with or followed by hemodynamic instability, glutamatergic neurotoxicity, edema formation or hemorrhage and glial scarring.[20] The reversible fluctuations of the blood supply ratio in the impacted SC region reflect the severity of symptoms accompanied by increasing pain.

Sub-acute conditions that worsen within days to a week point to oxidative stress, breakdown of the blood–SC barrier, activation of innate immunity, release of inflammatory mediators and ischemic apoptosis or

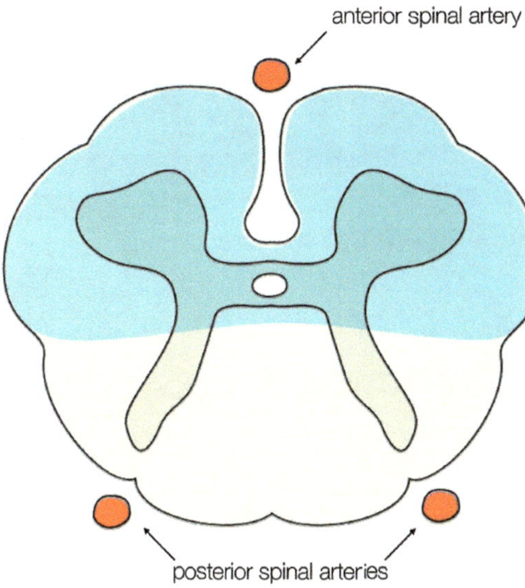

Figure 12.2 Main arteries of the SC.

benign necrosis. Symptoms like neuropathic pain and autonomic erectile dysfunction are observed.[21] There is a spontaneous and substantial improvement of symptoms that occurs in about 7% of patients (as with a spinal transient ischemic attack [TIA]).[22,23]

Delayed irreversible events are indicated by complete inflammation, demyelination, multiple lesion stabilization, necrosis and neuronal death.[24] Worsening conditions are bilateral paresis or paralysis and sensory loss below the level of lesion. Sphincter dysfunction is present with loss of bladder and bowel control.[25,26] Central SC ischemia syndrome, Brown–Séquard syndrome and complete transverse spinal stroke are less common.[27,28]

12.2.2 Biochemical Alterations in Acute and Chronic Conditions

There are some similarities in the mechanisms of brain and SC ischemia in acute and chronic conditions. However, it has been shown that 30–40% of SC mitochondria have lower respiratory activities during oxidative phosphorylation and uncoupling. The reduced respiratory activity of SC mitochondria is probably associated with lower amounts of the enzyme complexes responsible for electron transport and ATP synthesis. Furthermore, SC mitochondria have significantly lower H^+/Ca^{2+} ratios compared with brain mitochondria.[29] It seems that these peculiarities determine the lower possibilities of SC recovery after ischemia.

Metabolic dysfunction in the SC due to ischemia is associated with ionic imbalance (K^+ and Ca^{2+} influx), emission of excitatory amino acids and their derivatives, hyperglycolysis and lactate acidosis in areas of vascular

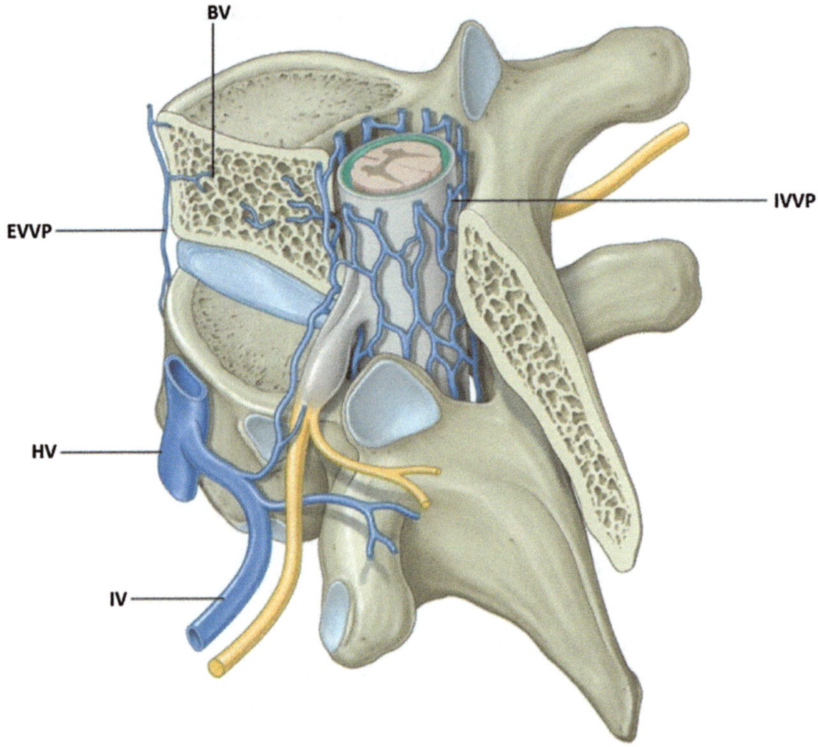

Figure 12.3 Veins of the SC. BV: basivertebral vena; EVVP: external vertebral venous plexus; HV: hemiazygos vein; IV: intervertebral vein; IVVP: internal vertebral venous plexus.

deficit or impact.[30] Ionic imbalance results in extracellular K^+ increases that initiate membrane depolarization and the release of the excitatory amino acid glutamate. Abundant glutamate concentrations activate α-amino-3-hydroxy-5-methyl-4-isoxazole propionic acid receptor (AMPAR) and kainate receptor (KAR) in spinal white matter[31–33] that are associated with Ca^{2+} influx, which persists during the first hours to days. Breached ionic homeostasis triggers increased glucose consumption and massive excitation followed by a wave of energy deficit and glycolysis with lactate production.[34] After the initial period of hyperglycolysis, cerebral glucose use is diminished. Depressed cerebral glucose metabolism may occur in the absence of overt clinical symptoms.[35] Elevated lactate levels can result in neuronal dysfunction by inducing acidosis, membrane damage, altered blood–SC barrier permeability and cerebral edema. Increased levels of lactate after SCI may leave neurons more vulnerable to a secondary ischemic injury.[34]

Current blood assays including routine complete blood count, fasting serum glucose, erythrocyte sedimentation rate, lipid panel for cholesterol, low-density lipoprotein cholesterol, high-density lipoprotein cholesterol, triglycerides and

Table 12.1 Pathophysiological types of SCI.

Type	Origin	Risk factors	Cases (%)	Ref.
Arterial	Vascular	Atherosclerosis, hypertension, diabetes mellitus, heart disease and other	~33	1, 2, 12, 19, 26, 39
	Surgery	Aortic pathologies (thoracoabdominal aneurysms, dissection, coarctation)	1–21	2, 6, 14, 15, 22, 24, 27, 45
	Dissection	Vertebral artery dissection	10–25	11
	Degenerative spine disease	Herniated disc, fibrocartilaginous embolism	~40 for herniated disc, unknown for fibrocartilaginous embolism	3, 14, 24, 27, 43
	Others	Spinal TIA, systemic hypotension, vasculitis, hypercoagulable conditions	Unknown	1, 12, 19, 25, 27, 40
Venous	Degenerative spine disease	Herniated disc	~60	14
	Arteriovenous malformations	Spinal dural arteriovenous fistula	60–80	16, 41, 44
	Others	Spinal TIA	Unknown	12, 13
Mixed	Trauma	Traumatic SCI	Up to 100	15, 18, 19

electrolytes are used to evaluate common causes of myelopathy.[36] Blood or cerebral spinal fluid (CSF) detections of excessive or deficient potassium in cases of flaccid quadriparesis, infectious causes (syphilis or viruses) using polymerase chain reaction and autoimmune assessment of erythrocyte sedimentation rate, antinuclear antibodies (Abs) and complement level assays might be considered in the broad differential diagnosis of myelopathy. Leukocytosis in blood suggests an infectious myelitis or other infectious cause of SC impairment.[36] To date, laboratory methods are directed at excluding other causes of myelopathy and do not "rule in" SC infarction.

12.2.3 Radiological Patterns of SC Impairments

Basic spinal MRI modalities (T1, T2-weighted or short-T1 inversion recovery) are used for suspected SC ischemia to identify or exclude lesions within the cord or cord vasculature.[37,38] Diffusion-weighted imaging is particularly sensitive to ischemic changes (showing areas of hyperintensities) and may become a standard addition to T1-, T2-weighted modalities.[39,40]

In cases when MRI is contraindicated (metal implants, medical instability, claustrophobia and/or obesity) or unavailable, then a computed tomography (CT) and enhanced CT myelography is employed for excluding cord compression, but not for diagnosing SC ischemia, due to its low sensitivity for detecting SC lesions.[41]

Trans-arterial spinal angiography (TASA) is a gold standard in the diagnosis and treatment of spinal arteriovenous malformation and is able to identify the correct location of the fistula site and the feeding artery and estimating the venous drainage.[42] However, this method is invasive and potentially time-consuming, requiring selective catheterization of a number of arteries supplying the SC.[43] Compared to TASA, spinal magnetic resonance angiography is preferable because of its noninvasive and time-saving characteristics.[43]

Current MRI technology overcomes peculiarities of lesion visualization in cervical *vs.* thoracic and lumbar regions with the aim of improving images, particularly for situations like SC swelling, as well as providing the possibility of post-contrast T1-weighted image enhancement that, to date, has common features in most ischemic, inflammatory, infectious or malignant processes.[44] Besides, MRI and functional assessment are limited in the diagnosis of subtle SC impairments, and they do not reflect the severity and consequences after a spinal stroke.[45]

There is still a medical need for accurate and sensitive biomarkers for SC infarction specific to the injured spinal region in acute and chronic phases of assessments.[46]

12.3 Potential Biomarkers of SC Impairments

12.3.1 Neuronal Markers in SCI

There are limited research data that have been presented on the assessment of neuronal biomarkers in SC impairment, specifically ischemia.

Levels of the astroglial marker protein S100β and neuron-specific enolase (NSE) detected in the serum and CSF of rats with induced moderate to severe SCI were significantly elevated compared to in those with mild SCI within 6 hours after the injury. Both biomarker concentrations did not appear to be different within graded groups (mild, moderate and severe), although they remained higher than those in the control group.[47] Analog elevations of S100β in the CSF of patients with SC infarction 6 hours after thoracic endovascular stenting have been demonstrated compared to those without an adverse event.[48] It was suggested that an increased concentration of S100β in the CSF is a marker of SC ischemia.[47]

In our feasibility comparative studies, the detection of S100β Abs in serum samples from patients with SC ischemia ($n = 26$) and SCI ($n = 17$) showed no increase compared to control groups of healthy persons ($n = 20$) and those with radiculopathy ($n = 32$; authors' unpublished data). These preliminary results suggest that S100β might not be a marker of a compromised blood–SC barrier.

Increased concentrations of NSE, glial fibrillary acidic protein (GFAP) and tau protein have been identified in the CSF after episodes of SC ischemia during cross-clamping and reperfusion (aortic aneurysm surgery), and elevated CSF concentrations of neurofilament light chain protein (NFL) were measured in patients who underwent surgery for lumbar disc herniation.[49] These biomarkers are not limited to SC infarction or SCI, and abundant amounts have been measured in the biological fluids of patients with TBI, intracerebral and subarachnoid hemorrhage and even persons without brain injury.[47]

12.3.2 Biomarkers of Inflammation

SC ischemia implies an initial acute ischemic insult followed by inflammation with blood–SC barrier dysfunction and the release of cytokines, free radicals and other inflammatory mediators that accelerate damage.[15] The study of thoracic aortic ischemia-reperfusion (TAR) showed interleukin (IL)-6 levels in SC tissue to be significantly increased at 24 hours compared to the sham group and the 6 hour TAR group ($p = 0.0001$ for 24 hours compared with sham). This indicates activation of local pro-inflammatory cytokine and chemokine responses in viable motor neurons in the lumbar and thoracic SC segments.[50] SCIs in the cervical and upper thoracic regions that are responsible for the autonomic nervous system frequently cause cerebrovascular disorders concomitant with visceral inflammation and acute elevation of C–C motif ligand 2 (CCL2) and CCL3 markers.[51]

It was reported[52] that free radical and cyclooxygenase-catalyzed oxidation of arachidonic acid is involved in SC ischemia. Enhanced oxidative injury and inflammatory response with a rapid increase of both 8-iso-prostaglandin F2α and 15-keto-dihydro-prostaglandin F2α in the biological fluids (plasma and urine) play special roles in SC ischemia.[52]

12.3.3 Neurotoxicity Markers for SC Ischemia and SCI

An exceptional feature of the SC vascularity resides in the various ratios of blood supply in the central system *vs.* that of the peripheral system: the cervical region has greater peripheral and central arterial blood supplies, while the thoracic region has a greater peripheral and lesser central supply, and the lumbar and upper sacral regions have a lesser peripheral and greater central supply.[15] The search for potential vascular biomarkers reflecting SC impairment to cervical, thoracic and lumbar regions and depicting the involvement of gray and white matter in hemodynamic infarction of the SC would be invaluable.

Synaptic transmission in SC depends on ionotropic glutamate receptors (GluR): *N*-methyl-D-aspartate receptor (NMDAR), AMPAR and KAR.[53] NMDARs were found in neurons and glia throughout the brain and SC.[54] Four subunits of AMPAR (GluR1–GluR4) were revealed at detectable levels in the lumbar region of the SC.[55] KARs (GluR5a) are preferentially expressed in the dorsal horn (and dorsal root ganglion) of the SC.[56]

Comparative experimental studies show that SC neurons are less vulnerable to activation of NMDAR and more sensitive to AMPA/KAR[33,55] input, especially in the cervical SC region containing phrenic (C3–C5 level) motor neurons.[57] It is suggested that NMDARs regulate arterial microvessels of gray matter and AMPARs/KARs could be associated with venous circulation in white matter.[58]

Spinal infarction might compromise the blood–SC barrier, facilitating GluR antigen release from the SC vasculature, which can activate the immune system.[59] The normal immune potential can be unleashed due to a leaking blood–SC barrier initiating the production of Abs to natural antigens. The latter being detected in serum is likely to be associated with chronic processes of SC impairment.

In our explorative research, Abs to the NR2 subunits of NMDARs were found in 14 out of 26 patients with chronic SC ischemia ($p=0.002$), confirming the primary vascular cause of the disease involving gray matter supplied by the central artery. At the same time, AMPAR Abs were elevated in 10 out of 17 patients with chronic traumatic SC injury ($p=0.02$). This possibly indicates the extent of white matter injury[31] and the compensatory long-term increases of AMPARs expression in motoneurons after SC injury.[57] There were MRI-confirmed lesion(s) in the cervical ($n=16$), thoracic ($n=8$), cervical/thoracic ($n=2$), lumbar ($n=4$) and thoracic/lumbar ($n=8$) areas of the SC.

It is necessary to note that in serum samples of some patients, Abs to more than one GluR have been detected. Five ($n=2$ with cervical lesions) out of 26 patients with SC ischemia and five ($n=3$ with cervical lesions) out of 17 SCI patients had abnormal concentrations of both NR2 and AMPAR Abs, suggesting the simultaneous implication of SC gray and white matter (GRACE Laboratories, Atlanta, GA, USA). Increased concentrations of KAR Abs (DRD Ltd, Moscow, Russia) were found in SCI patients ($n=9$, 1.50–3.05 ng mL^{-1}) and those with SC ischemia ($n=9$, 1.5–3.2 ng mL^{-1}). It is likely that the KAR response is for white matter venous circulation while NR2 locates to the arteries in the SC as well. At the same time, 13 patients (five with SCI and eight with SC ischemia) had both KAR and NR2 Abs elevated above the threshold of 1.5 ng mL^{-1}. In addition, 10 patients (seven with SCI and three with SC ischemia) had increased KAR Ab (1.5–3.0 ng mL^{-1}) and AMPA Ab (1.5–3.3 ng mL^{-1}) concentrations. This could indicate that arterial and venous blood circulation disturbances caused SC infarction. According to functional assessment, patients with elevated NR2 Abs had mild to moderate neurological symptoms (paraparesis) and gray matter lesions, while white matter impairments were associated with more severe neurological outcomes (tetraparesis to tetraplegia) accompanied by abnormal levels of AMPAR and/or KAR Abs.

In addition, we assessed the subtypes of GluR, which are more effective for SC ischemia and injuries. It was revealed that immune epitopes of NMDAR,

AMPAR and KAR are slightly different and belong to various subtypes of GluR that are not similar to those of the brain. There are some data indicating that NR3B and GluR3–4 (AMPA) are found mainly in motor neurons of the SC.[54] This should be carefully studied further in order to comprehend the major peculiarities of the arteriovenous circulation in the SC and CBF in the brain.

12.4 Therapy Options for SC Impairments

To date, the management of SC impairments remains suboptimal. There are a number of chemical agents under experimental research. Currently available treatment for SC ischemia and SCI is mostly based on symptomatic and pathogenic presentations depending on the acuteness of the symptom presentation.

12.4.1 Acute States

Limited treatment options to cure acute SC injuries of ischemic and traumatic origin are available (Table 12.2). It was recommended that thrombolysis be performed within 4.5 hours after the onset of ASA syndrome to give a better recovery.[60] Early surgical intervention for SC stabilization and decompression is utilized in acute SCI cases.[61] Endovascular embolization and obliteration of arterialized veins *via* microsurgery are treatment options for arteriovenous malformations (spinal dural fistula).[42] Endovascular treatment is offered to repair abnormal SC connections between an artery and a vein (arteriovenous fistula), demonstrating a significant improvement of neurological symptoms.[62]

A recent review of studies concerning methylprednisolone use in acute SCI showed little confidence in a long-term benefit in terms of improved motor recovery.[63,64]

12.4.2 Chronic Conditions

A small number of pharmacological agents have shown beneficial effects in subacute and chronic periods of SC ischemia. Limited studies reported that dextrorphan and other NMDAR inhibitors (Table 12.2) had protective effects in patients who suffered SC infarction and delayed ischemia due to SCI.[65] The administration of the opiate antagonist naloxone in the postoperative period (thoracoabdominal aortic aneurysm repair) attenuated glutamate neurotoxicity and resulted in a 30% decrease in the SC infarction rate.[65] The enhancement of collateral blood flow by use of vasodilators is another approach to SC post-ischemic injury reduction.[65] In our earlier studies, intravenous administration of euphyllin increased SC blood flow in cervical and lumbar regions.[14] Derivatives of the glycoprotein erythropoietin diminish lesion size by suppressing inflammation and restricting oxidative stress after SCI while aiding in structural and functional recuperation of the SC.[63]

Table 12.2 Therapeutic strategies for ischemic and traumatic SCIs.

Drug/procedure	Pharmacological properties	Effect	Ref.
In practice			
Thrombolysis (Alteplase)[a]	Recombinant tissue plasminogen activator	Stimulates the conversion of plasminogen to plasmin	59
Surgery[a]	—	Decompression of SC in injury/arteriovenous malformations	19, 41, 61
Methylprednisolone[a]	Corticosteroid	Cellular and lysosomal membrane stabilizer	62–64
Glutamate inhibitors: dextrorphan, riluzole, memantine, MK-801	NMDAR antagonists	Inhibit receptors' enzyme degradation; Ca^{2+} channel presynaptic membranes	62, 64
Naloxone hydrochloride	Opioid antagonist	Attenuates NMDA-mediated neurotoxicity	64
Prostaglandin E1, papaverine, euphyllin	Vasodilatators	Decrease SC perfusion pressure and increase CSF pressure	14, 64
Carbamylated erythropoietin (EPO), asialo-EPO	Cytokines	Suppression of inflammation, restriction of oxidative stress	62, 65
Research mode			
Mild hypothermia	—	Reduces NMDA channel activation	62, 64, 65
Hyperbaric oxygen	—	Increases blood oxygen, reduction of oxidative stress	67
Adenosine	A2-adenosine receptor agonist	Suppresses γ-aminobutyric acid	65
Nerve growth factor and neuropeptides	Neurotrophic factors	Stimulation of increase in neuronal metabolism	65
Cellular approach	Mesenchymal stem cells	Promotes neuronal regeneration; replaces neurons	62, 65

[a]Utilized mostly in acute states.

12.4.3 Trends in Treatment Optimization

SC protection during aortic surgery reconstruction including either enhanced oxygen supply or mild hypothermia is used to prevent SC ischemia.[65] Additionally, moderate hypothermia reduces the neurotoxicity and acidosis caused by ischemia 50-fold, allowing neurons and their circuits to remain viable for an extended time.[66] Combined use of hypothermia and adenosine A2A receptor agonists might increase neuroprotection during prolonged SC ischemia.[67] Hyperbaric oxygen treatment alone or simultaneously with administration of iloprost, a vasorelaxing agent, may improve neurological scores in SC ischemia–reperfusion.[68]

Administration of nerve growth factor promoted SC tissue repair following experimental traumatic SCI and ischemia.[63] Cell therapy by use of glial, neural stem and mesenchymal stem cells in different experimental models of SC impairments is considered a beneficial therapeutic tool for preventing the development of secondary injury.[63,66] Positive outcomes of cell therapy in SCI experimental models increased the possibility of successful translation to clinical practice by improving the functional recovery of patients with SC ischemia and SCI.[63]

12.5 Conclusion

SC ischemia is a rare but devastating disorder caused by trauma and a wide range of diseases directly or indirectly affecting the SC vascular system. The functional assessment and current neuroradiological methods are capable of recognizing severe SCIs; however, they are limited in terms of early diagnosis at the asymptomatic stage. Neuronal and inflammatory biomarkers that are suitable for cerebral ischemia, however, supply somewhat limited information on SC impairments and do not associate with vascular injury or depict region-specific changes. The high demand for specific vascular biomarkers for SC impairments associated with neurotoxicity in specific regions of the SC that correlate with the severity of ischemia still remains to be fulfilled.

Here, we present a pilot study of Abs to different subunits of glutamate receptors in patients with chronic SC ischemia and SCI, intending to find an association of their alteration with severity of disease and region of vascular damage. It seems that NMDAR association with arterial circulation in SC gray matter and the AMPAR/KAR role in the arteriovenous system of white matter should be extended to more specific GluR subtypes consisting of specific immune epitopes to cervical, thoracic and lumbar regions.

Technology that would allow monitoring of vascular biomarkers would potentially answer a number of questions regarding: (i) how the blood flow is altered; (ii) what is initially affected—arterial or venous system; and (iii) targeting the development of potential pharmacological agents for reversing the neurotoxicity. Targeted treatment of occluded arterial and venous circulations in the SC may have potential benefit in improving the blood supply to ischemic areas of the SC. Further optimization of neuroimaging of the SC combined with the implementation of GluR biomarkers could potentially meet requirements for the early diagnosis of SC ischemia, rendering it treatable and potentially preventing unfavorable outcomes (paraplegia and tetraplegia).

Abbreviations

Ab	antibodies
AMPAR	α-amino-3-hydroxy-5-methyl-4-isoxazole propionic receptors
ASA	anterior spinal artery
ATP	adenosine triphosphate
CBC	complete blood count
CBF	cerebral blood flow
CCL	C–C motif ligand
CT	computed tomography
DWI	diffusion-weighted imaging
GFAP	glial fibrillary acidic protein
GluR	glutamate receptors
HDL	high density lipoproteins
KAR	kainate receptors
LDL	low density lipoproteins
MCP	monocyte chemoattractant protein
MIP	macrophage inflammatory protein
MRI	magnetic resonance imaging
NFL	neurofilament light chain protein
NMDAR	*N*-methyl-D-aspartate receptors
NSE	neuron-specific enolase
PCR	polymerase chain reaction
PSA	posterior spinal artery
SC	spinal cord
SCI	spinal cord injury
STIR	short-TI inversion recovery
TAR	thoracic aortic ischemia-reperfusion
TASA	*trans*-arterial spinal angiography
TBI	traumatic brain injury
TIA	transient ischemic attack

References

1. S. Weidauer, M. Nichweiss, E. Hattingen and J. Berkefeld, *Neuroradiology*, 2015, **57**, 241.
2. S. Weidauer, M. Nichweiss, H. Lanfermann and F. E. Zanella, *Neuroradiology*, 2002, **44**, 851.
3. J. A. Nelson, C. Y. Ho and M. R. Golomb, *Pediatr. Neurol.*, 2016, **56**, 76.
4. J. R. Nance and M. R. Golomb, *Pediatr. Neurol.*, 2007, **36**, 209.

5. F. Gaillard. Acute Spinal Cord Ischemia Syndrome. (accessed on March 1 2017) https://radiopaedia.org/articles/acute-spinal-cord-ischaemia-syndrome.
6. A. Sivadasan, M. Alexander, A. K. Patil and S. Mani, *Ann. Indian Acad. Neurol.*, 2013, **16**, 190.
7. M. I. Vargas, J. Gariani and R. Sztajzel, *et al.*, *Am. J. Neuroradiol.*, 2015, **36**, 825.
8. J. Chen and P. Gailloud, *Neurology*, 2011, **77**, 1235.
9. T. F. Scott. Spinal Cord Infarction. (accessed on March 1 2017) http://emedicine.medscape.com/article/1164217-overview.
10. N. Vongveeranonchai, M. Zavahreh, D. Strbian and S. Sundararajan, *Stroke*, 2014, **45**, 203.
11. S. Richard, C. Abdallah and A. Chanson, *et al.*, *J. Spinal Cord Med.*, 2014, **37**, 233.
12. J. Novy, A. Carruzzo, P. Maeder and J. Bogosslavsky, *Arch. Neurol.*, 2006, **63**, 1113.
13. M. S. Hussain, A. Shuaib and Z. A. Siddiqi, *Neurology*, 2005, **64**, 761.
14. A. A. Skoromets, *in Spinal Angioneurology*, ed. A. A. Skoromets, A. P. Skoromets, T. A. Skoromets and T. P. Tissen, Medpress, Moscow, 2003, pp. 5–20 (in Russian).
15. N. M. Martirosyan, J. S. Feuerstein, N. Theodore, D. D. Cavalcanti, R. F. Spetzler and M. C. Preul, *J. Neurosurg. Spine*, 2011, **15**, 238.
16. S. K. Pillai, T. Subramaniam and G. G. Rao, *Br. Med. J. Case Rep.*, 2011, **8**, 43.
17. V. Bartanusz, M. Ziu, L. E. Wood and J. L. Caron, *J. Med. Case Rep.*, 2012, **6**, 314.
18. C. E. Hulsebosch, *Adv. Physiol. Educ.*, 2002, **26**, 238.
19. C. A. Oyinbo, *Acta Neurobiol. Exp.*, 2011, **71**, 281.
20. A. M. Granados, L. M. Garcia, C. A. Ortega and A. Lopez, *Rev. Colomb Radiol.*, 2011, **22**, 1.
21. S. Ram, A. Osman and V. N. Cassar-Pullicino, *et al.*, *Spinal Cord*, 2004, **42**, 481.
22. L. Restrepo and J. F. Guttin, *Tex Heart Inst. J.*, 2006, **33**, 74.
23. C. E. Robertson, R. D. Brown Jr., E. F. M. Wijdicks and A. A. Rabinstein, *Neurology*, 2012, **78**, 114.
24. R. George, *Heart India*, 2015, **3**, 61.
25. W. Kuker, M. Weller and U. Klose *et al.*, *J. Neurol.*, 2004, **251**, 818.
26. J. J. Wong, J. Dufton and S. A. Mior, *J. Can. Chiropr. Assoc.*, 2012, **56**, 58.
27. M. Mullen, M. L. McGarvey, S. E. Kasner and J. L. Wilterdink. Spinal Cord Infarction: Vascular Anatomy and Etiologies. (accessed on March 1 2017) http://hgp.gob.ec/index.html/uptodate/contents/UTD.htm?39/43/40631#.
28. M. Mullen, M. L. McGarvey, S. E. Kasner and J. L. Wilterdink. Spinal Cord Infarction: Prognosis and Treatment. (accessed on March 1 2017) http://hgp.gob.ec/index.html/uptodate/contents/mobipreview.htm?22/50/23333?source=related_link.

29. A. V. Panov, in *Biomarkers for Traumatic Brain Injury*, ed. S. A. Dambinova, R. L. Hayes and K. K. W. Wang, The Royal Society of Chemistry, Cambridge, 2012, pp. 107–109.
30. V. Belegu, M. Oudega, D. S. Gary and J. W. McDonald, *Neurosurg. Clin. North Am.*, 2007, **18**, 143.
31. G. K. Kanellopoulos, X. M. Xu, C. Y. Hsu, X. Lu, T. M. Sundt and N. T. Kouchoukos, *Stroke*, 2000, **31**, 1945.
32. S. D. Grossman, L. J. Rosenberg and J. R. Wrathall, *Exp. Neurol.*, 2001, **168**, 283.
33. A. Barbon, F. Fumagalli, L. Caracciolo, L. Madaschi, E. Lesma, C. Mora, S. Carelli, T. A. Slotkin, G. Racagni, A. M. Di Giulio, A. Gorio and S. Barlati, *J. Neurochem.*, 2010, **114**, 397.
34. J. E. Rinholm, N. B. Hamilton, N. Kessaris, W. D. Richardson, L. H. Bergersen and D. Attwell, *J. Neurosci.*, 2011, **31**, 538.
35. M. Bergseider, D. A. Hovda, S. M. Lee, D. F. Kelly, D. L. McArthur, P. M. Vespa, J. H. Lee, S. C. Huang, N. A. Martin, M. E. Phelps and D. P. Becker, *J. Neurotrauma*, 2000, **17**, 389.
36. M. Mullen, M. L. McGarvey, S. E. Kasner, J. L. Wilterdink. Spinal Cord Infarction: Clinical presentation and Diagnosis. (accessed on March 1 2017) http://hgp.gob.ec/index.html/uptodate/contents/mobipreview.htm?29/33/30230?source=related_link.
37. C. Masson, J. P. Pruvo, J. F. Meder and C. Cordonnier, *et al.*, *J. Neurol. Neurosurg. Psychiatry*, 2004, **75**, 1431.
38. R. G. Nogueira, R. Ferreria, P. E. Grant and S. E. Maier *et al.*, *Stroke*, 2012, **43**, 532.
39. M. M. Thurnher and R. Bammer, *Neuroradiology*, 2006, **48**, 795.
40. K. Nedeltchev, T. J. Loher, F. Stepper, M. Arnold, G. Schroth, H. P. Mattle and M. Sturzenegger, *Stroke*, 2004, **35**, 560.
41. C.-Y. Lien, C.-C. Lui, C.-H. Lu and W.-N. Chang, *Acta Neurol. Taiwan*, 2014, **23**, 29.
42. M. Amarouche, J. L. Hart and A. Siddiqui, *et al.*, *Am. J. Neuroradiol.*, 2015, **36**, 417.
43. T. P. Duprez, L. Danvoye, D. Hernalsteen and G. Cosnard, *et al.*, *Am. J. Neuroradiol.*, 2005, **26**, 496.
44. C. Amanieu, M. Hermier and N. Peyron, *et al.*, *Diagn. Interv. Imaging*, 2014, **95**, 897.
45. O. Hdiji, N. Bouzidi, M. Damak and C. Mhiri, *J. Med. Case Rep.*, 2016, **10**, 99.
46. M. H. Pow, A. J. F. Hosman and J. J. van Middendorp, *et al.*, *Spinal Cord*, 2009, **47**, 519.
47. G. B. Brunnekreef, R. H. Heijmen, W. B. Gerristen, M. A. Shepens, H. T. ter Beek and E. P. van Dongen, *Eur. J. Vasc. Endovasc. Surg.*, 2007, **34**, 169.
48. A. Winnerkvist, R. E. Anderson, L. O. Hansson, L. Rosengren, A. E. Estrera, T. T. Huynh, E. E. Porat and H. J. Safi, *Eur. J. Cardiothorac. Surg.*, 2007, **31**, 637.

49. H. Albadawi, J. W. Chen, R. Oklu, Y. Wu, G. Wojtkiewicz, B. Pulli, J. D. Milner, R. P. Cambria and M. T. Watkins, *Radiology*, 2017, **282**, 202.
50. E. M. Besecker, G. M. Deiter, N. Pironi, T. K. Cooper and G. M. Holmes, *Am. J. Physiol. Regul. Integr. Comp. Physiol.*, 2017, **312**, 146.
51. S. Basu, A. Hellberg and A. T. Ulus, *et al.*, *FEBS Lett.*, 2001, **508**, 36.
52. C. Matute, M. Domercq, A. Perez-Samartin and B. R. Ransom, *Stroke*, 2013, **44**, 1204.
53. A. Contractor and G. T. Swanson, in *The Glutamate Receptors*, ed. R. W. Gereau IV and G. T. Swanson, Humana Press, LLS, 2008, pp. 101–127.
54. M. P. Hefferan, K. Kucharova, K. Kinjo, O. Kakinohana, G. Sekerkova, S. Nakamura, T. Fuchigami, Z. Tomori, T. L. Yaksh, N. Kurtz and M. Marsala, *J. Neurosci*, 2007, **27**, 11179.
55. J. L. Rozas, in *Kainate Receptors. Novel Signaling Insights*, ed. A. Rodriguez-Moreno and T. S. Sihra, Springer Science + Business Media, LLS, Landes Bioscience, 2011, 69–76.
56. C. B. Mantilla, J. P. Bailey, W.-Z. Zhan and G. C. Sieck, *Exp. Neurol.*, 2012, **234**, 191.
57. S. A. Dambinova, J. D. Mullins and T. Gennarelli, in *NEUROTRAUMA: A Comprehensive Book On Brain Injury and Spinal Cord Injury*, ed. K. K. Wang, Oxford University Press, New York, 2017.
58. J. D. Weissman, G. A. Khunteev, R. Heath and S. A. Dambinova, *J. Neurol. Sci.*, 2011, **300**, 97.
59. K. I. Muller, L. H. Steffensen and S. H. Johnsen, *Br. Med. J. Case Rep.*, 2012, **10**, 1136.
60. L. Grassner and D. Maier, *Neural. Regen. Res.*, 2016, **11**, 1928.
61. J. M. C. van Dijk, K. G. TerBrugge, R. A. Willinsky, R. I. Farb and M. C. Wallace, *Stroke*, 2002, **33**, 1578.
62. M. A. Anwar, T. S. Al Shehabi and A. H. Eid, *Front. Cell. Neurosci.*, 2016, **10**, 98.
63. N. Evaniew, E. P. Belley-Cote, N. Fallah, V. K. Noonan, C. S. Rivers and M. F. Dvorak, *J. Neurotrauma*, 2016, **33**, 468.
64. T. Kunihara, S. Kubota, S. Wakasa, N. Shiya and Y. Matsui, in *Diagnosis and Treatment of Abdominal and Thoracic Aortic Aneurysms Including the Ascending Aorta and the Aortic Arch*, ed. R. Grundmann, InTech, 2011, pp. 198–201.
65. D. P. Kuffler, *Ther. Clin. Risk Manag.*, 2012, **8**, 185.
66. P. E. Parrino, I. L. Kron, S. D. Ross, K. S. Shockey, M. J. Fisher, J. R. Gaughen Jr., D. F. Kallmes, J. A. Kern and C. J. Tribble, *J. Vasc. Surg.*, 2000, **32**, 171.
67. G. Ilhan, M. Aksun, B. Ozpak, T. Gunes, S. Bozok, M. E. Durakoglugil, L. Tumkaya and A. Gurbuz, *Eur. J. Cardiothorac. Surg.*, 2013, **44**, 332.

CHAPTER 13

Advanced Approaches in Stem Cell Therapy for Stroke and Traumatic Brain Injury

A. GORODINSKY

Grace Laboratories, LLC, P.O. Box 33797, Atlanta, GA 30033, USA
Email: alexgorodinsky@hotmail.com

13.1 Introduction

Although the therapeutic approaches during the acute phases of stroke and traumatic brain injury (TBI) are very different (reviewed in Chapter 1), the pathophysiology of stroke and TBI presents many common problems associated with inflammation, apoptosis, glial scarring, *etc.* (Chapter 2). Emerging cell therapy, particularly involving stem cells (SCs), offers several approaches to compensate for the insufficient endogenous cell replacement and regeneration mechanisms,[1] including exogenous neurogenesis (allogeneic cell replacement), neuroprotection, stimulation of endogenous neurogenesis, modulation of inflammatory and immune response and stimulation of angiogenesis. In this review, the origins and potential sources of SCs, the experimental data accumulated in preclinical studies and clinical trials of SC therapy in stroke and TBI will be discussed.

Drug Discovery Series No. 62
Acute Brain Impairment: Scientific Discoveries and Translational Research
Edited by Philip V. Peplow, Svetlana A. Dambinova, Thomas A. Gennarelli and Bridget Martinez

Published by the Royal Society of Chemistry, www.rsc.org

13.2 Origins and Sources of SCs

The term "stem cell" was introduced in 1868 by German biologist Ernst Haeckel, and for the next 100 years, SC biology was mainly focused on the germ cells and their role in embryonic development, as well as on hematopoietic SCs.[2] The mammalian brain was traditionally considered as an organ lacking any regenerative capacity,[1] but the first data pointing to the existence of a postnatal neurogenesis surfaced in 1965, when differentiation of "immature" cells into neurons was reported for the granular layer of the dentate gyrus (DG) of the hippocampus.[3] By the end of the 20th century, SCs were identified in many tissues of the adult body,[4] and the brain was no exception: neural SCs (NSCs) were found primarily in the DG and subventricular zone (SVZ) surrounding the lateral ventricle, but also in the striatum, neocortex, substantia nigra and spinal cord (reviewed in Gage[5]). At the same time, grafting of embryonic neural tissue into the brains of patients with Parkinson's disease (PD) has already been introduced into clinical practice.[6] With the advances of SC biology and the discovery of different types of SCs, the concept of neurotransplantation as a possible cure for a wide range of central nervous system (CNS) disorders including stroke and TBI, gained strength and evolved from supplying a ready-made supplement to an impaired brain function (*e.g.*, dopaminergic neurons in PD) to transplantation of SCs potentially capable of tissue-specific proliferation and rebuilding of the damaged brain structures.

Table 13.1 presents different types of human SCs, which were either used for neurotransplantation in experimental models of stroke and TBI and/or clinical practice. Several aspects are of a particular importance (references are included in Table 13.1):

- Potency (*i.e.*, an ability to differentiate into different cell lineages). Pluripotent SCs, such as embryonic SCs (ESCs) or induced pluripotent SCs (iPSCs), can differentiate into any type of adult tissue, while multipotent SCs, such as NSCs, have a repertoire mostly limited to a specific (in this particular case, nervous) tissue. In this review, the use of neural progenitors (NPs) will be discussed. These are proliferative cells that are found, for example, in the SVZ, but unlike multipotential NSCs that can continuously proliferate, NPs have a limited self-renewal capacity and can be either multipotential, bipotential or unipotential.[7]
- A possibility of patient-specific therapy (*i.e.*, personalized medicine provided by autologous transplants; *e.g.*, of mesenchymal SCs [MSCs] or iPSCs and autologous iPSC-derived cell lines).
- Availability. When a considerable quantity of the SCs immediately available for transplantation is a critical factor, allogenic SC lines, which can be delivered as off-the-shelf products, have a clear advantage over patient-specific cell preparations, such as autologous MSCs or iPSCs, the accumulation of necessary quantities of which may require days or weeks of tissue culture.

Table 13.1 Types and sources of human stem cells.[a]

Stem cell	Origin	Potency	Off-the-shelf availability	Tumorigenic potential	Ref.
Autologous stem cells					
MNCs	BM	Multipotent	No	None	11
MSCs	BM	Multipotent	No	None	12
MAPC	BM	Multipotent	No	None	13
EPCs	BM	Multipotent	No	None	14
MUSECs	BM	Pluripotent	No	None	15
ADSC	AT	Multipotent	No	None	16
iPSCs	Adult fibroblasts	Pluripotent	No	High	17
Allogenic stem cells					
NSCs (primary cultures)	FB	Multipotent	No	None	18–20
HB1.F3	FB	Multipotent	Yes	None/TBD	21, 22
NSI-566RSC	Fetal spinal cord	Multipotent	Yes	None/TBD	23
CTX0E03	FB	Multipotent	Yes	None/TBD	24
ESCs	Embryo	Pluripotent	Yes	High	25–28
NT2N	Embryonal carcinoma	Multipotent	Yes	None/TBD	29, 30
NSCs (stable lines)	ESCs	Multipotent	Yes	None/TBD	31
hpNSCs	Oocytes	Pluripotent	Yes	High	32
ISC-hpNSC®	hpNSCs	Multipotent	Yes	None/TBD	33
iPSCs	Adult tissues	Pluripotent	Yes	High	17, 34–36
iNs	iPSCs	Multipotent	Yes	Medium	37
MSCs	UCB	Multipotent	Yes	None	38
	BM	Multipotent	Yes	None	38

[a]AT: adipose tissue; ADSC: adipose tissue-derived stromal cell; BDNF: brain-derived neurotrophic factor; BM: bone marrow; BMPC: bone marrow precursor cell; EPC: endothelial progenitor cell; ESC: embryonic stem cell; FB: fetal brain; iN: induced neuronal cell; iPSC: induced pluripotent stem cell; MAPC: multipotent adult progenitor cell; MNC: mononuclear cell; MSC: mesenchymal stem cell; MUSEC: multilineage-differentiating stress-enduring cell; NSC: neural stem cell; TBD: to be determined.

- Potential tumorigenicity of genetically modified cell lines as well as naïve ESCs or iPSCs.
- Ethical issues associated mostly with the use of human fetal tissues and human ESCs (hESCs; the derivation of which involves destruction of a human embryo). Lately, several ways of bypassing the stage of destroying a live human embryo and thus avoiding such ethical issues were suggested and tried, such as altered nuclear transfer,[8] deriving ESCs from a blastomere (a single cell from an eight-cell blastocyst)[9] or producing hESCs from parthenogenetically activated unfertilized oocytes.[10]

13.3 Transplantation of SCs in Animal Models of Stroke and TBI

Preclinical studies of SC therapy address a wide range of issues, including the effect of SCs on recovery of sensorimotor and cognitive deficits after experimental stroke and TBI of different severity levels, survival, proliferation and differentiation of SCs *in vivo*, molecular mechanisms of action and potential anti-inflammatory activity, stimulation of endogenous neurogenesis, *etc.* Although almost all of the SC types discussed earlier (Section 13.2) were tested in experimental studies, NSCs from different sources and MSCs are currently considered to have the greatest clinical potential in the cell therapy of CNS disorders.[39,40] It has been estimated that MSCs and NSCs produce a significant improvement in functional outcome in 44.4% and 24.4% of studies performed on experimental models of stroke, respectively.[41] Most of the animal studies of SC transplantation in stroke are focused on ischemic stroke (IS) models; far fewer reports are found on the intracerebral hemorrhage (ICH).

While the original idea of SC neurotransplantation was to replace the damaged brain cells by taking advantage of the differential potential of the grafted cells, the results of animal studies credit their therapeutic effect mostly to an indirect, paracrine action on the endogenous regenerative processes. The data presented in Tables 13.2 and 13.3 (the animal models used in these studies are discussed in detail in Chapter 4) point to neurotrophic, neuroprotective and anti-inflammatory mechanisms of action of NSCs and MSCs, rather than these SCs being a direct source of replacement of the damaged neural cells. Microglia, as the major contributors to the inflammatory response after stroke and TBI, are often viewed as major "early response" targets for the transplanted SCs, MSCs in particular.[42]

In the majority of experimental studies, SCs of various types are grafted into the mammalian brain in the form of a cell suspension, but using three-dimensional (3D) scaffolds (which were shown to be beneficial for cell transplant survival and proliferation[43]) is becoming more popular, particularly when the goal is to specifically induce neuronal differentiation and prolong the survival of grafted NSCs and NPs (reviewed in Cohen and Jensen[44]). In a recently published study, intracerebral transplantation of polymeric substrate-supported human iPSC-derived neurons improved the graft survival rate 38-fold and allowed for delivery of multiple neuronal subtypes.[45] While improving the survival and proliferation rate can significantly increase the value of neuronal differentiation of transplanted NSCs for long-term positive therapeutic effects in stroke and TBI, a direct involvement of MSCs in neural regeneration is viewed with more skepticism. Neural induction of MSCs was described more than a decade ago,[46] and since then, considerable experimental data have accumulated showing that MSCs can express some neuronal and glial cell markers after grafting into a mammalian brain.[47] Although a terminal neural transdifferentiation of MSCs can be achieved *in vitro*,[48] to the best of our knowledge, no published data have shown differentiation of naïve MSCs into functional neurons *in vivo*.

Table 13.2 Experimental studies of stem cell therapy of stroke.[a]

Type of SC/line	Source	Model	ROA	SMF	Comments	Ref.
NSCs	Rat FB	pMCAO	IC	Yes	Improved SMF, reduced atrophy and migration to the injured site (if grafted elsewhere) observed only if animals housed in EE	20, 49, 50
NSCs	Human FB	tMCAO	IC	Yes	Differentiation (mostly in neurons) observed in migrating cells	51, 52
NSCs HB1.F3	Human FB	tMCAO	IC	Yes	Overexpression of VEGF increased survival of the transplanted cells and their effect on the structural and functional recovery; effects may be due to intervention of neuroinflammatory system	53–56
NSCs NSI-566RSC	Human FSC	tMCAO	IC	Yes	8 week observation: evidence of neuronal differentiation	23
NSCs NT2N	Human embryonal carcinoma	tMCAO	IC	Yes	Long-lasting SMF improvements, survived in the host brain for 14 months without neoplastic conversion	29, 30
NSCs CTX0E03	Human FB	tMCAO	IC	Yes	Long-term (up to 12 weeks) observations; SMF improvements accompanied by a high rate of proliferating microglia	57, 58
NPs ENStem-A[b]	Human ESCs	tMCAO	IC	Yes	Migration of NPs to the contralateral (damaged) area accompanied by decreased apoptosis and gliosis	59
NPs ENStem-A[b]	Human ESCs	HI	IA	Yes	Survival and migration of NPs to the cortex and hippocampus significantly elevated if pre-treated with BDNF	60
iPSCs	Mouse fibroblasts	tMCAO	IC	No	Implanted iPSCs differentiated into neurons and neuroblasts, but did not improve SMF and developed tumors—larger in MCAO animals than in sham-operated controls	61
iPSCs	Rat fibroblasts	ICH	IC	Yes	SMF improvement and survival rate was accompanied by attenuation of cerebral inflammatory reaction and neuronal injuries	62

NPs	Human iPSCs	tMCAO	IC	Yes	Protective actions of the graft observed early and were possibly due to trophic effects, which preceded maturation and functional integration of the transplanted NPs	63
NSCs	Human iPSCs	tMCAO	IC	Yes	Survival and neuronal differentiation of the graft observed over a period of up to 4 months; SMF improvement attributed to enhanced endogenous plasticity, lower microglia/macrophage activity and neuronal damage; differentiation of the graft could be supported by release of differentiation factors from co-injected microspheres	64–66
MSCs	Different sources	Different models	IV, IA, or IC	Yes	Injected IV or IA early after the brain injury or IC in later stages, MSCs migrate to the injury site and improve SMF by secreting paracrine factors, which support endogenous angiogenesis, neurogenesis and synapse formation and attenuate neuronal death by suppressing activated microglia	reviewed in 67–71
MUSECs	Human BM	tMCAO SALS	IC	Yes	MUSECs migrate to the peri-lesion area, differentiate into mature neurons, facilitate neuronal reconstruction and promote long-term recovery	72, 73
SB623	Human MSCs	tMCAO	IC	Yes	SB623 line was made by transfection of human MSCs with the intracellular domain of Notch-1 transmembrane receptor protein	74
MultiStem	Human MAPC	tMCAO	IC	Yes	IC grafting of MultiStem did not require immunosuppression; proposed action: secreting of trophic modulators, which inhibit neuronal cell death and inflammation, promote angiogenesis and indirectly contribute to endogenous neurogenesis	75

[a]BDNF: brain-derived neurotrophic factor; BM: bone marrow; EE: enhanced environment; FB: fetal brain; FPI: fluid percussion injury; FSC: fetal spinal cord; HI: hypoxic–ischemic stroke model; IA: intra-arterial; IC: intracerebral; iPSC: induced pluripotent stem cell; IV: intravenous; MAPC: multipotent adult progenitor cell; MCAO: intraluminal middle cerebral artery occlusion (t: transient, p: permanent); MSC: mesenchymal stem cell; MUSEC: multilineage-differentiating stress-enduring cell; NP: neural progenitor; NSC: neural stem cell; ROA: route of administration; SALS: subacute lacunar stroke model; SC: stem cell; SMF: sensorimotor functions; VEGF: vascular endothelial growth factor.
[b]Produced by EMD Millipore (Billerica, MA, USA).

Table 13.3 Experimental studies of stem cell therapy of traumatic brain injury.[a]

Type of SC/line	Source	Model	ROA	SMF	Comments	Ref.
NSCs	Rat FB	FPI	IC	Yes	Survival of the graft significantly improved when neurotransplantation was combined with NGF infusion	76
NSCs	Rat FB	FPI	IA	Yes	Reduced glial scarring and improvements in SMF observed only if neurotransplantation performed shortly after TBI	77
NSCs	Human FB	CCI	IC	Yes	Proliferation and differentiation into neurons and astrocytes (not oligodendrocytes) increased through a 6 week observation period	78
NSCs	Human FB	FPI	IC	Yes	Some NSCs grafted into ipsilateral hippocampus survived, differentiated into GDNF-secreting neurons and produced improvement of CF	79
NSCs C17.2	Mouse neonatal brain	CCI, FPI	IC	Yes	Long-term survival of the graft and improvement of CF observed only with C17.2 cells genetically engineered to overexpress GDNF	80, 81
NSCs	Rat FB	CCI	IC	Yes	Survival of the graft and its effects on SMF were significantly potentiated if the cells overexpressed BDNF	82
NSCs K048	Human FB	CCI	IC	ND	Enhanced neural repair though modulating of the host immune response: elevated expression of anti-inflammatory interleukin-4 receptor-α and diminished expression of pro-inflammatory interferon-γ receptor-β	83
NSCs NT2N	Human embryonal carcinoma	FPI	IC	No	CF (but not SMF) were improved if NT2N cells overexpressing NGF	84–86

mESCs	Mouse embryo	FPI	IC	Yes	Caused tumor formation in 20% of the experimental animals	87
Neurons	mESCs	CCI	IC	Yes	Both mESC-derived GABAergic neurons and astrocytes infiltrated the injured brain, but only neurons improved SMF	88
iPSCs	Rat fibroblasts	CCI	IC	Yes	iPSCs alone produced SMF, but not CF recovery; full CF restoration was achieved when the rats were maintained in EE	89
NPs	Mouse iPSCs	CCI	IC	Yes	Young (P14) rats were used to model juvenile post-traumatic stress disorder; impaired by TBI, social behavior was markedly improved by NPs pre-treated with hypoxic preconditioning	90
MSCs	Different sources	Different models	IV, IA or IC	Yes	The functional benefits of the MSCs are associated with their anti-inflammatory and neuroprotective effects, as well as with stimulation of endogenous neurogenesis, which was shown to be activated by TBI	Reviewed in 91–93 67, 94, 95
MSCs	Human BM	FPI	IV	Yes	*Secretome* collected from MSCs improved SMF and CF, increased the number of newly forming neurons and decreased neuronal apoptosis and lesion size; all effects were significantly enhanced by hypoxic preconditioning of MSCs	96

[a] AD-cSVF: adipose-derived cellular stromal vascular fraction; ADSC: adipose tissue-derived stromal cell; BDNF: brain-derived neurotrophic factor; BM: bone marrow; CCI: cortical contusion injury; CF: cognitive function; EE: enhanced environment; ESC: embryonic stem cell (m: mouse, h: human); FB: fetal brain; FPI: fluid percussion injury; GDNF: glial cell-derived neurotrophic factor; IA: intra-arterial; IC: intracerebral; iPSC: induced pluripotent stem cell; IV: intravenous; MSC: mesenchymal stem cell; ND: not addressed in the study; NGF: nerve growth factor; NSC: neural stem cell; ROA: route of administration; SC: stem cell; SMF: sensorimotor functions; TBI: traumatic brain injury.

13.4 SC Therapy in Stroke and TBI: Clinical Trials

The first clinical trials of SC therapy in stroke and TBI began in the late 1990s following the promising results of early experimental studies, but it was in the second decade of this century that the growing interest in MSCs and accumulating preclinical data boosted the number of clinical trials in this area. Although in the majority of the clinical trials reviewed below autologous MSCs were used, the limited quantity of these cells that could be obtained without additional expansion may not fit the needs of a rapid intervention, particularly in IS. Stable allogenic cell lines offered as off-the-shelf products by a growing number of companies provide an alternative solution to this problem. Several of such cell lines have already been tested on animal models (reviewed in earlier sections) and are now undergoing clinical trials for the treatment of stroke and TBI.

13.4.1 Stroke

Most of the experimental studies of SC therapy of stroke showed some degree of functional improvement (Section 13.3), usually attributed to the mechanisms of SC action that have been addressed, to SC-stimulated endogenous neurogenesis or to immunomodulation and neuroprotection, rather than to a structural and functional contribution of the SC-originated neuronal growth. As pointed out in a comprehensive review,[97] functional replacement of the damaged neurons by the grafted SCs would require the production of lesion-specific neuronal phenotypes, along with case-tailored structural and trophic support. Thus, while indirect, supportive mechanisms of the therapeutic action of SCs are finding their application in clinical practice, direct cellular replacement and restoration of damaged brain structures from the graft material are still at a comparatively early stage of development, which would require advancements in the field of new biomaterials, as well as long-term experimental studies.[98] The most recent guidelines from the "Stem Cells as an Emerging Paradigm in Stroke" (STEPS) meetings (launched under the auspices of the National Institutes of Health [NIH] and Food and Drug Administration [FDA] in 2007), along with a number of recommendations on clinical trial design, point to a need for a better definition and more experimental data on chronic stroke, noting the comparatively low volume of animal model research of chronic stroke in comparison to such studies in acute stroke.[99]

The first clinical trials were focused on safety and feasibility with varying sources of SCs, methods of their preparations, routes of administration, doses and types of stroke. The results of the registered and/or published clinical trials in IS are summarized in Table 13.4. In most of the trials (28), autologous MSCs or mononuclear cells (MNCs) are/were used, being isolated from the patients' bone marrow (BM; 23), but also from adipose tissue

Table 13.4 Clinical trials of stem cells therapy in patients with ischemic stroke.[a]

ID, year, reference, sponsor	Location	Cell type/source	ROA	Time after stroke	Number of patients	Purpose Phase	Status/results
NSCs							
N/A, 1999,[105] Layton BioScience, Inc.	Pittsburgh, PA, USA	LBS neurons Clonal line from human embryonal carcinoma	IC	1–6 years	12	Safety, feasibility Phase I	Completed/no adverse effects, some neurological improvement
N/A[b], 2004,[106,107] Layton BioScience, Inc.	Pittsburgh, PA, USA	LBS neurons Clonal line from human embryonal carcinoma	IC	1–6 years	18	Safety, feasibility Phase II	Completed/no adverse effects, some improvement but less robust than in Phase I; four cases—marked improvement in CF
NCT01151124, 2010–2016,[108] ReNeuron, Ltd	Glasgow, UK	hNSCs CTX0E03 Clonal line from human FB	IC	6–60 months	13	Safety Phase I	Completed/no adverse effects, improved neurological function
NCT02117635, 2014–2017, ReNeuron, Ltd	Birmingham, UK, among others	hNSCs CTX0E03 Clonal line from human FB	IC	2–3 months	23	Efficacy Phase II	Ongoing, not recruiting
N/A, 2000–2004[110]	Novosibirsk, Russia	Fetal SCs (suspension of NSCs and HSPCs)	IT	4–24 months	10	Safety, efficacy	Completed/no serious complications, significant functional improvements
N/A, 2013–2016, Neuralstem, Inc.	Beijing, China	hNSCs—NSI-566 FSC	IC	4–24 months	118	Safety, efficacy Phases I/II	Ongoing

Table 13.4 (*Continued*)

ID, year, reference, sponsor	Location	Cell type/source	ROA	Time after stroke	Number of patients	Purpose Phase	Status/results
Autologous MSCs and MSC-like SCs							
NCT01501773, 2006–2007[111]	New Delhi, India, among others	MNCs Autologous BM	IV	7–30 days	11	Safety, feasibility Phase I	Completed/safe and feasible Larger study needed to examine efficacy
NCT02425670, 2011–2014[112]	New Delhi, India, among others	MSCs Autologous BM	IV	7–30 days	120	Safety, feasibility Phase II	Completed/safe but no beneficial effect on stroke outcome
NCT02245698, 2008–2017, Neurogen Brain and Spine Institute	Navi Mumbai, India	MNCs Autologous BM	IT	N/A Subacute/ chronic IS	200	Safety, feasibility Phase II	Ongoing, recruiting
N/A, 2011[113]	New Delhi, India	MSCs Autologous BM	IV	3–12 months	12	Safety, feasibility and efficacy Phases I, II	Completed/safe and feasible Further trials needed for efficacy
N/A, 2012[114]	New Delhi, India	MSCs, MNCs Autologous BM	IV	3–12 months	40	Safety, feasibility and efficacy Phases I, II	Completed/safe, some functional improvement
N/A, 2005[115]	Suwon, South Korea	MSCs Autologous BM	IV	Up to 1 year	30	Feasibility, efficacy and safety Phases I, II	Completed/safe, feasible, some functional improvement

N/A, 2010[116]	Suwon, South Korea	MSCs Autologous BM	IV	Less than 7 days	52	Long-term safety and efficacy Phase II	Completed/with a 5 year follow-up: safe, feasible, functional improvement
NCT01716481, 2012–2016[100]	Seoul, South Korea	MSCs preconditioned with autologous serum Autologous BM	IV	Acute and chronic patients	60	Efficacy Phase III	Ongoing or completed
NCT01468064, 2011–2017	Guangzhou, China	MSCs and EPCs Autologous BM	IV	Less than 7 days	20	Safety Phase I	Ongoing, recruiting
NCT02564328[c], 2014–2017	Guangzhou, China	MSCs Autologous BM	IV	6–60 months	40	Safety, efficacy Phase I	Ongoing, recruiting
NCT02605707, 2014–2017	Guangzhou, China	EPCs Autologous blood	IV	6–60 months	30	Safety, efficacy Phases I, II	Ongoing, recruiting
NCT00950521, 2009–2011[102]	Taichung, Taiwan	CD34 + HSPC Autologous blood combined with subcutaneous G-CSF	IC	6–60 months	30	Efficacy Phase II	Completed/safe, no serious adverse effects, some functional improvements
NCT02813512, 2016–2017, Gwo Xi Stem Cell Applied Technology Co., Ltd	Hsinchu, Taiwan	ADSCs Autologous AT	IC	6 months to 10 years	N/A	Safety, efficacy Phase I	Not yet recruiting
NCT01028794, 2008–2013[117]	Kobe, Japan	MNCs Autologous BM	IV	7–10 days	12	Safety, feasibility Phases I/IIa	Completed/safe and feasible
N/A, 2011[118]	Sapporo, Japan	MSCs, expanded in autologous serum Autologous BM	IV	36–133 days	12	Safety, feasibility Phase I	Completed/safe and feasible

Table 13.4 (*Continued*)

ID, year, reference, sponsor	Location	Cell type/source	ROA	Time after stroke	Number of patients	Purpose Phase	Status/results
N/A, 2016[119]	Sapporo, Japan	MSCs, expanded in autologous serum Autologous BM	IV	Less than 40 days	N/A	Feasibility, safety and efficacy Phase III	Ongoing/none
N/A, 2016[101]	Sapporo, Japan	MSCs HUNS001-01 cells Autologous BM	IC	N/A	≥6	Safety, feasibility Phase I	Ongoing/none
NCT00761982, 2008–2011[120]	Seville, Spain	CD34+ MNCs Autologous BM	IA	5–9 days	20	Safety, efficacy Phases I, II	Completed/safe and feasible but no apparent improvement
NCT02290483, 2015–2020	Seville, Spain	MNCs Autologous BM	IA	Less than 7 days	76	Safety, efficacy Phase II	Ongoing, recruiting
N/A, 2010[121]	Rio de Janeiro, Brazil	MNCs Autologous BM	IA	59–82 days	6	Safety, feasibility Phase I	Completed/safe and feasible
N/A, 2009[122,123]	Havana, Cuba	MSCs Autologous BM	IC	1–10 years	5	Safety, tolerability Phase I	Completed/safe, excellent tolerance, some functional improvement; one case of significant cognitive improvement
NCT00875654, 2009–2016	Grenoble, France	MSCs Autologous BM	IV	Less than 6 weeks	30	Feasibility, tolerability Phase II	Ongoing, not recruiting/none

NCT01461720, 2012–2016	Kuala Lumpur, Malaysia	MSCs Autologous BM	IV	2 weeks to 2 months	50	Efficacy Phase II	Ongoing, recruiting
NCT00535197, 2007–2012[124]	London, UK	CD34 + MNCs Autologous BM	IC	Less than 7 days	5	Safety, tolerability Phases I, II	Completed/safe, some functional improvement
NCT00859014, 2009–2014[125]	Houston, TX, USA	MNCs Autologous BM	IV	24–72 hours	25	Safety, feasibility Phase I	Completed/safe and feasible
Allogenic MSCs and MSC-derived SCs							
NCT01436487, 2011–2015,[126] Athersys, Inc.	Cleveland, OH, USA, among others; UK (33 sites)	MultiStem®, human MAPC-derived cells	IV	24–36 hours	134	Safety, efficacy Phase II	Completed/safe and feasible Pre-specified efficacy endpoint not achieved Some improvement at 1 year point
NCT02961504, Healios K.K., Athersys, Inc.	Japan	HLCM051 MultiStem®, human MAPC-derived cells	IV	<36 hours	220	Safety, efficacy Phases II, III	Not yet started
NCT01922908, 2016–2018	Houston, TX, USA	MSCs Allogenic BM	IV	3–10 days	48	Safety, feasibility Phases I, II	Not yet recruiting
NCT01297413, 2011–2016, Stemedica Cell Technologies, Inc.	San Diego, CA, USA	MSCs Allogenic BM	IV	More than 6 months	38	Safety, tolerability and preliminary efficacy Phases I, II	Active, not recruiting

Table 13.4 (*Continued*)

ID, year, reference, sponsor	Location	Cell type/source	ROA	Time after stroke	Number of patients	Purpose Phase	Status/results
NCT02397018, 2015–2017	Durham, NC, USA	UCB (ABO/Rh matched, unfractionated)	IV	3–10 days	10	Safety Phase I	Active, not recruiting
NCT02580019, 2016–2017	Beijing, China	MSCs Allogenic UCB	IV	Within 3 months	N/A	Safety, efficacy Phase II	Not yet recruiting
NCT03004976, 2017–2020	Durham, NC, USA, among others	UCB (ABO/Rh matched, unfractionated)	IV	3–10 days	100	Efficacy Phase II	Not yet recruiting
NCT01287936, 2011–2015,[127] SanBio, Inc.	Birmingham, AL, USA, among others	SB623 Modified allogenic MSCs	IC	6–60 months	18	Safety Phases I, IIa	Completed/safe, significant clinical outcome improvement
NCT02448641, 2016–2017, SanBio, Inc.	Stanford, CA, USA, among others	SB623 Modified allogenic MSCs	IC	6–60 months	156	Safety, efficacy Phase IIb	Ongoing, recruiting
NCT02378974, 2015–2017, CHABiotech Co., Ltd	Seongnam, South Korea	Cordstem-ST UCB	IV	Within 7 days	18	Safety and potential therapeutic effects Phases I, II	Ongoing, not recruiting

[a]AD-cSVF: adipose-derived cellular stromal vascular fraction; ADSC: adipose tissue-derived stromal cell; BM: bone marrow; CF: cognitive functions; EPC: endothelial progenitor cell; ESC: embryonic stem cell (m: mouse, h: human); FB: fetal brain; FSC: fetal spinal cord; G-CSF: granulocyte colony-stimulating factor; HSPC: hematopoietic stem and progenitor cell; IA: intra-arterial; IC: intracerebral; ICH: internal cerebral hemorrhage; iPSC: induced pluripotent stem cell; IS: ischemic stroke; IT: intrathecal; IV: intravenous; MAPC: multipotent adult progenitor cell; MNC: mononuclear cell; MSC: mesenchymal stem cell; NP: neural progenitor; NSC: neural stem cell; ROA: route of administration; SC: stem cell; SMF: sensorimotor functions; UCB: umbilical cord blood.
[b]Trial included patients with the basal ganglia stroke resulting from either IS or ICH.
[c]Trial NCT02564328 includes both IS and ICH patients.

(AT; 2, also ref. 128) and blood (3). Several technical modifications were tested, such as preconditioning of MSCs with autologous serum,[100] expansion in the presence of allogeneic human platelet lysate[101] or the combination of SC transplantation with subcutaneous injections of granulocyte colony-stimulating factor (G-CSF).[102] Intravenous or intra-arterial infusions are the preferred routes of administration of autologous MSCs, which seems more relevant to acute stroke cases when the SC migration to the brain might be stimulated by the IS-induced up-regulation of chemoattractants,[103] while IC injections are mostly employed in chronic cases. The majority of the trials proved MSCs to be safe, and subsequent trials with larger recruitment have been launched. The reported efficiency of the completed trials, however, still varied from mild functional improvement to none at all (reviewed in Shichinohe and Houkin[101] and Bang[104]). More Phase II/III trials with efficacy as the primary outcome are ongoing and/or recruiting (Table 13.4).

The first clinical trial of neurotransplantation of a stable cell line (LBS neurons, produced from a human embryonal carcinoma-derived NT2N line by Layton BioScience, Inc.) in IS was conducted in 1999. Phase I included 12 patients with chronic IS; IC infusion of LBS neurons showed no adverse effects and some neurological improvement.[105] Phase II included 18 patients with basal ganglia stroke resulting from both IS and ICH. The safety of the procedure was confirmed, but functional improvement was less obvious than in the earlier study, possibly because of a different choice of participants.[106] A remarkable cognitive improvement, however, was observed in four patients.[107] Recently, the results of the first clinical trial of a drug product (CTX-DP) made from a human fetal brain-derived NSC line CTX0E03 (Table 13.2) by ReNeuron, Ltd (UK), were published: implantation of the NSCs into the ipsilateral putamen of the patients with chronic IS produced no adverse effects and resulted in some improvements in neurological functions.[108] Efficacy of these cells is the primary outcome of the ongoing Phase II study (Table 13.4; reviewed in Borlongan[103]). The safety and efficacy of NSI-566 cells derived from human fetal spinal cord and produced by Neuralstem, Inc. (USA), are undergoing Phase I/II trials in China (Table 13.4).

As mentioned earlier, off-the-shelf allogenic MSCs and MSC-derived stable cell lines are becoming more and more popular. According to Athersys, Inc. (USA), which produces one such cell line, the potential market for this product could be $15–20 billion annually.[109] The first allogenic MSC-derived cell line, SB623, was genetically engineered from human MSCs and produced by SanBio, Inc. (USA). After successful preclinical tests (Table 13.2), the safety and feasibility of these cells in IS have been proved (NCT01287936), and its efficacy is being tested in an ongoing Phase IIb trial (NCT02448641). Clinical trials of two genetically engineered lines, CTX-DP and SB623, are reviewed in Borlongan.[103] MultiStem®, human MAPC-derived cells produced by Athersys, Inc. (USA), proved to be safe and feasible

Table 13.5 Clinical trials of stem cells therapy in patients with intracerebral hemorrhage.[a]

ID, year, reference, sponsor	Location	Cell type/source	ROA	Time after stroke	Number of patients	Purpose phase	Status/results
N/A, 2007–2014[130]	Siping, China	MSCs Allogenic UCB Autologous BM	IC	2 weeks	24	Safety, efficacy Phase I	Completed/both BM and UCB MSC-treated patients showed an improved functional outcome after 5 years
N/A, 2012[129]	Liaocheng, China	MSCs Autologous BM	IC	5–6 days	100	To assess the clinical effect of MSC transplantation	Completed/MSCs reduced neurological impairment and improved daily activities
NCT02564328[b], 2014–2017	Guangzhou, China	MSCs Autologous BM	IV	6–60 months	40	Safety, efficacy Phase I	Ongoing, recruiting
NCT02283879, 2015–2017, Shenzhen Hornetcorn Bio-technology Company, Ltd	Guangzhou, China	MSCs Allogenic UCB	IV	3–60 months	20	Safety, efficacy Phase I	Ongoing, not recruiting

[a]BM: bone marrow; IC: intracerebral; ICH: internal cerebral hemorrhage; IS: ischemic stroke; IV: intravenous; MSC: mesenchymal stem cell; ROA: route of administration; UCB: umbilical cord blood.
[b]Trial NCT02564328 includes both IS and ICH patients.

in a large Phase II trial (conducted in 33 UK and US hospitals); a Phase III trial aimed at MultiStem® efficiency is planned in Japan and is sponsored by the Healios K.K. company (trial NCT02961504). The safety and a potential therapeutic effect of Cordstem-ST cells produced from umbilical cord blood by a Korean company are being tested in a Phase I/II trial (NCT02378974). Allogenic MSC-derived lines are also produced by Hornetcorn Bio-technology Company, Ltd (China), Gwo Xi Stem Cell Applied Technology Co., Ltd. (Taiwan), Stemedica Cell Technologies, Inc. (USA), and by a number of hospital-based centers such as Neurogen Brain and Spine Institute (India).

Results of the preclinical studies of ICH and perspectives on their transfer into clinical practice are discussed in a recent detailed review.[68] Despite the fact that ICH accounts for 6.5–9.6% of all stroke cases around the world,[2] only four clinical trials in this area could be located, all of which are from China (Table 13.5). Two trials have already been completed; IC implantation of autologous MSCs[129] or autologous or allogenic MSCs[130] shortly after ICH did not cause any adverse effects and improved functional outcome. In one of the trials, the positive effects of MSC treatment were observed for the duration of a 5 year-long study.[130] In two ongoing trials, the safety and efficacy of intravenous infusion of autologous (NCT02564328) or allogenic (NCT02283879) MSCs are being tested in patients with longer than a 3 month history of ICH (Table 13.5).

13.4.2 Traumatic Brain Injury

Preclinical studies and the results of the first clinical trials of SC therapy in TBI are reviewed in Dekmak *et al.*[92] We could locate only ten clinical trials, eight using autologous MNCs and MSCs, one using fetal humans NSCs and hematopoietic stem and progenitor cells and one using the allogenic human MSC-derived cell line SB623 (Table 13.6). The safety and feasibility of intravenous infusion of MNCs within the first 2 days after TBI in children and adults were shown in Phase I trials conducted in Houston, TX, USA (trials NCT00254722 and NCT01575470).[131–133] Both trials have progressed to Phase II (ongoing trials NCT01851083 and NCT02525432).

Healeon Medical, Inc. (USA), developed an effective protocol for the isolation of an autologous adipose-derived cellular stromal vascular fraction, which will be tested in a large, currently recruiting Phase I trial (NCT02959294). Patients are also being recruited for a Phase I trial of genetically modified, allogenic MSC-derived SB623 cell treatment of TBI with longer than a 1 year history (NCT02416492).

Table 13.6 Clinical trials of stem cell therapy in patients with traumatic brain injury.[a]

ID, year, reference, sponsor	Location	Cell type/ source	ROA	Time after TBI	Number of patients	Purpose Phase	Status/results
Autologous MSCs and MSC-like SCs							
NCT00254722, 2006–2009[131,132]	Houston, TX, USA	BMPCs Autologous BM	IV	Less than 36 hours	10 children 5–14 years old	Safety, efficacy Phase I	Completed/safe, improves functional outcome
NCT01851083, 2013–2018	Houston, TX, USA	MNCs Autologous BM	IV	Less than 48 hours	50 children 5–17 years old	Safety, biological activity Phase II	Enrolling participants by invitation only
NCT01575470, 2012–2015[133]	Houston, TX, USA	MNCs Autologous BM	IV	Less than 24 hours	25	Safety, efficacy Phases I, II	Completed/safe and feasible
NCT02525432, 2013–2018	Houston, TX, USA	MNCs Autologous BM	IV	Less than 48 hours	55	Effect on brain structure and neurocognitive and functional outcomes Phase IIb	Enrolling participants by invitation only
NCT02959294, 2016–2021, Healeon Medical, Inc.	Stevensville, MT, USA	AD-cSVF Autologous AT	IV	At least 1 month	200	Efficacy in concussions, mild TBI	Recruiting

N/A, 2010–2014[134]	Mumbai, India	MNCs Autologous BM	IT	N/A	14	Safety, efficacy	Completed/safe, some functional improvement
NCT02028104, 2010–2016	Mumbai, India	MNCs Autologous BM	IT	N/A	43	Safety, efficacy Phase I	Completed/N/A
N/A, 2013[135]	Yichang, China	MSCs Autologous BM	IT	More than 1 month	97	Safety, feasibility Phase I	Completed/safe and feasible, some functional improvement
Allogenic MSCs and MSC-like SCs							
NCT02416492, 2016–2017, SanBio, Inc.	Los Angeles, CA, USA, among others	SB623 Modified allogenic MSCs	IC	More than 12 months	52	Safety, efficacy Phase II	Recruiting
Human fetal NSCs							
N/A, 2001–2004[136]	Novosibirsk, Russia	Fetal SCs (suspension of NSCs and HSPCs)	IT	5–8 weeks	38	Safety, efficacy in severe TBI patients	Completed/no serious complications, significant functional improvements

[a]AD-cSVF: adipose-derived cellular stromal vascular fraction; BM: bone marrow; BMPC: bone marrow precursor cell; HSPC: hematopoietic stem and progenitor cell; IC: intracerebral; IV: intravenous; MNC: mononuclear cell; MSC: mesenchymal stem cell; NSC: neural stem cell; ROA: route of administration; SC: stem cell; TBI: traumatic brain injury.

13.5 Conclusion

The pathophysiology of stroke and TBI presents many common problems, including neuronal cell death, and the original idea of SC transplantation therapy for these disorders was to replace the dying cells with grafted ones. Neural cell transplantations in stroke and TBI animal models started with primary cultures of the fetal brain and neuronal cell lines produced from embryonic tissues. The experimental research in this area was stimulated by better understanding of the mechanisms of endogenous neurogenesis and ischemic brain injury, as well as the rapid development of SC biology with such landmarks as the isolation of ESCs and later iPSCs, the discovery of tissue-specific progenitors and the isolation and characterization of MNCs and their more refined components, such as MSCs, endothelial progenitor cells, or multilineage-differentiating stress-enduring cells. Gradually, it became apparent that functional and structural recovery from experimental stroke and TBI was mostly due to the paracrine effects of the graft rather than its structural contribution to the repair of brain tissue. Direct cellular replacement and restoration of the damaged brain structure is still at a comparatively early stage of development, while transplantation of SCs has found its way from experimental research to clinical trials due to the proven therapeutic effect of their stimulation of endogenous neurogenesis, immunomodulation and neuroprotection. The largest numbers of clinical trials are conducted in IS and with autologous MSCs and MSC-like stromal cells isolated mostly from BM, but also from AT and blood. Allogenic MSCs and off-the-shelf NSCs and MSC-derived cell lines are being produced by a growing number of companies and, along with other benefits, this brings the advantage of much larger quantities of SCs that are immediately available for transplantation, which can be critical for a successful treatment of acute stroke and TBI. Future perspectives are associated with a better understanding of the graft–host interactions in the injured brain, long-term survival of transplanted cells and their direct structural aid to endogenous neurogenesis, refinement and standardization of the clinical protocols and advances in the combination therapy of stroke and TBI.

References

1. H. Okano, *J. Neurosci Res.*, 2002, **69**(6), 698.
2. M. Ramalho-Santos and H. Willenbring, *Cell Stem Cell*, 2007, **1**(1), 35.
3. J. Altman and G. D. Das, *J. Comp. Neurol.*, 1965, **124**(3), 319.
4. G. de Haan and G. Van Zant, *J. Am. Aging Assoc.*, 2002, **25**(2), 79.
5. F. H. Gage, *Science*, 2000, **287**(5457), 1433.
6. O. Lindvall and P. Hagell, *Neurosurg. Focus*, 2002, **13**(5), e2.
7. M. G. Clausi, E. Kumari and S. W. Levison, *Neural. Regen. Res.*, 2016, **11**(1), 45.
8. A. Meissner and R. Jaenisch, *Nature*, 2006, **439**(7073), 212.

9. Y. Chung, I. Klimanskaya, S. Becker, J. Marh, S. J. Lu, J. Johnson, L. Meisner and R. Lanza, *Nature*, 2006, **439**(7073), 216.
10. E. S. Revazova, N. A. Turovets, O. D. Kochetkova, L. B. Kindarova, L. N. Kuzmichev, J. D. Janus and M. V. Pryzhkova, *Cloning Stem Cell*, 2007, **9**(3), 432.
11. J. J. Bara, R. G. Richards, M. Alini and M. J. Stoddart, *Stem Cells*, 2014, **32**(7), 1713.
12. A. M. DiMarino, A. I. Caplan and T. L. Bonfield, *Front. Immunol.*, 2013, **4**, 201.
13. Y. Jiang, B. Vaessen, T. Lenvik, M. Blackstad, M. Reyes and C. M. Verfaillie, *Exp. Hematol.*, 2002, **30**(8), 896.
14. Y. Fan, F. Shen, T. Frenzel, W. Zhu, J. Ye, J. Liu, Y. Chen, H. Su, W. L. Young and G. Y. Yang, *Ann. Neurol.*, 2010, **67**(4), 488.
15. Y. Kuroda, M. Kitada, S. Wakao, K. Nishikawa, Y. Tanimura, H. Makinoshima, M. Goda, H. Akashi, A. Inutsuka, A. Niwa, T. Shigemoto, Y. Nabeshima, T. Nakahata, Y. Nabeshima, Y. Fujiyoshi and M. Dezawa, *Proc. Natl. Acad. Sci. U. S. A.*, 2010, **107**(19), 8639.
16. D. A. De Ugarte, Z. Alfonso, P. A. Zuk, A. Elbarbary, M. Zhu, P. Ashjian, P. Benhaim, M. H. Hedrick and J. K. Fraser, *Immunol. Lett.*, 2003, **89**(2–3), 267.
17. K. Takahashi, K. Tanabe, M. Ohnuki, M. Narita, T. Ichisaka, K. Tomoda and S. Yamanaka, *Cell*, 2007, **131**(5), 861.
18. T. Ishii and K. Eto, *World J. Stem Cells*, 2014, **6**(4), 404.
19. C. J. Boer, *J. Neurol.*, 1994, **242**(1), 1.
20. H. Nishino and C. V. Borlongan, *Prog. Brain Res.*, 2000, **127**, 461.
21. G. Cocks, N. Romanyuk, T. Amemori, P. Jendelova, O. Forostyak, A. R. Jeffries, L. Perfect, S. Thuret, G. Dayanithi, E. Sykova and J. Price, *Stem Cell Res. Ther.*, 2013, **4**(3), 69.
22. S. U. Kim, *Neuropathology*, 2004, **24**(3), 159.
23. N. Tajiri, D. M. Quach, Y. Kaneko, S. Wu, D. Lee, T. Lam, K. L. Hayama, T. G. Hazel, K. Johe, M. C. Wu and C. V. Borlongan, *PLoS One*, 2014, **9**(3), e91408.
24. J. Sinden, K. Pollock and P. Stroemer, Pat. US 20060067918 A1, 2006.
25. F. Erdo, C. Buhrle, J. Blunk, M. Hoehn, Y. Xia, B. Fleischmann, M. Föcking, E. Küstermann, E. Kolossov, J. Hescheler, K. A. Hossmann and T. Trapp, *J. Cereb. Blood Flow Metab.*, 2003, **23**(7), 780.
26. G. M. Keller, *Curr. Opin. Cell Biol.*, 1995, **7**(6), 862.
27. J. A. Thomson, J. Itskovitz-Eldor, S. S. Shapiro, M. A. Waknitz, J. J. Swiergiel, V. S. Marshall and J. M. Jones, *Science*, 1998, **282**(5391), 1145.
28. M. M. Daadi and G. K. Steinberg, *Regen. Med.*, 2009, **4**(2), 251.
29. C. V. Borlongan, Y. Tajima, J. Q. Trojanowski, V. M. Lee and P. R. Sanberg, *Exp. Neurol.*, 1998, **149**(2), 310.
30. S. R. Kleppner, K. A. Robinson, J. Q. Trojanowski and V. M. Lee, *J. Comp. Neurol.*, 1995, **357**(4), 618.
31. D. Drury-Stewart, M. Song, O. Mohamad, Y. Guo, X. Gu, D. Chen and L. Wei, *Stem Cell Res. Ther.*, 2013, **4**(4), 93.

32. E. S. Revazova, N. A. Turovets, O. D. Kochetkova, L. B. Kindarova, L. N. Kuzmichev, J. D. Janus and M. V. Pryzhkova, *Cloning Stem Cell*, 2007, **9**(3), 432.
33. R. Gonzalez, I. Garitaonandia, M. Poustovoitov, T. Abramihina, C. McEntire, B. Culp, J. Attwood, A. Noskov, T. Christiansen-Weber, M. Khater, S. Mora-Castilla, C. To, A. Crain, G. Sherman, A. Semechkin, L. C. Laurent, J. D. Elsworth, J. Sladek, E. Y. Snyder, D. Redmond, Jr. and R. A. Kern, *Cell Transplant.*, 2016, **25**, 1945–1966.
34. H. Okano, M. Nakamura, K. Yoshida, Y. Okada, O. Tsuji, S. Nori, E. Ikeda, S. Yamanaka and K. Miura, *Circ. Res.*, 2013, **112**(3), 523.
35. M. G. Cefalo, A. Carai, E. Miele, A. Po, E. Ferretti, A. Mastronuzzi and I. M. Germano, *Stem Cells Int.*, 2016, **2016**, 4869071.
36. O. Tsuji, K. Miura, Y. Okada, K. Fujiyoshi, M. Mukaino, N. Nagoshi, K. Kitamura, G. Kumagai, M. Nishino, S. Tomisato, H. Higashi, T. Nagai, H. Katoh, K. Kohda, Y. Matsuzaki, M. Yuzaki, E. Ikeda, Y. Toyama, M. Nakamura, S. Yamanaka and H. Okano, *Proc. Natl. Acad. Sci. U. S. A.*, 2010, **107**(28), 12704.
37. Y. Zhang, C. Pak, Y. Han, H. Ahlenius, Z. Zhang, S. Chanda, S. Marro, C. Patzke, C. Acuna, J. Covy, W. Xu, N. Yang, T. Danko, L. Chen, M. Wernig and T. C. Südhof, *Neuron*, 2013, **78**(5), 785.
38. S. Kern, H. Eichler, J. Stoeve, H. Klüter and K. Bieback, *Stem Cells*, 2006, **24**(5), 1294.
39. P. A. Walker, M. T. Harting, S. K. Shah, M.-C. Day, R. El Khoury, S. I. Savitz, J. Baumgartner and C. S. Cox, Jr., *Stem Cells Int.*, 2010, **2010**, 369578.
40. S. A. Goldman, *Cell Stem Cell*, 2016, **18**(2), 174.
41. W. K. Leong, M. D. Lewis and S. A. Koblar, *Stem Cells*, 2013, **31**(6), 1040.
42. S. I. Savitz and C. S. Cox, Jr., *Stem Cells*, 2016, **34**(3), 537.
43. M. I. El-Sherbiny and M. H. Yacoub, *Glob. Cardiol. Sci. Pract.*, 2013, **2013**(3), 316.
44. L. K. Cohen and M. B. Jensen, *Arch. Neurosci.*, 2015, **2**(4), e25364.
45. A. L. Carlson, N. K. Bennett, N. L. Francis, A. Halikere, S. Clarke, J. C. Moore, R. P. Hart, K. Paradiso, M. Wernig, J. Kohn, Z. P. Pang and P. V. Moghe, *Nat. Commun.*, 2016, **7**, 10862.
46. X. R. Ortiz-Gonzalez, C. D. Keene, C. M. Verfaillie and W. C. Low, *Curr. Neurovasc. Res.*, 2004, **1**(3), 207.
47. D. Sun, *Neural Regen. Res.*, 2016, **11**(1), 18.
48. A. Aguilera-Castrejon, H. Pasantes-Morales, J. J. Montesinos, L. V. Cortés-Medina, M. E. Castro-Manrreza, H. Mayani and G. Ramos-Mandujano, *Neurochem. Res.*, 2016, DOI: 10.1007/s11064-016-2086-7.
49. B. Mattsson, J. C. Sørensen, J. Zimmer and B. B. Johansson, *Stroke*, 1997, **28**(6), 1225.
50. Y. Wang, Z. Zhao, S. V. Rege, M. Wang, G. Si, Y. Zhou, S. Wang, J. H. Griffin, S. A. Goldman and B. V. Zlokovic, *Nat. Med.*, 2016, **22**(9), 1050.

51. S. Kelly, T. M. Bliss, A. K. Shah, G. H. Sun, M. Ma, W. C. Foo, J. Masel, M. A. Yenari, I. L. Weissman, N. Uchida, T. Palmer and G. K. Steinberg, *Proc. Natl. Acad. Sci. U. S. A.*, 2004, **101**(32), 11839.
52. V. Darsalia, T. Kallur and Z. Kokaia, *Eur. J. Neurosci.*, 2007, **26**(3), 605.
53. K. Chu, M. Kim, S. H. Chae, S. W. Jeong, K. S. Kang, K. H. Jung, J. Kim, Y. J. Kim, L. Kang, S. U. Kim and B. W. Yoon, *Neurosci. Res.*, 2004, **50**(4), 459.
54. S. W. Jeong, K. Chu, K. H. Jung, S. U. Kim, M. Kim and J. K. Roh, *Stroke*, 2003, **34**(9), 2258.
55. H. J. Lee, K. S. Kim, I. H. Park and S. U. Kim, *PLoS One*, 2007, **2**(1), e156.
56. T. Watanabe, A. Nagai, A. M. Sheikh, S. Mitaki, K. Wakabayashi, S. U. Kim, S. Kobayashi and S. Yamaguchi, *Brain Res.*, 2016, **1631**, 194.
57. K. Pollock, P. Stroemer, S. Patel, L. Stevanato, A. Hope, E. Miljan, Z. Dong, H. Hodges, J. Price and J. D. Sinden, *Exp. Neurol.*, 2006, **199**(1), 143.
58. Z. Hassani, J. O'Reilly, Y. Pearse, P. Stroemer, E. Tang, J. Sinden, J. Price and S. Thuret, *PLoS One*, 2012, **7**(11), e50444.
59. D. J. Chang, S.-H. Oh, N. Lee, C. Choi, I. Jeon, H. S. Kim, D. A. Shin, S. E. Lee, D. Kim and J. Song, *Exp. Mol. Med.*, 2013, **45**, e53.
60. S. Rosenblum, T. N. Smith, N. Wang, J. Y. Chua, E. Westbroek, K. Wang and R. Guzman, *Cell Transplant.*, 2015, **24**(12), 2449.
61. H. Kawai, T. Yamashita, Y. Ohta, K. Deguchi, S. Nagotani, X. Zhang, Y. Ikeda, T. Matsuura and K. Abe, *J. Cereb. Blood Flow Metab.*, 2010, **30**(8), 1487.
62. J. Qin, X. Ma, H. Qi, B. Song, Y. Wang, X. Wen, Q. M. Wang, S. Sun, Y. Li, R. Zhang, X. Liu, H. Hou, G. Gong and Y. Xu, *PLoS One*, 2015, **10**(6), e0129881.
63. J. Polentes, P. Jendelova, M. Cailleret, H. Braun, N. Romanyuk, P. Tropel, M. Brenot, V. Itier, C. Seminatore, K. Baldauf, K. Turnovcova, D. Jirak, M. Teletin, J. Côme, J. Tournois, K. Reymann, E. Sykova, S. Viville and B. Onteniente, *Cell Transplant.*, 2012, **21**(12), 2587.
64. K. Oki, J. Tatarishvili, J. Wood, P. Koch, S. Wattananit, Y. Mine, E. Monni, D. Tornero, H. Ahlenius, J. Ladewig, O. Brüstle, O. Lindvall and Z. Kokaia, *Stem Cells*, 2012, **30**(6), 1120.
65. J. Tatarishvili, K. Oki, E. Monni, P. Koch, T. Memanishvili, A. M. Buga, V. Verma, A. Popa-Wagner, O. Brüstle, O. Lindvall and Z. Kokaia, *Restor. Neurol. Neurosci.*, 2014, **32**(4), 547.
66. T. Memanishvili, N. Kupatadze, D. Tugushi, R. Katsarava, S. Wattananit, N. Hara, D. Tornero and Z. Kokaia, *Biomed. Mater.*, 2016, **11**(2), 025011.
67. A. Rolfe and D. Sun in *Brain Neurotrauma: Molecular, Neuropsychological, and Rehabilitation Aspects*, ed. F. H. Kobeissy, CRC Press/Taylor & Francis, Boca Raton (FL), 2015, 587–596.
68. P. H. Rosado-de-Castro, F. G. de Carvalho, G. R. de Freitas, R. Mendez-Otero and P. M. Pimentel-Coelho, *Stem Cells Int.*, 2016, **2016**, 4617983.

69. Y. Hu, N. Liu, P. Zhang, C. Pan, Y. Zhang, Y. Tang, H. Deng, M. Aimaiti, Y. Zhang, H. Zhou, G. Wu and Z. Tang, *Mol. Neurobiol.*, 2016, **53**(8), 5269.
70. Y. C. Hsuan, C. H. Lin, C. P. Chang and M. T. Lin, *Brain Behav.*, 2016, **6**(10), e00526.
71. Y. Hu, N. Liu, P. Zhang, C. Pan, Y. Zhang, Y. Tang, H. Deng, M. Aimaiti, Y. Zhang, H. Zhou, G. Wu and Z. Tang, *Mol. Neurobiol.*, 2016, **53**(8), 5269.
72. T. Yamauchi, Y. Kuroda, T. Morita, H. Shichinohe, K. Houkin, M. Dezawa and S. Kuroda, *PLoS One*, 2015, **10**(3), e0116009.
73. H. Uchida, K. Niizuma, Y. Kushida, S. Wakao, T. Tominaga, C. V. Borlongan and M. Dezawa, *Stroke*, 2017, **48**(2), 428.
74. T. Yasuhara, N. Matsukawa, K. Hara, M. Maki, M. M. Ali, S. J. Yu, E. Bae, G. Yu, L. Xu, M. McGrogan, K. Bankiewicz, C. Case and C. V. Borlongan, *Stem Cells Dev.*, 2009, **18**(10), 1501.
75. R. W. Mays, C. V. Borlongan, T. Yasuhara, K. Hara, M. Maki, J. E. Carroll, R. J. Deans and D. C. Hess, *J. Exp. Stroke Transl. Med.*, 2010, **3**(1), 34.
76. G. Sinson, M. Voddi and T. K. McIntosh, *J. Neurosurg.*, 1996, **84**(4), 655.
77. H. Soares and T. K. McIntosh, *J. Neural Transplant. Plast.*, 1991, **2**(3–4), 207.
78. A. Wennersten, X. Meier, S. Holmin, L. Wahlberg and T. Mathiesen, *J. Neurosurg.*, 2004, **100**(1), 88.
79. J. Gao, D. S. Prough, D. J. McAdoo, J. J. Grady, M. O. Parsley, L. Ma, Y. I. Tarasenko and P. Wu, *Exp. Neurol.*, 2006, **201**(2), 281.
80. P. Riess, C. Zhang, K. E. Saatman, H. L. Laurer, L. G. Longhi, R. Raghupathi, P. M. Lenzlinger, J. Lifshitz, J. Boockvar, E. Neugebauer, E. Y. Snyder and T. K. McIntosh, *Neurosurgery*, 2002, **51**(4), 1043.
81. A. Bakshi, S. Shimiz, C. A. Keck, S. Cho, D. G. LeBold, D. Morales, A. E. Snyder, D. J. Watson and T. K. McIntosh, *Eur. J. Neurosci.*, 2006, **23**(8), 2119.
82. H. Ma, B. Yu, L. Kong, Y. Zhang and Y. Shi, *Neurochem. Res.*, 2012, **37**(1), 69.
83. J. Gao, R. J. Grill, T. J. Dunn, S. Bedi, J. A. Labastida, R. A. Hetz, H. Xue, J. R. Thonhoff, D. S. DeWitt, D. S. Prough, C. S. Cox, Jr. and P. Wu, *Cell Transplant.*, 2016, **25**(10), 1863.
84. M. F. Philips, J. K. Muir, K. E. Saatman, R. Raghupathi, V. M. Lee, J. Q. Trojanowski and T. K. McIntosh, *J. Neurosurg*, 1999, **90**(1), 116.
85. D. J. Watson, L. Longhi, E. B. Lee, C. T. Fulp, S. Fujimoto, N. C. Royo, M. A. Passini, J. Q. Trojanowski, V. M. Lee, T. K. McIntosh and J. H. Wolfe, *J. Neuropathol. Exp. Neurol.*, 2003, **62**(4), 368.
86. L. Longhi, D. J. Watson, K. E. Saatman, H. J. Thompson, C. Zhang, S. Fujimoto, N. Royo, D. Castelbuono, R. Raghupathi, J. Q. Trojanowski, V. M. Lee, J. H. Wolfe, N. Stocchetti and T. K. McIntosh, *J. Neurotrauma*, 2004, **21**(12), 1723.
87. P. Riess, M. Molcanyi, K. Bentz, M. Maegele, C. Simanski, C. Carlitscheck, A. Schneider, J. Hescheler, B. Bouillon, U. Schäfer and E. Neugebauer, *J. Neurotrauma*, 2007, **24**(1), 216.

88. G. D. Becerra, L. M. Tatko, E. S. Pak, A. K. Murashov and M. R. Hoane, *Behav. Brain Res.*, 2007, **179**(1), 118.
89. J. Dunkerson, K. E. Moritz, J. Young, T. Pionk, K. Fink, J. Rossignol, G. Dunbar and J. S. Smith, *Restor. Neurol. Neurosci.*, 2014, **32**(5), 675.
90. Z. Z. Wei, J. H. Lee, Y. Zhang, Y. B. Zhu, T. C. Deveau, X. Gu, M. M. Winter, J. Li, L. Wei and S. P. Yu, *Cell Transplant*, 2016, **25**(5), 79.
91. W. Peng, J. Sun, C. Sheng, Z. Wang, Y. Wang, C. Zhang and R. Fan, *Stem Cell Res. Ther.*, 2015, **6**, 47.
92. A. Dekmak, S. Mantash, A. Shaito, A. Toutonji, N. Ramadan, H. Ghazale, N. Kassem, H. Darwish and K. Zibara, *Behav. Brain Res.*, 2016, **S0166–4328**(16), 3134.
93. M. A. Matthay, S. Pati and J. W. Lee, *Stem Cells*, 2017, **35**(2), 316.
94. A. Mahmood, D. Lu, C. Qu, A. Goussev and M. Chopp, *Neurosurgery*, 2005, **57**(5), 1026.
95. H.-J. Kim, J.-H. Lee and S. H. Kim, *J. Neurotrauma*, 2010, **27**(1), 131.
96. C.-P. Chang, C.-C. Chio, C.-U. Cheong, C.-M. Chao, B.-C. Cheng and M.-T. Lin, *Clin. Sci.*, 2013, **124**(3), 165.
97. M. Dihné, H. P. Hartung and R. J. Seitz, *Stroke*, 2011, **42**(8), 2342.
98. M. Janowski, D. C. Wagner and J. Boltze, *Stroke*, 2015, **46**(8), 2354.
99. S. I. Savitz, S. C. Cramer, L. Wechsler and STEPS 3 Consortium (32 Collaborators), *Stroke*, 2014, **45**(2), 634.
100. S. J. Kim, G. J. Moon, W. H. Chang, Y. H. Kim, O. Y. Bang and STARTING-2 collaborators (19), *Trials*, 2013, **14**, 317.
101. H. Shichinohe and K. Houkin, *Neurol. Med. Chir.*, 2016, **56**(10), 59.
102. D. C. Chen, S. Z. Lin, J. R. Fan, C. H. Lin, W. Lee, C. C. Lin, Y. J. Liu, C. H. Tsai, J. C. Chen, D. Y. Cho, C. C. Lee and W. C. Shyu, *Cell Transplant.*, 2014, **23**(12), 1599.
103. C. V. Borlongan, *Mol. Ther.*, 2016, **24**(10), 1710.
104. O. Y. Bang, *J. Clin. Neurol.*, 2016, **12**(1), 14.
105. C.-P. Chang, C.-C. Chio, C.-U. Cheong, C.-M. Chao, B.-C. Cheng and M.-T. Lin, *Clin. Sci.*, 2013, **124**(3), 165.
106. D. Kondziolka, L. Wechsler, S. Goldstein, C. Meltzer, K. R. Thulborn, J. Gebel, P. Jannetta, S. DeCesare, E. M. Elder, M. McGrogan, M. A. Reitman and L. Bynum, *Neurology*, 2000, **55**(4), 565.
107. C. S. Stilley, C. M. Ryan, D. Kondziolka, A. Bender, S. DeCesare and L. Wechsler, *Neurology*, 2004, **63**(7), 1320.
108. D. Kalladka, J. Sinden, K. Pollock, C. Haig, J. McLean, W. Smith, A. McConnachie, C. Santosh, P. M. Bath, L. Dunn and K. W. Muir, *Lancet*, 2016, **388**(10046), 787.
109. Athersys, Inc. (company's website): http://www.athersys.com/neurologicalII.cfm.
110. S. S. Rabinovich, V. I. Seledtsov, N. V. Banul, O. V. Poveshchenko, V. V. Senyukov, S. V. Astrakov, D. M. Samarin and V.Ya. Taraban, *Kletochnye Tekhnologii v Biologii i Meditsine*, 2005, **1**(1), 25, [Article in Russian].
111. K. Prasad, A. Sharma, A. Garg, S. Mohanty, S. Bhatnagar, S. Johri, K. K. Singh, V. Nair, R. S. Sarkar, S. P. Gorthi, K. M. Hassan,

S. Prabhakar, N. Marwaha, N. Khandelwal, U. K. Misra, J. Kalita, S. Nityanand and InveST Study Group, *Stroke*, 2014, **45**(12), 3618.
112. A. Bhasin, M. V. Srivastava, S. S. Kumaran, S. Mohanty, R. Bhatia, S. Bose, S. Gaikwad, A. Garg and B. Airan, *Cerebrovasc. Dis. Extra.*, 2011, **1**(1), 93.
113. A. Bhasin, M. V. Srivastava, S. Mohanty, R. Bhatia, S. S. Kumaran and S. Bose, *Clin. Neurol. Neurosurg.*, 2013, **115**(7), 1003.
114. A. Bhasin, M. V. Srivastava, S. Mohanty, R. Bhatia, S. S. Kumaran and S. Bose, *Clin. Neurol. Neurosurg.*, 2013, **115**(7), 1003.
115. O. Y. Bang, J. S. Lee, P. H. Lee and G. Lee, *Ann. Neurol.*, 2005, **57**(6), 874.
116. J. S. Lee, J. M. Hong, G. J. Moon, P. H. Lee, Y. H. Ahn and O. Y. Bang, *Stem Cells*, 2010, **28**(6), 1099.
117. A. Taguchi, C. Sakai, T. Soma, Y. Kasahara, D. M. Stern, K. Kajimoto, M. Ihara, T. Daimon, K. Yamahara, K. Doi, N. Kohara, H. Nishimura, T. Matsuyama, H. Naritomi, N. Sakai and K. Nagatsuka, *Stem Cells Dev.*, 2015, **24**(19), 2207.
118. O. Honmou, K. Houkin, T. Matsunaga, Y. Niitsu, S. Ishiai, R. Onodera, S. G. Waxman and J. D. Kocsis, *Brain*, 2011, **134**(Pt 6), 1790.
119. O. Honmou, *Nihon Rinsho*, 2016, **74**(4), 649, [Article in Japanese].
120. A. Moniche, A. Gonzalez, J. R. Gonzalez-Marcos, M. Carmona, P. Piñero, I. Espigado, D. Garcia-Solis, A. Cayuela, J. Montaner, C. Boada, A. Rosell, M. D. Jimenez, A. Mayol and A. Gil-Peralta, *Stroke*, 2012, **43**(8), 2242.
121. L. M. Barbosa da Fonseca, B. Gutfilen, P. H. Rosado de Castro, V. Battistella, R. C. S. Goldenberg, T. Kasai-Brunswick, C. L. R. Chagas, E. Wajnberg, A. Maiolino, S. Salles Xavier, C. Andre, R. Mendez-Otero and G. R. de Freitas, *Exp. Neurol.*, 2010, **221**(1), 122.
122. C. Suárez-Monteagudo, P. Hernández-Ramírez, L. Alvarez-González, I. García-Maeso, K. de la Cuétara-Bernal, L. Castillo-Díaz, M. L. Bringas-Vega, G. Martínez-Aching, L. M. Morales-Chacón, M. M. Báez-Martín, C. Sánchez-Catasús, M. Carballo-Barreda, R. Rodríguez-Rojas, L. Gómez-Fernández, E. Alberti-Amador, C. Macías-Abraham, E. D. Balea, L. C. Rosales, L. Del Valle Pérez, B. B. Ferrer, R. M. González and J. A. Bergado, *Restor. Neurol. Neurosci.*, 2009, **27**(3), 151.
123. M. L. Bringas, C. Suarez, C. Sanchez, M. Alvarez, P. Valdes, S. Salazar, D. Chongo and M. Jahanshahi, *Br. Med. J. Case Rep.*, 2011, **2011**, bcr0320113944.
124. S. Banerjee, P. Bentley, M. Hamady, S. Marley, J. Davis, A. Shlebak, J. Nicholls, D. A. Williamson, S. L. Jensen, M. Gordon, N. Habib and J. Chataway, *Stem Cells Transl. Med.*, 2014, **3**(11), 1322.
125. S. I. Savitz, V. Misra, M. Kasam, H. Juneja, C. S. Cox, Jr., S. Alderman, I. Aisiku, S. Kar, A. Gee and J. C. Grotta, *Ann Neurol.*, 2011, **70**(1), 59.
126. D. C. Hess, C. A. Sila, A. J. Furlan, L. R. Wechsler, J. A. Switzer and R. W. Mays, *Int. J. Stroke*, 2014, **9**(3), 381.
127. G. K. Steinberg, D. Kondziolka, L. R. Wechsler, L. D. Lunsford, M. L. Coburn, J. B. Billigen, A. S. Kim, J. N. Johnson, D. Bates, B. King,

C. Case, M. McGrogan, E. W. Yankee and N. E. Schwartz, *Stroke*, 2016, **47**(7), 1817.
128. E. Díez-Tejedor, M. Gutiérrez-Fernández, P. Martínez-Sánchez, B. Rodríguez-Frutos, G. Ruiz-Ares, M. L. Lara and B. F. Gimeno, *J. Stroke Cerebrovasc. Dis.*, 2014, **23**(10), 2694.
129. Z. M. Li, Z. T. Zhang, C. J. Guo, F. Y. Geng, F. Qiang and L. X. Wang, *Clin. Neurol. Neurosurg.*, 2013, **115**(1), 72.
130. Z. Chang, G. Mao, L. Sun, Q. Ao, Y. Gu and Y. Liu, *Exp. Ther. Med.*, 2016, **12**(6), 3535.
131. C. S. Cox, Jr., J. E. Baumgartner, M. T. Harting, L. L. Worth, P. A. Walker, S. K. Shah, L. Ewing-Cobbs, K. M. Hasan, M. C. Day, D. Lee, F. Jimenez and A. Gee, *Neurosurgery*, 2011, **68**(3), 588.
132. G. P. Liao, M. T. Harting, R. A. Hetz, P. A. Walker, S. K. Shah, C. J. Corkins, T. G. Hughes, F. Jimenez, S. C. Kosmach, M. C. Day, K. Tsao, D. A. Lee, L. L. Worth, J. E. Baumgartner and C. S. Cox, Jr., *Pediatr. Crit. Care. Med.*, 2015, **16**(3), 245.
133. C. S. Cox, Jr., R. A. Hetz, G. P. Liao, B. M. Aertker, L. Ewing-Cobbs, J. Juranek, S. I. Savitz, M. L. Jackson, A. N. Romanowska-Pawliczek, F. Triolo, P. K. Dash, C. Pedroza, D. A. Lee, L. Worth, I. P. Aisiku, H. A. Choi, J. B. Holcomb and R. S. Kitagawa, *Stem Cells*, 2016, DOI: 10.1002/stem.2538.
134. A. Sharma, H. Sane, P. Kulkarni, J. Yadav, N. Gokulchandran, H. Biju and P. Badhe, *Springerplus*, 2015, **4**, 26.
135. C. Tian, X. Wang, X. Wang, L. Wang, X. Wang, S. Wu and Z. Wan, *Exp. Clin. Transplant.*, 2013, **11**(2), 176.
136. V. I. Seledtsov, S. S. Rabinovich, O. V. Parlyuk, M. Y. Kafanova, S. V. Astrakov, G. V. Seledtsova, D. M. Samarin and O. V. Poveschenko, *Biomed. Pharmacother.*, 2005, **59**(7), 415.

CHAPTER 14

Testing and Mechanisms of Neuroprotective Agents for Cerebral Ischemic Injury in Clinical Trials, 2010–2016

S. L. GRUPKE* AND J. F. FRASER

University of Kentucky, Department of Neurosurgery, 800 Rose St, Lexington, KY 40515, USA
*Email: stephen.grupke@uky.edu

14.1 Introduction

Ischemic stroke affects approximately 15 million people worldwide each year, and it accounts for approximately 6.2 million deaths.[1] Stroke is also a leading cause of adult disability, second worldwide only to dementia. In the United States, one in four stroke survivors needs assistance with activities of daily living 6 months from onset.[2] Despite the global impact that this disease has, pharmacological interventions for acute ischemic stroke remain limited. Currently, the only drug with demonstrated efficacy in the setting of acute stroke is tissue plasminogen activator (tPA),[3] which, along with mechanical reperfusion *via* endovascular thrombectomy, aims to limit brain injury by re-establishing blood flow to ischemic tissue. Neuroprotective agents have been studied as a second strategy for preserving tissue and mitigating the damage incurred by stroke. The rationale behind this discipline is that ischemic tissue in a stroke exists on a gradient of viability.

Drug Discovery Series No. 62
Acute Brain Impairment: Scientific Discoveries and Translational Research
Edited by Philip V. Peplow, Svetlana A. Dambinova, Thomas A. Gennarelli and Bridget Martinez

Published by the Royal Society of Chemistry, www.rsc.org

Normal cerebral blood flow is about 50 mL 100 g $tissue^{-1}$ $minute^{-1}$, and neurons can survive in a relatively quiescent state at a perfusion as low as 23 mL 100 g $tissue^{-1}$ $minute^{-1}$. At the onset of a stroke, downstream blood flow is limited, availability of oxygen and substrates to the tissue are diminished, clearance of metabolic byproducts is reduced, and cells go into metabolic failure. Blood flow of less than 12 mL 100 g $tissue^{-1}$ $minute^{-1}$ is rapidly fatal to neurons and quickly results in an irreversible infarct.[4] In the spectrum between the living quiescent and irreversibly dead cells, however, lies the ischemic penumbra, a region of tissue that is at danger of progressing to complete infarct, but may be potentially salvageable. Programmed apoptotic pathways have been activated in this area and may continue to lead ultimately to cell death even after reperfusion of the tissue. The discipline of neuroprotection aims to identify areas in the apoptotic biochemical cascade that can be exploited with pharmacologic agents, halting cell death pathways and saving neurons and, by extension, neurologic function.

At the onset of an ischemic insult, inadequate availability of energy substrates leads to failure of the ATP-dependent plasma membrane ion pumps needed to maintain polarization. This results in a loss of ion concentration gradients across cell membranes, as calcium and sodium ions flow freely into the neuron, leading to depolarization of the cell. Excitatory neurotransmitters (primarily glutamate) are released en masse. Following that, there is a large focal excitatory response, leading to further neuron depolarization *via* *N*-methyl-D-aspartate (NMDA) and α-amino-3-hydroxy-5-methyl-4-isoxazoleproprionic acid (AMPA) receptor activation, exacerbating and perpetuating this cycle. An increase in intracellular calcium ions (arising from both free-flowing ions from the extracellular milieu and the release of mitochondrial stores) leads to activation of calcium-dependent caspases, promoting cell breakdown and apoptosis. In the process, reactive oxidative species are released, leading to further membrane degradation by lipid peroxidation. An inflammatory response is mounted, and neutrophils are recruited from the circulation by way of cytokine release, making their way into the ischemic tissue through a blood–brain barrier that has been weakened through breakdown of endothelial tight junctions. These neutrophils release more reactive oxidative species, and the cell degradation cycle continues (Figure 14.1).[5]

14.2 Neuroprotective Agents

14.2.1 Magnesium

Magnesium sulfate ($MgSO_4$) has many characteristics that would make it an effective neuroprotective agent. Ionic magnesium is a voltage-dependent blocker of the ion pore associated with the NMDA receptor.[6] It is also a direct blocker of several calcium channels, it has been shown to decrease spontaneous presynaptic release of glutamate,[7] and it enhances mitochondrial

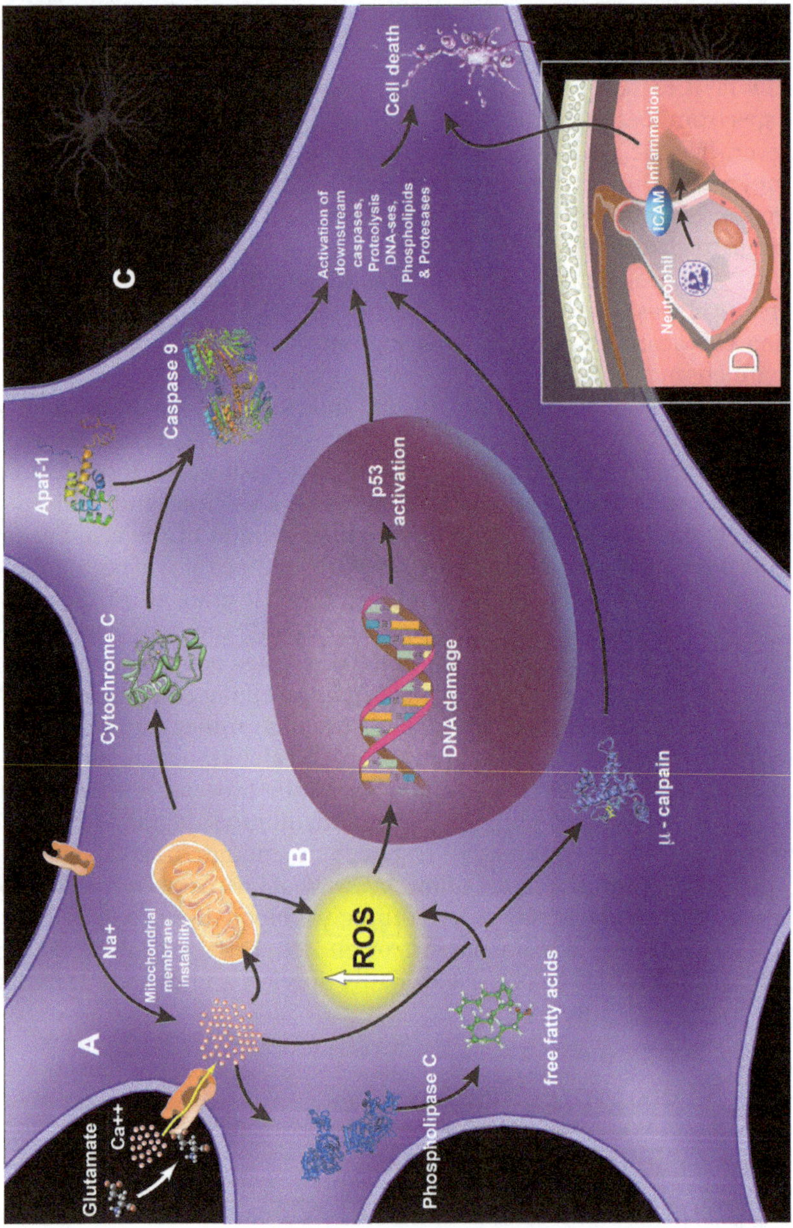
Glutamate
Ca++
Phospholipase C
A
Na+
Mitochondrial membrane instability
free fatty acids
ROS
B
μ - calpain
Cytochrome C
DNA damage
p53 activation
Apaf-1
Caspase 9
C
Activation of downstream caspases, Proteolysis DNA-ses, Phospholipids & Protesases
Cell death
D
Neutrophil
ICAM
Inflammation

sequestration of intracellular calcium.[8] Together, these characteristics make it potentially useful in blunting glutamate-mediated excitotoxicity. It is inexpensive, easy to administer, and widely available due to its routine use in hypomagnesemia, pre-eclampsia, cardiac arrhythmia, and other pathologies.[9] This widespread use also allows for ample study of its side-effect profile, which has proven to be tolerable and treatable. It is excreted through the kidneys with a half-life of about 4 hours, and overdose can be detected clinically by checking for the loss of deep tendon reflexes.

Multiple studies in animals using different models for stroke have demonstrated the efficacy of $MgSO_4$ as a neuroprotective agent, showing a reduction in infarct size in a dose-dependent fashion when magnesium was given along with a surgically induced stroke.[10–12] Small pilot studies in humans showed a modest reduction of long-term death and disability associated with magnesium administration upon ictus.[13,14] The first large-scale trial examining the efficacy of $MgSO_4$ in humans was reported in 2004: the Intravenous Magnesium Efficacy in Stroke (IMAGES) trial[15] was a massive multicenter, double-blind, placebo-controlled study looking at 2589 patients, one of the largest samples for a neuroprotective agent trial. These patients were randomized within 12 hours of the onset of an acute ischemic stroke to receive either an intravenous bolus of 16 mmol $MgSO_4$ with an infusion of 65 mmol over the next 24 hours, or a matching saline placebo. Patients had to have an ischemic stroke with a "last-known normal" within 12 hours of intervention, be at least 18 years old and conscious, and have limb weakness defined as a score of 1 or higher on the National Institutes for Health Stroke Scale (NIHSS) for limb strength in one of their limbs. Brain imaging confirming the diagnosis was performed within 7 days of stroke onset. Patients were excluded for pregnancy, renal impairment, or contraindication for magnesium administration. The primary outcome of the trial was ultimately "global outcome", encompassing death and disability as a

Figure 14.1 Illustration of cell signaling cascades and mechanisms of ischemia targeted by clinically evaluated neuroprotective agents. Glutamate-mediated sodium and calcium channel activation leads to depolarization and activation of cascades leading to apoptosis, targeted by inhibitors of ligand binding or direct blockade of the channels (A). Production of reactive oxidative species (ROS), (B) can lead to DNA damage and trigger downstream cell death mechanisms. Administration of neurotrophic and vasoactive agents may modify the extracellular matrix (C) and has shown promise in the treatment of cerebral infarcts in animals. Finally, neutrophil-mediated inflammatory changes have been associated with brain injury in ischemia (D), and antibodies designed to target cell surface receptors on neutrophils and endothelial cells have been trialed in order to prevent infiltration of these cells into the ischemic bed and subsequent downstream oxidative damage.[75] Reprinted from *Clinical Neurology and Neurosurgery*, Volume 129, S. L. Grupke, J. Hall, M. Dobbs *et al.* Understanding history, and not repeating it. Neuroprotection for acute ischemic stroke: from review to preview, 1–9, Copyright 2015, with permission from Elsevier.

Barthel Index (BI)[16] less than 95 and a modified Rankin Score (mRS)[17] greater than 1. Secondary endpoints were death and disability examined independently based on BI and mRS. No statistically significant difference was seen at 90 days between groups in the primary or any of the secondary endpoints. The pre-specified subgroup analysis did not show any difference between patients treated within 6 hours of ictus and those treated 6–12 hours out.

The authors of the IMAGES trial explain the negative results of the study as possibly stemming from the relatively late "time to treat". The average timespan between stroke onset and drug administration was 7 hours, and by this time it was postulated that viable penumbra may be reduced to a very small volume and differences in tissue salvage attributable to neuroprotective drugs would be blunted. From this stemmed the rationale for the Field Administration of Stroke Therapy—Magnesium (FAST-MAG) trial reported in 2015,[18] which aimed to reduce the amount of time from onset of stroke symptoms to drug treatment by implementing a protocol for administration by first-responders. In this multicenter, randomized, double-blind, placebo-controlled study, patients with a stroke (defined clinically) within 2 hours were randomized to receive intravenous magnesium or saline placebo, similar to the IMAGES trial. Here, however, emergency medical technicians and receiving hospitals were trained in field administration of $MgSO_4$ as per the study protocol, ensuring administration of the drug in a relatively short time window. The primary outcome in this study was the degree of disability measured on the mRS 3 months from stroke ictus, with secondary measures at 90 days being BI, NIHSS, and Glasgow Outcome Scale (GOS). A total of 1700 patients were enrolled over the course of 7 years. Impressively, the group was able to demonstrate the feasibility of implementing a system for the rapid administration of medication in the setting of acute ischemic stroke: the average time from stroke onset to initiation of study drug infusion was 45 minutes. Discouragingly, even with this system in place, there was no statistically significant difference noted in 90 day disability outcomes on the mRS between treatment and placebo groups. No differences were seen in any of the secondary outcome measures using dichotomized standard outcome instruments to evaluate for excellent recovery, minimal/no disability, functional independence, and neurological deficit. While the clinical results of FAST-MAG were neutral, the systems goals accomplished by this group were important for future neuroprotective agent studies. As mentioned above, they proved the feasibility of the field administration of the medication to reduce the time to treat. They developed a system by which the patient was rapidly screened for stroke by the emergency provider using a validated rapid clinical assessment;[19] this was then supplemented with a call to an enrolling physician who discussed the case with the paramedic and the patient (or their proxy). This significantly reduced the rate of enrollment of patients with stroke-mimicking conditions, a known obstacle that can complicate interpretation of data in neuroprotective drug studies. The protocol also utilized a phone system for constant immediate access to knowledgeable physician consent elicitors. This allowed the patient or their proxy to ask questions and give consent for study enrollment in parallel with the treatment being performed by the paramedics, ensuring no delay in the patient's care.

14.2.2 Minocycline

Minocycline is a semi-synthetic antibiotic derived from tetracycline, most commonly used in patients as a treatment for a wide range of bacterial infections such as acne vulgaris, bacterial meningitis, or gonorrhea and chlamydia urethritis. Its bacteriostatic actions stem from its ability to block bacterial ribosomal subunits from synthesizing proteins. It is also used in the treatment of mild rheumatoid arthritis as it has anti-inflammatory characteristics.

Along with these actions, minocycline has shown promise as a neuroprotective agent by demonstrating clinical benefit in animal models of neurodegeneration mimicking multiple sclerosis, Huntington's disease, and amyotrophic lateral sclerosis (ALS).[20–22] These actions are likely broad, as minocycline has been shown to inhibit many key steps in different branches of the programmed cell death pathways. It decreases the activity of several caspases implicated in apoptosis, though not directly. It has been shown to be protective in the setting of NMDA-induced toxicity in cultured mouse neurons by reducing the mitochondrial release of cytochrome c. This mechanism was then verified in an *in vivo* setting in a mouse model of ALS,[23] a model in which cytochrome c release and subsequent apoptosis is correlated with progression of disease.[24] Minocycline has demonstrated the ability to protect cultured rat cerebellar granule cells from reactive oxidative species by blocking phosphorylation of p38 mitogen-activated protein (MAP) kinase.[25] It plays a role in the inflammatory cascade, as shown by its role in the treatment of rheumatoid arthritis, affecting cytokine production by inhibiting tumor necrosis factor-α (TNF-α) and interferon-γ production in activated T cells. It also is known to block matrix metalloproteinase (MMP) activity, altering the composition of the extracellular milieu and reducing infiltration of inflammatory cells into the local environment.[26]

In animal studies looking at minocycline's ability to protect the ischemic penumbra during a transient middle cerebral artery occlusion (MCAO), pretreatment with minocycline and treatment up to 4 hours after ictus showed marked reductions in the volumes of infarcted cortex.[27] This benefit was associated with the biochemical finding of decreased microglial interleukin-1β, a pro-inflammatory cytokine, and decreased cyclooxygenase-2 (COX-2), an inducible enzyme that is involved in the production of pro-inflammatory prostaglandins. In 2007, an open-label, evaluator-blinded study looking at 152 patients was reported demonstrating encouraging findings for minocycline's potential as a neuroprotective agent in humans. In this study, Lampl and colleagues[28] looked at adult patients with an NIHSS >5 presenting within 6–24 hours from onset of stroke symptoms. Patients were randomized to receive 200 mg minocycline orally once a day for 5 days or a placebo. Patient exams were recorded as BI, mRS, and NIHSS, with NIHSS findings categorized as complete improvement (0–1), mild (2–7), moderate (8–14), and severe (>15). A physician blinded to treatment assignment scored the patients on days 0, 7, 30, and 90.

Average time to treat in this study was 12.6 hours, and clinical benefit could be seen at day 7 *via* BI, mRS, and NIHSS measurements. These clinical improvements were durable to 90 day follow-up, with NIHSS of 1.6 ± 1.9 in the treatment group and 6.5 ± 3.8 for the patients receiving placebo ($p < 0.0001$). These findings engendered hope and optimism in a discipline that had been uniformly met with failure when trying to translate bench-top biochemical and animal behavioral findings to the patient's bedside; however, larger double-blinded studies would have to be performed to confirm these data.

The Minocycline to Improve Neurologic Outcome in Stroke (MINOS) author group published their Phase I data in 2010.[29] Sixty patients were enrolled, and in this study, patients with acute ischemic stroke within 6 hours of onset with NIHSS >1 were given an intravenous loading dose of minocycline and were assigned to tiers of 3.0, 4.5, 6.0, and 10.0 mg kg^{-1} daily divided into two doses per day. Continual reassessment was used to determine the maximum tolerated dose for each patient. Three deaths occurred during hospitalization, two by malignant cerebral edema and one by extension of the initial infarct. Two more deaths occurred before the 90 day follow-up, one by gastric cancer and one by exacerbation of congestive heart failure. One dose-limiting toxicity occurred in an 88 year-old patient with elevated liver enzymes; all other adverse effects were mild, self-limiting, and not related to dose. In total, 60% of the patients received tPA, and there were no instances of severe intracerebral hemorrhage. The authors posit that this may be due to minocycline's inhibitory action towards MMP.

Minocycline has been found to be promising in an open-label, evaluator-blinded study with a relatively small sample size, and it has been shown to be tolerable at experimentally therapeutic doses, feasible to administer, and safe to give in conjunction with tPA. Despite this, large prospective trials are still needed. The Neuroprotection with Minocycline Therapy for Acute Stroke Recovery Trial (NeuMast; ClinicalTrials.gov #NCT00930020) proposed to do that using parameters similar to Lampl *et al.*: oral minocycline administered 3–48 hours after ictus with a primary endpoint of mRS at 90 days. This study has been reported as terminated due to futility at interim analysis.[30]

14.2.3 Lovastatin

Lovastatin is a statin drug that is clinically used to lower serum cholesterol levels due to its ability to block 3-hydroxy-3-methylglutaryl coenzyme A reductase, a limiting step in the synthesis of cholesterol. There is evidence that the statin's beneficial effect in the setting of central nervous system injury is not completely attributable to its cholesterol-lowering properties. In animal models for multiple sclerosis, statin therapy reduced relapse frequency and led to improved functional outcomes by blocking the T-helper cell type 1 immune response.[31] Statin use in rat transient MCAO stroke models demonstrated a decrease in the size of infarcts in treated animals relative to those receiving placebo, with attenuated glutamate and improved cerebral blood flow in the infarcted regions of treated rats.[32] Mice treated with a statin for 2 weeks after an induced stroke showed improved functional recovery with an immunohistochemically demonstrated increase in brain-derived neurotrophic factor and vascular endothelial growth factor in the ischemic penumbra, suggesting a neurotrophic and angiogenic component to the statin's mechanism of neuroprotection.[33]

A meta-analysis of eight randomized controlled trials compared 625 total patients receiving statin therapy (any statin, any dose) to placebo or no treatment administered within 2 weeks of an acute ischemic stroke. There were no deaths from ischemic stroke or from adverse drug events in the studies that reported these data, though there were insufficient data available to make a statement about safety or efficacy of statins in acute ischemic stroke patients.[34]

The Neuroprotection with Statin Therapy for Acute Recovery Trial (NeuSTART) Phase I study was published in 2009.[35] Based on the successes of past animal studies, this non-randomized study was a dose-escalation and dose-finding study in which acute ischemic stroke patients were treated within 24 hours of ictus with high-dose lovastatin. Patients receiving tPA were excluded because of concerns of increased risk of hemorrhagic conversion given statin-related platelet aggregation inhibition. Patients were administered a tiered dose orally, divided into four doses per day, with 20 mg given orally once per day for 27 days afterward to avoid withdrawal effects. The primary endpoint chosen was hepatic or muscular toxicity based on the side-effect profile of statins in prior studies. Hepatic toxicity was defined as development of jaundice, coagulopathy, or aspartate aminotransferase/alanine aminotransferase levels more than three-times the upper limit of normal. Muscle toxicity was defined as creatine kinase levels more than ten-times the upper limit of normal, or more than five-times the upper limit of normal with muscle pain or weakness not explained by the patient's stroke. The investigators did not look at clinical improvement; however, this initial study showed that lovastatin treatment at levels above those approved by the FDA (8 $mg\,kg^{-1}\,day^{-1}$) for 3 days were tolerable and feasible. The authors also measured serum levels of inflammatory biomarkers, and interestingly there was a significant decrease in TNF receptor 1, but not TNF-α. Platelet aggregation tests also demonstrated no effect on platelet function related to high-dose lovastatin.

Currently, researchers have completed enrollment but have not completed collection of outcome data for a Phase 2 trial examining placebo *vs.* standard-dose lovastatin *vs.* 3 days of high-dose lovastatin (640 $mg\,day^{-1}$). Primary outcomes again will be hepatic or muscular toxicity defined as before, with secondary outcomes being clinical status at 90 days as measured with the NIHSS, BI, and mRS. Again, laboratory data on serum lipid levels and inflammatory markers will be collected (ClinicalTrials.gov identifier #NCT01976936).

14.2.4 Citalopram

Citalopram is a selective serotonin reuptake inhibitor (SSRI), a class of drugs used for the treatment of major depression and anxiety. They have been implicated in neuroprotection in animal models: a rat MCAO/reperfusion stroke model with fluoxetine administration within 6 hours of stroke had significantly reduced stroke volume compared to non-treated controls, with comparable improvement on motor, sensory, and balance tests.[36] Fluoxetine treatment in these animals was found to be associated with inhibited NF-κB activity both *in vivo* and in microglial cultures, ultimately repressing downstream expression of genes for TNF-α and COX-2, pointing to an anti-inflammatory mechanism of neuroprotection.

SSRIs are frequently used to treat post-stroke depression, and as such there were enough data for a large meta-analysis of post-stroke patients receiving these drugs. Mead *et al.* looked at 52 trials with 4060 patients receiving any SSRI at any dose after an acute ischemic stroke. A large degree of heterogeneity between studies led to the conclusion that SSRIs generally produced modest benefits with regard to dependency, disability, neurologic deficits, and anxiety, though there were not enough data to declare one SSRI superior or to make a statement regarding time to treat as it related to outcomes.[37]

TALOS is a multicenter, randomized, double-blind, placebo-controlled study assessing citalopram treatment *vs.* placebo in patients with acute ischemic stroke. Patients must be adults with their first time ever ischemic stroke and be randomized into a treatment group within 7 days of their stroke onset. Treatment arms will get 20 mg citalopram orally once per day or a comparable placebo. Follow-up will be by phone or in clinic at 2 weeks and 1, 3, and 6 months. Primary outcome measures are mRS at 6 months and vascular death. Secondary outcome measures are post-stroke depression, BI, and mini mental status exam scores. In total, 642 patients have been enrolled in the study. The study is closed for enrollment, but still tabulating follow-up data.[38]

14.2.5 Donepezil

Donepezil is a centrally acting, reversible acetylcholinesterase inhibitor, commonly used as a therapy in Alzheimer's disease patients to improve cognition and behavior under the premise that Alzheimer's disease is a progressive neurodegenerative disease resulting in the loss of central cholinergic neurons. Inhibiting the breakdown of acetylcholine (ACh) means there is more of this transmitter available, thus mitigating the symptoms (but not modifying the course) of the disease. Post-stroke dementia, aphasia, and cognitive deficits without dementia are common, and donepezil has been used to treat these conditions to some success.[39–41] It is known that pretreatment with donepezil can reduce glutamate-induced toxicity and subsequent apoptosis in cultured rat cortical neurons in a manner dependent on nicotinic ACh receptors.[42] In rats given a permanent suture occlusion of the middle cerebral artery, treatment with donepezil 2 hours before or 1 hour after showed marked improvement in the volume of infarcted tissue, and this benefit was eliminated with concurrent administration of a nicotinic ACh receptor blocker.[43] The Mayo Acute Stroke Trial for Enhancing Recovery (MASTER) published Phase IIa data in 2011 from their single-arm, multicenter clinical trial looking at adults treated within 24 hours of ischemic stroke onset with donepezil 5 mg day^{-1} for 30 days and then 10 mg day^{-1} for 60 days. The aim of this study was to ensure the safety, tolerability, and efficacy of this intervention in this patient population. The study enrolled 33 adults, none of which had any serious adverse events related to drug administration. Three patients discontinued use and three patients asked for de-escalation to 5 mg day^{-1} secondary to intolerable side effects. Fifteen patients had an NIHSS $\leq$1 at their 90 day follow-up. The investigators conclude that as donepezil was safe, tolerated well, and produced a percentage of patients with good outcomes greater than historic controls. It warrants further study in a large-scale investigation.[44,45]

14.2.6 Citicoline

Citicoline is an exogenous form of cytadine-5′-diphosphocholine, a precursor to phosphatidylcholine. This structural component of the plasma membrane is broken down into fatty acids and free radicals during an ischemic event. Citicoline is known to have neuroprotective qualities, acting as a free radical scavenger and a membrane stabilizer. It has demonstrated efficacy in animal studies, decreasing the volume of infarct in MCAO models compared to vehicle administration.[46] It was demonstrated to be safe and well tolerated, lending itself to being studied extensively as a potential neuroprotective adjunct to standard stroke therapy. Many initial trials produced conflicting or modest results with regard to the agent's efficacy,[47–49] though a meta-analysis of several double-blinded, randomized, placebo-controlled trials pooled 1372 patients receiving various doses of citicoline orally after moderate to severe stroke and evaluated patients based on a global outcome measure derived from NIHSS, GOS, BI, and mRS as suggested by the National Institute of Neurological Disease and Stroke (NINDS).[50] This meta-analysis showed that treatment with citicoline led to a statistically significant increase in the likelihood of a favorable outcome at 3 months of follow-up.[51]

The International Citicoline Trial on acUte Stroke (ICTUS) aimed to definitively settle the debate on citicoline's efficacy. This massive, international, double-blinded, randomized, placebo-controlled study enrolled 2298 patients over 5 years. Patients received 1000 mg intravenous citicoline twice a day for 3 days then two 500 mg tablets orally (or dissolved and per nasogastric tube) twice a day. Patients in the placebo arm received a comparable vehicle. The primary endpoint, as in the aforementioned meta-analysis, was the NINDS global test based on BI, NIHSS, and mRS. Secondary measures were outcome on individual assessment instruments, change in NIHSS from baseline at 3 months, and dichotomized mRS. Ultimately, the benefits brought about with citicoline in the meta-analysis were not re-demonstrated, as there were no differences in primary or secondary outcome measures between the groups. As a testament to citicoline's well-documented safety profile, there was also no difference in adverse events between the groups. The authors postulated that in the interim 10 years between the studies, best medical management for stroke had advanced to the point that any modest benefits seen with this drug were buffered and no longer statistically significant.

14.2.7 Tocotrienols

Tocotrienols are naturally occurring members of the vitamin E family along with tocopherols. There are four variants, and they can be found in palm oils and various other seeds and grains. They act as anti-oxidants, scavenging reactive oxidative species and free radicals. Along with this ability, this group of substances has been shown to inhibit glutamate-mediated cell death in cultured mouse hippocampal cells with administration up to 6 hours after insult by blocking c-Src kinase, an early species in the apoptosis cascade.[52] In human studies, 121 Malaysian volunteers with white matter hypodensities on screening brain magnetic resonance imaging (MRI) were divided into treatment and placebo groups. These patients were followed every 3 months for 2 years and had extensive laboratory tests performed, including serum cholesterol and triglycerides, fasting glucose, hepatic enzymes, and inflammatory markers. Follow-up MRIs were performed at 1 and 2 years and compared to baseline. The primary endpoint was change in white matter lesion volume or load on MRI; secondary endpoints were improvements in markers associated with cardiovascular risks (C-reactive protein, lipid profile, *etc.*). Interestingly, the treatment cohort had a nearly unchanged volume of white matter lesions, while the placebo group had a steady increase in overall load. At 2 years, the groups had a statistically significant difference radiographically. For laboratory values, there was no difference between groups. Adverse events were minimal, with five patients in the treatment group reporting mild self-limited diarrhea upon treatment initiation.[53]

Given these known neuroprotective qualities, their reassuring safety profile, and their mass availability, tocotrienols make an attractive candidate for use in acute ischemic stroke. Currently, the Stroke and Tocotrienol: Unique Role in Neuroprotection (SATURN) trial is enrolling patients that present between 1 and 10 days after a stroke for randomization to placebo *vs.* 200 mg tocotrienols orally twice a day, with an estimated completion date of December 2018 (ClinicalTrials.gov identifier #NCT02263924).

14.2.8 Cyclosporine

Cyclosporine is a lipophilic cyclic peptide drug originally isolated from fungi, used most commonly to treat immune-mediated diseases. The drug binds to the intracellular receptor cyclophilin and then binds to and inhibits calcineurin, a calcium-dependent phosphatase.[54] As a result, downstream transcription factors are inhibited and the expression of many inflammatory cytokines (such as interleukin-2) is decreased. Cyclosporine blocks mitochondrial permeability transition pores and inhibits cytochrome c release, thus inhibiting apoptotic pathways. As this mechanism is highly implicated in post-ischemia/reperfusion-associated programmed cell death within the penumbra, cyclosporine may have potential as a neuroprotective agent. Indeed, like many other candidate drugs, cyclosporine use has demonstrated histological and functional benefits in animal models of stroke.[55–58] It also has shown benefit in the setting of myocardial ischemia after coronary artery angioplasty and stenting, an analogous type of reperfusion injury in a different organ system.[59]

In the Cyclosporine in Acute Ischemic Stroke trial, Nighoghossian *et al.* looked at patients with acute ischemic stroke who were eligible for intravenous tPA thrombolysis (presenting within 4.5 hours, NIHSS between 6 and 18) and compared infarct size on MRI between a cohort treated with cyclosporine along with tPA with patients receiving a placebo vehicle at time of thrombolysis. Patient selection was randomized. An MRI with a T2 fluid-attenuated inversion recovery sequence was obtained immediately and again at between 2 and 6 weeks from ictus to measure the final stroke volume. This value was the primary endpoint of the study. Adverse events were defined as symptomatic hematoma, neurological decline (decrease in NIHSS of 4 points or more), or systemic complications. A total of 127 patients were randomized, and 110 met the primary endpoint. Eleven patients died (six in the treatment arm and five in the placebo group), two required pacemaker placement and could not get an MRI, and the rest withdrew from the study. The recanalization rate with tPA was comparable between groups (65% treatment *vs.* 59% placebo). Primary analysis showed a 21% reduction in the size of the infarct in the cyclosporine-treated patients on follow-up; however, this did not meet statistical significance. The clinical assessment at 1 month also showed no difference in the amount of functional independence between cohorts.[60]

14.2.9 Cutamesine (SA4503)

Cutamesine is an agonist of the σ-1 receptor, an endoplasmic reticulum protein that modulates calcium signaling and is implicated in inhibition of microglial activation through interference with calcium ion influx.[61] These receptors act as chaperones, stabilizing various cellular proteins, and as such they potentially provide neuroprotection at many different biochemical pathways. They may facilitate neurotrophic pathways, as activation with cutamesine has been linked to increased secretion of brain-derived neurotrophic factor in cultured neuroblastoma cells.[62] They may help prevent apoptosis by increasing B cell lymphoma-2 (Bcl-2) protein in the setting of oxidative stress.[63] Ligand binding to this receptor has been shown experimentally to reduce downstream inflammatory markers such as TNF-α and interleukin-10, and administration of an agonist of this receptor 24 hours after experimentally induced stroke in rats significantly diminished infarct size.[64] In 2014, Urfer *et al.* published their Phase II clinical trial data in which 60 patients with acute ischemic stroke with NIHSS $\geq$4 were randomized at between 48 and 72 hours from symptom onset to receive 1 mg cutamesine per day orally, 3 mg day^{-1}, or placebo for 28 days. The patients were followed until 56 days post-stroke, with the primary outcome measure being the safety and tolerability of cutamesine and the secondary outcome measures being changes in functional status as determined by assessments at baseline, 28 days, and 56 days. One patient in the 3 mg day^{-1} cohort died 6 days after randomization due to hemorrhagic conversion of the stroke, another had an intracerebral hemorrhage the day after randomization, and one had progression of the stroke. One patient in the 1 mg day^{-1} cohort had respiratory failure, and no serious adverse events were reported in the placebo group. These serious adverse events were not felt to be related to the study drug. Improvement in the NIHSS between assessments showed a dose-dependent trend toward better outcomes in the treatment groups; however, these data did not meet statistical significance. The study concluded that this drug is safe and tolerable in this patient population, and further study is needed to assess efficacy.[65]

14.2.10 Intra-arterial Administration of Neuroprotective Drugs as an Adjunct to Mechanical Thrombectomy—Verapamil and Beyond

Up to this point, no drug has met the rigorous scientific criteria for safety and efficacy as a neuroprotective agent in acute ischemic stroke. The previously described drugs are either still being studied or have met the same fate as dozens of other drugs examined over the last few decades: the drug works reliably and reproducibly in cultured cells or slice preparations and in animal models of stroke, but when the drug is examined for efficacy in human stroke patients, it shows no benefit over standard-of-care therapy. There have been many explanations posited for this. Humans, unlike laboratory animals, are a much more diverse test population. Lab animals have a high degree of genetic homogeneity, and animal models are not often tested with the varying constellation of comorbidities with which stroke patients present. Strokes in the laboratory are created in a controlled fashion, allowing for a more consistent stroke territory and a well-defined time of ictus. The experimental drugs can be given at a uniform time from stroke onset, whereas in human trials, initial treatment administration can vary from minutes to days. In a pathology where minutes can mean the difference to the survival of millions of neurons, this is significant.

In many animal models of stroke, an intracranial artery is obstructed for a predetermined amount of time and then the obstruction is removed, allowing for reperfusion. Until recently, that would have been a markedly different situation from the typical patient experience. Most patients are not candidates for pharmacological thrombolysis with tPA due to a narrow treatment window and several contraindications. Even in studies selecting for patients receiving thrombolytics like the Cyclosporine in Acute Ischemic Stroke trial above, only about 60% of the patients had a physiologic reperfusion, as tPA tends to be less efficacious in large vessel occlusions and cardiogenic strokes. A very large proportion of the patients treated in these studies have a vascular obstruction immediately proximal to the region of brain we are intending to treat. Combine the heterogeneous nature of stroke patient populations with the fact that even if there is a beneficial effect of the treatment drug, it may only be getting to the intended target in a fraction of the patients, and it is clear how a modest treatment effect may become diluted and appear insignificant.

The advent of routine mechanical thrombectomy may be an equalizing factor. Prior to 2015, mechanical thrombectomy was a niche intervention with narrow treatment windows, which, like tPA, meant it was not used in a majority of stroke patients. After several large prospective studies clearly demonstrated this intervention's superiority over conventional best medical care, thrombectomy has become the standard of care in patients with an emergent large vessel occlusion.[66–70] After performing the thrombectomy, the neurointerventionalist is left with a reperfused penumbra and a catheter in the parent vessel through which to directly administer a neuroprotective agent (or agents). This "equalizes" some of the patient population's heterogeneity. Bioavailability is less of a question as the drug is administered directly into the brain parenchyma downstream from the occlusion. By virtue of being a candidate for thrombectomy, the patients would all have emergent large vessel occlusion strokes with documentable reperfusion, and they would all be treated within a relatively narrow time window. There is minimal added mechanistic risk to drug administration, because the parent vessel is selected already for the delivery of standard-of-care treatment.

Verapamil is an L-type calcium channel blocker that makes an attractive candidate for neuroprotection trials with intra-arterial administration after thrombectomy. It has the same mechanism of action as previously trialed potentially neuroprotective agents, and it has a well-described history of safe intra-arterial delivery, as intra-arterial infusion of this drug is a currently used treatment for cerebral vasospasm.

Verapamil was tested on an animal model designed to simulate large vessel occlusion with subsequent recanalization, with intra-arterial drug administration immediately after. The mouse is sedated and its carotid bifurcation and middle cerebral artery are exposed. The distal external carotid artery if permanently ligated, while the internal carotid and middle cerebral arteries are clamped for an hour. The vessels are then unclamped, allowed to reperfuse for 5 minutes, then the study drug or a control vehicle is infused through an angio-catheter in the external carotid artery proximal to the ligation. The animals have their wounds sutured and are allowed to wake up, after which they are tested behaviorally for a week. Mice that received verapamil did significantly better on behavioral motor tests than the control group. After a week, the animals were sacrificed and histologic preparations to assess stroke volume showed a significantly smaller infarct in the treated group.[71] Furthermore, intra-arterial verapamil improved the survival of mature neurons in the target region and reduced cellular apoptosis. Finally, the experiments were blinded and randomized in order to mimic a clinical trial.

The Superselective Administration of Verapamil during Recanalization in Acute Ischemic Stroke (SAVER-I) trial is a recently completed Phase I trial demonstrating the safety and feasibility of intra-arterial verapamil administration after mechanical thrombectomy. The primary endpoint was symptomatic cerebral hemorrhage, with the secondary endpoint being mRS at 30 days. No patients suffered an intracerebral hemorrhage or died as a result of the procedure (ClinicalTrials.gov identifier #NCT02235558, article in press).

This study is the first Phase I trial to take post-thrombectomy intra-arterial neuroprotective drug testing from the laboratory to the angiography suite. A second trial is already proposed to give intra-arterial magnesium and verapamil in combination (MAVARIC study; ClinicalTrials.gov identifier #NCT02912663[72]). In addition, two other trials are planned to give intravenous drugs before/during thrombectomy.[73,74] This represents a potentially exciting time for the discipline of neuroprotection. If these studies are able to demonstrate clinical benefits in patients analogous to those seen in animal studies, then previously trialed drugs may have a new life in future studies, and intra-arterial administration of a neuroprotective drug or cocktail of drugs after thrombectomy may one day be a standard of care in acute stroke management.

References

1. Global Health Estimates 2015: Deaths by Cause, Age and Sex, Estimates for 2000–2015. Geneva: World Health Organization; 2015. Available from: http://www.who.int/healthinfo/global_burden_disease/en/. Accessed December 2016.
2. A. S. Go, D. Mozaffarian, V. L. Roger, E. J. Benjamin, J. D. Berry and M. J. Blaha, *et al.*, Heart disease and stroke statistics–2014 update: a report from the American Heart Association, *Circulation.*, 2014, **129**(3), e28.
3. National Institute of Neurological D. Stroke, rt PASSG. Tissue plasminogen activator for acute ischemic stroke, *N. Engl. J. Med.*, 1995, **333**(24), 1581.
4. A. M. Harper, Physiology of Cerebral Bloodflow, *Br. J Anaesth.*, 1965, **37**, 225.

5. B. C. White, J. M. Sullivan, D. J. DeGracia, B. J. O'Neil, R. W. Neumar and L. I. Grossman, *et al.*, Brain ischemia and reperfusion: molecular mechanisms of neuronal injury, *J. Neurol. Sci.*, 2000, **179**(S1–2), 1.
6. L. Nowak, P. Bregestovski, P. Ascher, A. Herbet and A. Prochiantz, Magnesium gates glutamate-activated channels in mouse central neurones, *Nature*, 1984, **307**(5950), 462.
7. D. A. Smith, J. H. Connick and T. W. Stone, Effect of changing extracellular levels of magnesium on spontaneous activity and glutamate release in the mouse neocortical slice, *Br. J. Pharmacol.*, 1989, **97**(2), 475.
8. M. Favaron and P. Bernardi, Tissue-specific modulation of the mitochondrial calcium uniporter by magnesium ions, *FEBS Lett.*, 1985, **183**(2), 260.
9. L. Dube and J. C. Granry, The therapeutic use of magnesium in anesthesiology, intensive care and emergency medicine: a review, *Can. J. Anaesth.*, 2003, **50**(7), 732.
10. Y. Yang, Q. Li, F. Ahmad and A. Shuaib, Survival and histological evaluation of therapeutic window of post-ischemia treatment with magnesium sulfate in embolic stroke model of rat, *Neurosci. Lett.*, 2000, **285**(2), 119.
11. M. B. Marinov, K. S. Harbaugh, P. J. Hoopes, H. J. Pikus and R. E. Harbaugh, Neuroprotective effects of preischemia intraarterial magnesium sulfate in reversible focal cerebral ischemia, *J. Neurosurg.*, 1996, **85**(1), 117.
12. Y. Izumi, S. Roussel, E. Pinard and J. Seylaz, Reduction of infarct volume by magnesium after middle cerebral artery occlusion in rats, *J. Cereb. Blood Flow Metab.*, 1991, **11**(6), 1025.
13. K. W. Muir and K. R. Lees, A randomized, double-blind, placebo-controlled pilot trial of intravenous magnesium sulfate in acute stroke, *Ann. N. Y. Acad. Sci.*, 1995, **765**, 315.
14. Y. Lampl, R. Gilad, D. Geva, Y. Eshel and M. Sadeh, Intravenous administration of magnesium sulfate in acute stroke: a randomized double-blind study, *Clin. Neuropharmacol.*, 2001, **24**(1), 11.
15. K. W. Muir, K. R. Lees, I. Ford and S. Davis, Intravenous Magnesium Efficacy in Stroke Study I. Magnesium for acute stroke (Intravenous Magnesium Efficacy in Stroke trial): randomised controlled trial, *Lancet*, 2004, **363**(9407), 439.
16. F. I. Mahoney and D. W. Barthel, Functional Evaluation: The Barthel Index, *Md. State Med. J.*, 1965, **14**, 61.
17. J. Rankin, Cerebral vascular accidents in patients over the age of 60. II. Prognosis, *Scott. Med. J.*, 1957, **2**(5), 200.
18. J. L. Saver, S. Starkman, M. Eckstein, S. J. Stratton, F. D. Pratt and S. Hamilton, *et al.*, Prehospital use of magnesium sulfate as neuroprotection in acute stroke, *N. Engl. J. Med.*, 2015, **372**(6), 528.
19. C. S. Kidwell, S. Starkman, M. Eckstein, K. Weems and J. L. Saver, Identifying stroke in the field. Prospective validation of the Los Angeles prehospital stroke screen (LAPSS), *Stroke*, 2000, **31**(1), 71.

20. N. Popovic, A. Schubart, B. D. Goetz, S. C. Zhang, C. Linington and I. D. Duncan, Inhibition of autoimmune encephalomyelitis by a tetracycline, *Ann. Neurol.*, 2002, **51**(2), 215.
21. V. Brundula, N. B. Rewcastle, L. M. Metz, C. C. Bernard and V. W. Yong, Targeting leukocyte MMPs and transmigration: minocycline as a potential therapy for multiple sclerosis, *Brain*, 2002, **125**(Pt 6), 1297.
22. Y. Du, Z. Ma, S. Lin, R. C. Dodel, F. Gao and K. R. Bales, *et al.*, Minocycline prevents nigrostriatal dopaminergic neurodegeneration in the MPTP model of Parkinson's disease, *Proc. Natl. Acad. Sci. U. S. A.*, 2001, **98**(25), 14669.
23. S. Zhu, I. G. Stavrovskaya, M. Drozda, B. Y. Kim, V. Ona and M. Li, *et al.*, Minocycline inhibits cytochrome c release and delays progression of amyotrophic lateral sclerosis in mice, *Nature*, 2002, **417**(6884), 74.
24. P. C. Wong, C. A. Pardo, D. R. Borchelt, M. K. Lee, N. G. Copeland and N. A. Jenkins, *et al.*, An adverse property of a familial ALS-linked SOD1 mutation causes motor neuron disease characterized by vacuolar degeneration of mitochondria, *Neuron*, 1995, **14**(6), 1105.
25. S. Lin, Y. Zhang, R. Dodel, M. R. Farlow, S. M. Paul and Y. Du, Minocycline blocks nitric oxide-induced neurotoxicity by inhibition p38 MAP kinase in rat cerebellar granule neurons, *Neurosci. Lett.*, 2001, **315**(1–2), 61.
26. M. E. Ryan, A. Usman, N. S. Ramamurthy, L. M. Golub and R. A. Greenwald, Excessive matrix metalloproteinase activity in diabetes: inhibition by tetracycline analogues with zinc reactivity, *Curr. Med. Chem.*, 2001, **8**(3), 305.
27. J. Yrjanheikki, T. Tikka, R. Keinanen, G. Goldsteins, P. H. Chan and J. Koistinaho, A tetracycline derivative, minocycline, reduces inflammation and protects against focal cerebral ischemia with a wide therapeutic window, *Proc. Natl. Acad. Sci. U. S. A.*, 1999, **96**(23), 13496.
28. Y. Lampl, M. Boaz, R. Gilad, M. Lorberboym, R. Dabby and A. Rapoport, *et al.*, Minocycline treatment in acute stroke: an open-label, evaluator-blinded study, *Neurology*, 2007, **69**(14), 1404.
29. S. C. Fagan, J. L. Waller, F. T. Nichols, D. J. Edwards, L. C. Pettigrew and W. M. Clark, *et al.*, Minocycline to improve neurologic outcome in stroke (MINOS): a dose-finding study, *Stroke*, 2010, **41**(10), 2283.
30. R. Singh, Neuroprotection With Minocycline Therapy for Acute Stroke Recovery Trial (NeuMAST) clinicaltrials.gov2012. Available from: https://clinicaltrials.gov/ct2/show/NCT00930020. Accessed December 2016.
31. O. Aktas, S. Waiczies, A. Smorodchenko, J. Dorr, B. Seeger and T. Prozorovski, *et al.*, Treatment of relapsing paralysis in experimental encephalomyelitis by targeting Th1 cells through atorvastatin, *J. Exp. Med.*, 2003, **197**(6), 725.
32. C. Berger, F. Xia, M. H. Maurer and S. Schwab, Neuroprotection by pravastatin in acute ischemic stroke in rats, *Brain Res. Rev.*, 2008, **58**(1), 48.

33. J. Chen, C. Zhang, H. Jiang, Y. Li, L. Zhang and A. Robin, *et al.*, Atorvastatin induction of VEGF and BDNF promotes brain plasticity after stroke in mice, *J. Cereb. Blood Flow Metab.*, 2005, **25**(2), 281.
34. A. Squizzato, E. Romualdi, F. Dentali and W. Ageno, Statins for acute ischemic stroke, *Cochrane Database Syst Rev.*, 2011, (8), CD007551.
35. M. S. Elkind, R. L. Sacco, R. B. Macarthur, E. Peerschke, G. Neils and H. Andrews, *et al.*, High-dose lovastatin for acute ischemic stroke: results of the phase I dose escalation neuroprotection with statin therapy for acute recovery trial (NeuSTART), *Cerebrovasc. Dis.*, 2009, **28**(3), 266.
36. C. M. Lim, S. W. Kim, J. Y. Park, C. Kim, S. H. Yoon and J. K. Lee, Fluoxetine affords robust neuroprotection in the postischemic brain via its anti-inflammatory effect, *J. Neurosci. Res.*, 2009, **87**(4), 1037.
37. G. E. Mead, C. F. Hsieh, R. Lee, M. A. Kutlubaev, A. Claxton and G. J. Hankey, *et al.*, Selective serotonin reuptake inhibitors (SSRIs) for stroke recovery, *Cochrane Database Syst. Rev.*, 2012, **11**, CD009286.
38. K. L. Kraglund, J. K. Mortensen, E. L. Grove, S. P. Johnsen and G. Andersen, TALOS: a multicenter, randomized, double-blind, placebo-controlled trial to test the effects of citalopram in patients with acute stroke, *Int. J. Stroke*, 2015, **10**(6), 985.
39. M. L. Berthier, C. Green, C. Higueras, I. Fernandez, J. Hinojosa and M. C. Martin, A randomized, placebo-controlled study of donepezil in poststroke aphasia, *Neurology*, 2006, **67**(9), 1687.
40. M. L. Berthier, J. Hinojosa, C. Martin Mdel and I. Fernandez, Open-label study of donepezil in chronic poststroke aphasia, *Neurology*, 2003, **60**(7), 1218.
41. E. M. Whyte, E. J. Lenze, M. Butters, E. Skidmore, K. Koenig and M. A. Dew, *et al.*, An open-label pilot study of acetylcholinesterase inhibitors to promote functional recovery in elderly cognitively impaired stroke patients, *Cerebrovasc. Dis.*, 2008, **26**(3), 317.
42. Y. Takada, A. Yonezawa, T. Kume, H. Katsuki, S. Kaneko and H. Sugimoto, *et al.*, Nicotinic acetylcholine receptor-mediated neuroprotection by donepezil against glutamate neurotoxicity in rat cortical neurons, *J. Pharmacol. Exp. Ther.*, 2003, **306**(2), 772.
43. M. Fujiki, H. Kobayashi, S. Uchida, R. Inoue and K. Ishii, Neuroprotective effect of donepezil, a nicotinic acetylcholine-receptor activator, on cerebral infarction in rats, *Brain Res.*, 2005, **1043**(1–2), 236.
44. J. F. Meschia, R. B. McNeil, K. M. Barrett, T. G. Brott, N. R. Graff-Radford and R. D. Brown Jr., Mayo Acute Stroke Trial for Enhancing Recovery (MASTER) protocol, *J. Stroke Cerebrovasc. Dis.*, 2010, **19**(4), 299.
45. K. M. Barrett, T. G. Brott, R. D. Brown Jr., R. E. Carter, J. R. Geske and N. R. Graff-Radford, *et al.*, Enhancing recovery after acute ischemic stroke with donepezil as an adjuvant therapy to standard medical care: results of a phase IIA clinical trial, *J. Stroke Cerebrovasc. Dis.*, 2011, **20**(3), 177.

46. A. Shuaib, Y. Yang and Q. Li, Evaluating the efficacy of citicoline in embolic ischemic stroke in rats: neuroprotective effects when used alone or in combination with urokinase, *Exp. Neurol.*, 2000, **161**(2), 733.
47. W. M. Clark, S. J. Warach, L. C. Pettigrew, R. E. Gammans and L. A. Sabounjian, A randomized dose-response trial of citicoline in acute ischemic stroke patients. Citicoline Stroke Study Group, *Neurology*, 1997, **49**(3), 671.
48. W. M. Clark, B. J. Williams, K. A. Selzer, R. M. Zweifler, L. A. Sabounjian and R. E. Gammans, A randomized efficacy trial of citicoline in patients with acute ischemic stroke, *Stroke*, 1999, **30**(12), 2592.
49. W. M. Clark, L. R. Wechsler, L. A. Sabounjian and U. E. Schwiderski, Citicoline Stroke Study G. A phase III randomized efficacy trial of 2000 mg citicoline inacute ischemic stroke patients, *Neurology*, 2001, **57**(9), 1595.
50. B. C. Tilley, J. Marler, N. L. Geller, M. Lu, J. Legler and T. Brott, *et al.*, Use of a global test for multiple outcomes in stroke trials with application to the National Institute of Neurological Disorders and Stroke t-PA Stroke Trial, *Stroke*, 1996, **27**(11), 2136.
51. A. Davalos, J. Castillo, J. Alvarez-Sabin, J. J. Secades, J. Mercadal and S. Lopez, *et al.*, Oral citicoline in acute ischemic stroke: an individual patient data pooling analysis of clinical trials, *Stroke*, 2002, **33**(12), 2850.
52. C. K. Sen, S. Khanna, S. Roy and L. Packer, Molecular basis of vitamin E action. Tocotrienol potently inhibits glutamate-induced pp60(c-Src) kinase activation and death of HT4 neuronal cells, *J. Biol. Chem.*, 2000, **275**(17), 13049.
53. Y. Gopalan, I. L. Shuaib, E. Magosso, M. A. Ansari, M. R. Abu Bakar and J. W. Wong, *et al.*, Clinical investigation of the protective effects of palm vitamin E tocotrienols on brain white matter, *Stroke*, 2014, **45**(5), 1422.
54. G. Wiederrecht, E. Lam, S. Hung, M. Martin and N. Sigal, The mechanism of action of FK-506 and cyclosporin A, *Ann. N. Y. Acad. Sci.*, 1993, **696**, 9.
55. H. Uchino, R. Minamikawa-Tachino, T. Kristian, G. Perkins, M. Narazaki and B. K. Siesjo, *et al.*, Differential neuroprotection by cyclosporin A and FK506 following ischemia corresponds with differing abilities to inhibit calcineurin and the mitochondrial permeability transition, *Neurobiol. Dis.*, 2002, **10**(3), 219.
56. H. Uchino, E. Elmer, K. Uchino, P. A. Li, Q. P. He and M. L. Smith, *et al.*, Amelioration by cyclosporin A of brain damage in transient forebrain ischemia in the rat, *Brain Res.*, 1998, **812**(1–2), 216.
57. T. Yoshimoto and B. K. Siesjo, Posttreatment with the immunosuppressant cyclosporin A in transient focal ischemia, *Brain Res.*, 1999, **839**(2), 283.
58. S. Matsumoto, A. Isshiki, Y. Watanabe and T. Wieloch, Restricted clinical efficacy of cyclosporin A on rat transient middle cerebral artery occlusion, *Life Sci.*, 2002, **72**(4–5), 591.

59. C. Piot, P. Croisille, P. Staat, H. Thibault, G. Rioufol and N. Mewton, *et al.*, Effect of cyclosporine on reperfusion injury in acute myocardial infarction, *N. Engl. J. Med.*, 2008, **359**(5), 473.
60. N. Nighoghossian, Y. Berthezene, L. Mechtouff, L. Derex, T. H. Cho and T. Ritzenthaler, *et al.*, Cyclosporine in acute ischemic stroke, *Neurology.*, 2015, **84**(22), 2216.
61. A. A. Hall, Y. Herrera, C. T. Ajmo Jr., J. Cuevas and K. R. Pennypacker, Sigma receptors suppress multiple aspects of microglial activation, *Glia*, 2009, **57**(7), 744.
62. M. Fujimoto, T. Hayashi, R. Urfer, S. Mita and T. P. Su, Sigma-1 receptor chaperones regulate the secretion of brain-derived neurotrophic factor, *Synapse*, 2012, **66**(7), 630.
63. J. Meunier and T. Hayashi, Sigma-1 receptors regulate Bcl-2 expression by reactive oxygen species-dependent transcriptional regulation of nuclear factor kappaB, *J. Pharmacol. Exp. Ther.*, 2010, **332**(2), 388.
64. C. T. Ajmo Jr., D. O. Vernon, L. Collier, K. R. Pennypacker and J. Cuevas, Sigma receptor activation reduces infarct size at 24 hours after permanent middle cerebral artery occlusion in rats, *Curr. Neurovasc. Res.*, 2006, **3**(2), 89.
65. R. Urfer, H. J. Moebius, D. Skoloudik, E. Santamarina, W. Sato and S. Mita, *et al.*, Phase II trial of the Sigma-1 receptor agonist cutamesine (SA4503) for recovery enhancement after acute ischemic stroke, *Stroke*, 2014, **45**(11), 3304.
66. O. A. Berkhemer, P. S. Fransen, D. Beumer, L. A. van den Berg, H. F. Lingsma and A. J. Yoo, *et al.*, A randomized trial of intraarterial treatment for acute ischemic stroke, *N. Engl. J. Med.*, 2015, **372**(1), 11.
67. M. Goyal, A. M. Demchuk, B. K. Menon, M. Eesa, J. L. Rempel and J. Thornton, *et al.*, Randomized assessment of rapid endovascular treatment of ischemic stroke, *N. Engl. J. Med.*, 2015, **372**(11), 1019.
68. J. L. Saver, M. Goyal, A. Bonafe, H. C. Diener, E. I. Levy and V. M. Pereira, *et al.*, Stent-retriever thrombectomy after intravenous t-PA vs. t-PA alone in stroke, *N. Engl. J. Med.*, 2015, **372**(24), 2285.
69. B. C. Campbell, P. J. Mitchell and E.-I. Investigators, Endovascular therapy for ischemic stroke, *N. Engl. J. Med.*, 2015, **372**(24), 2365.
70. T. G. Jovin, A. Chamorro, E. Cobo, M. A. de Miquel, C. A. Molina and A. Rovira, *et al.*, Thrombectomy within 8 hours after symptom onset in ischemic stroke, *N. Engl. J. Med.*, 2015, **372**(24), 2296.
71. M. E. Maniskas, J. M. Roberts, I. Aron, J. F. Fraser and G. J. Bix, Stroke neuroprotection revisited: Intra-arterial verapamil is profoundly neuroprotective in experimental acute ischemic stroke, *J. Cereb. Blood Flow Metab*, 2016, **36**(4), 721.
72. J. F. Fraser, Magnesium And Verapamil After Recanalization in Ischemia of the Cerebrum: a Clinical and Translational Study. (MAVARIC) clinicaltrials.gov2017. Available from: https://clinicaltrials.gov/ct2/show/NCT02912663. Accessed December 2016.

73. M. D. Hill, Safety and Efficacy of NA-1 in Subjects Undergoing Endovascular Thrombectomy forStroke (ESCAPE-NA1) (ESCAPE-NA1) 2017. Available from: https://clinicaltrials.gov/ct2/show/NCT02930018. Accessed December 2016.
74. J. M. Hong, Efficacy and Safety of Neu2000KWL in Treating Acute Ischemic Stroke Receiving Endovascular Therapy (ENIS I) clinicaltrials.gov. Available from: https://clinicaltrials.gov/ct2/show/NCT02831088. Accessed December 2016.
75. S. Grupke, J. Hall, M. Dobbs, G. J. Bix and J. F. Fraser, Understanding history, and not repeating it. Neuroprotection for acute ischemic stroke: from review to preview, *Clin. Neurol. Neurosurg.*, 2015, **129**, 1.

CHAPTER 15

Difficulties in Clinical Trials to Treat Traumatic Brain Injury and Stroke

ALFONSO LAGARES*[a] AND MARÍA DE TOLEDO[b]

[a] Department of Neurosurgery, Hospital 12 de Octubre de Madrid, Instituto de Investigación imas12, Universidad Complutense de Madrid, Madrid, Spain; [b] Department of Neurology, Hospital Universitario La Princesa, Madrid, Spain
*Email: algadoc@yahoo.com

15.1 Introduction

Traumatic brain injury (TBI) and stroke are the main causes of acquired brain damage and are major causes of death and disability.[1] Both can affect all ages, and though TBI used to be more frequent in younger patients than stroke, this is not the case any more in many developed countries.[2,3] The costs produced by both illnesses are enormous in our societies. Both illnesses are able to transform an active life into a completely dependent one, causing familiar disruptions, loss of activity and enormous and long-lasting suffering. The human and economic costs associated with these two illnesses have determined the search for a better and quicker treatment of their consequences. In recent years, there have been dramatic changes in the western world regarding better treatment for both illnesses. Major improvements in management have seen the development of better acute care, with the evolution of trauma care systems determining quicker and better pre-hospital care.[4] Implementation of code stroke focused on early

Drug Discovery Series No. 62
Acute Brain Impairment: Scientific Discoveries and Translational Research
Edited by Philip V. Peplow, Svetlana A. Dambinova, Thomas A. Gennarelli and Bridget Martinez

Published by the Royal Society of Chemistry, www.rsc.org

revascularization has dramatically changed the initial care of stroke, but even after all the efforts done to obtain early revascularization in the first few hours, many patients will still have an ischemic brain injury. The better knowledge acquired from both experimental and clinical data regarding the importance of time of action in decreasing the effects of brain injury, the sequence of events that brain injury triggers and the effect of the maintenance of systemic physiological parameters in order to prevent secondary insults have determined the change in the pre-hospital and acute hospital management of these conditions. There have been advances in neurocritical care and rehabilitation that have helped to improve outcomes. Both conditions have been extensively studied, their clinical management studied thoroughly and standardized with the help of the development of clinical guidelines that are updated regularly. The use of these guidelines has increased during recent decades, contributing to a more standardized treatment of these conditions throughout the developed world. Stroke treatment has been completely changed with the development of treatments and devices capable of restoring blood flow through brain arteries. In contrast to all these advances in the clinical management of these patients, and despite huge investment from governments and drug companies, no neuroprotective drug therapy has been translated from animal research to humans.

Scientists have revealed many pathologic molecular processes that cause progressive damage after brain injury caused by both illnesses, suggesting multiple potential drug targets and different therapeutic windows. Many of these have proven effective in animal models. However, there is a lack of translation of any compound found effective in animals into humans. Though there are differences between stroke and TBI, there are also similarities between them relating to the difficulties in developing and conducting a finally successful clinical trial. In this chapter, we will try to provide reasons for the failure of past trials and give some ideas on how to improve the design of future trials.

15.2 Why Translation Has Failed: Reasons for Failure in the Translation from Animal to Human Studies

The traditional steps taken in developing new treatments begin with the identification in the laboratory of a potentially implicated molecular mechanism. Then a candidate drug is obtained, developed or chosen from the existing pharmacology and tested in an animal model looking for efficacy, dosing and toxicology. After these tests, assumptions are made relating to the existence of an identical mechanism in the human, sometimes supported by some clinical evidence, mainly based in indirect data. Assumptions are also made relating to pharmacology, distribution and dosing in humans suffering from the condition. After these assumptions are made, the clinical stage of testing begins, with Phase I testing toxicology and Phase II and III for dose escalation, safety and efficacy. In all these phases, there can be mistaken assumptions that could lead to the final result of a lack of efficacy in ill humans of a drug tested with success in animals.

15.2.1 Animal Models

Although animal models for TBI have been established in different species, most TBI investigation in animals has been performed in rodents. This is due to several reasons including simplicity of surgery and management and reduced costs, but also ethical reasons. The use of rodents allows the possibility of using knockout animals where the gene of a protein being studied can be ablated, or transgenic mice where the studied protein is overexpressed. The same happens regarding stroke investigation in animals.

Several models have been developed both for stroke and TBI. However, there are several questions regarding their reliability and their translation to the actual human situation.

15.2.1.1 Are Animal Models Really Comparable to the Clinical Situation?

Animal models have been developed to induce TBI in animals.[5–7] These include static loading with slow forces gradually applied to the head, impact of an object or fluid to the head or sudden detention of the head during motion not necessarily requiring contact with an object. These models produce a form of primary lesion, which can be either focal or diffuse, depending on the model. In that way, they will not exactly reproduce human trauma, in which a combination of focal and diffuse injury is nearly always encountered. Also, the similarity of response of a lissencephalic rodent brain—which is completely different in shape, morphology and proportion of constituents (white/grey matter and cerebrospinal fluid) to that of a human brain—under a similar trauma is difficult to imagine. Furthermore, TBI models in rodents normally do not produce disturbances of the level of consciousness, something that frequently occurs in human TBI.

Regarding experimental design, stroke models pose a similar problem.[8,9] Though there are different models of complete occlusion, focal ischemia or embolic ischemia, most preclinical studies have focused on middle cerebral artery occlusion as the model of choice. However, most clinical trials on neuroprotection in stroke have admitted patients with different types and severities of ischemia. Also, the effect of reperfusion has not been well studied in animals. Most humans now undergo clinical reperfusion, and this has not yet been taken into account in animal testing.

There are differences not only regarding the mechanism or primary pathophysiology, but also in the subject charachteristics.[6,10] Gender has not always been taken into account, and both stroke and TBI show differences in incidence and outcome regarding gender in humans. Also, age and co-morbidities are important. Most patients with stroke and TBI are mature and therefore can have important comorbidities that could affect the mechanism of action of a compound or the outcome. These factors are diabetes, alcohol and drug consumption and the use of other concomitant medication. These factors are not normally taken into account when trying a new compound in an animal model.

On the other hand, stroke and especially TBI are not only illnesses of the brain, but for humans they are systemic processes in which fundamental biological variables are affected that are also important for the outcome. The inclusion of these variables and their monitoring should also be accomplished in preclinical testing.

Neurocritical care has changed the management of human TBI and stroke by the implementation of management algorithms that include neuromonitoring and the use of sedatives and paralytics and other drugs to improve cerebral perfusion pressure and to diminish intracranial pressure. All of these factors have very seldom been taken into account in animal testing.

15.2.1.2 Variability in Testing Among Labs

The use of a specific model can change dramatically from lab to lab. Though there are guidelines or clear descriptions regarding how to perform such tests, the actual effect on the animal can be different due to variability in the experimental setting and in the experimental methodology or due to the strain, weight or sex of the animals. Also, the management of blood pressure, temperature, anaesthetic agents or other drugs used during the experiment can be completely different from lab to lab, making comparisons of the results difficult.

15.2.1.3 Animal Testing Outcome Assessment

The analysis of the size of the lesion has been used most frequently to assess the damage produced by the model and also the response or benefit achieved with the use of a certain compound. There has been variability among different labs in the methodology used for assessing this damaged volume. There are intrinsic problems in stroke testing, as oedema can be a confounder in determining the real size of the infarction. Also, this is not comparable to the human setting, in which the outcome analysis after a trial is normally performed using a clinical scale. Infrequently, behavioural testing has been used for the assessment of the response to a drug. The methodology used has been also variable. Assessment of functional outcome in the rodent is time consuming and challenging due to the fact that rodents develop learned compensatory behaviours that could be identified as recovery.

15.2.2 Assumptions Made in Translation: Time of Administration, Mechanism, Dosing and Publication Bias

Most drugs tested in animals have been used before or a short time after the brain insult.[9,11] The time window has varied in different clinical trials, most probably due to the fact that it is impossible to determine which patients will suffer a stroke or TBI, and there are also ethical considerations regarding the administration of the drug before informed consent is achieved. This is especially important in TBI and stroke clinical trials in which consent has to be obtained by a proxy, due to the frequently altered state of consciousness that patients present. This can lead to delays of hours until consent is obtained.

Most clinical trials in humans regarding neuroprotection have been conducted with only limited knowledge about the possible implication in humans of the target mechanism that was previously effective in animals. Also, most clinical trials have been conducted to assess drugs with effects limited to a simple or very few possible targeting mechanisms (single bullet strategy). Most neuroprotectors are based on their action on neurons and grey matter, leaving white matter without protection. However, most strokes and TBI affect white matter as well as grey matter, and in similar proportions.[12] Therefore, white matter and the mechanisms involved in the damage of glia and axons should also be targeted.

Most clinical trials have been conducted without strong evidence of the pharmacology of the drug in humans, the availability of the drug in brain tissue or the actual mechanisms or the effects on them in humans. Dosing has sometimes been insufficiently calculated. A good example of dosing problems is the failed clinical trial on progesterone in human TBI.[13,14] Also, the nature of the dose–response curve has often been assumed to have a sigmoidal nature, in which pushing the dose higher, up to the point when the response tends to level off, can enhance the response to a drug. However, numerous investigators have reported a U-shaped dose–response curve in different potentially neuroprotective compounds, in which there is an optimal response zone where protection is maximized, with less protection at lower and higher doses, and toxicity increasing in the latter.[15]

Negative results are very infrequently published. This could have led to the selection of drugs to be tested based only in positive information, thereby failing to show the other circumstances, such as type of head injury or stroke model, animal management or animal sex or age, in which the drug was not as effective or clearly ineffective. This publication bias could have led to the wrong selection of patients in the inclusion criteria of clinical trials.

15.2.3 Clinical Trials and Their Problems

The failure of clinical trials to prove benefit in humans of drugs already showing promising results in animals is not only restricted to the field of brain injury (ischemic or traumatic), but also includes many other fields, like sepsis, cancer and cardiology. These failures determine a huge loss of resources both for governments and pharmaceutical industries. However, while there have been several reviews regarding the need to take actions to prevent these failures, little has changed in the design of new clinical trials to improve their results.[13,14,16–22] In this section, we aim to discuss some of the elements in which improvement of design and conduct of clinical trials in humans could be focused in order to improve their results.

15.2.3.1 *Inclusion and Exclusion Criteria: Heterogeneity of the Chosen Samples*

Human illnesses are highly heterogeneous. As discussed previously, most evidence from preclinical animal trials has been based on the effectiveness of a drug in a certain model of stroke or TBI. However, the mechanisms, the timing of those mechanisms and the heterogeneity of the patients who present these illnesses regarding age, genetics and comorbidities are huge obstacles to establishing the effectiveness of a new drug.

Inclusion criteria regarding head injury have been too general in the clinical trials performed so far.[4,23] In general, inclusion policies have determined the inclusion of patients suffering very different kinds of trauma, including different mechanisms, the presence or absence of secondary insults and the presence or absence of significant systemic trauma. In general, inclusion in clinical trials of TBI has been determined by its severity, as measured by means of the Glasgow Coma Scale (GCS), the presence of intracranial pathology in the initial computed tomography (CT) or the need for intracranial pressure monitoring. Patients admitted with too severe head injuries in which survival for more than 24 hours was deemed improbable by the investigators have been excluded. Nevertheless, these kinds of inclusion criteria determine the inclusion of a very heterogeneous sample of patients, in terms of the kind of lesion, focal or diffuse, location of the lesions and prognosis. In closed head injury, it is quite difficult to ascertain the type and extent of injury at the diagnosis with the techniques used for the general management of the patients. The GCS is no longer a reliable indicator of injury severity, as there are many factors that prevent a precise estimation of the GCS in most TBI cases nowadays, including sedation or paralysis used as a therapeutic measure during prehospital management. Detailed screenings to eliminate fluctuations in GCS scores during the inclusion period are left to the criteria of the investigator, making the decision to enrol a patient based on the GCS difficult and sometimes arbitrary. A non-contrast CT scan of the head is the imaging modality used to determine the extent of intracranial damage in TBI patients. However, CT is only capable of determining the presence or absence of focal damage, but fails to identify lesions related to diffuse injury in many instances. The combination of imaging techniques (magnetic resonance [MR] and CT) could help to determine the location and extent of traumatic injury. This will incur costs of time and delaying inclusion, hampering the effect of drugs with a short time window of action.

A critical confounding factor in TBI studies is the presence of systemic injuries. Head trauma is infrequently isolated and therefore the presence of severe traumatic injuries to other organs or severe orthopaedic trauma can affect the final outcome analysis.[24] The refinement of inclusion criteria so as not to include severe systemic injury in TBI trials could, however, determine a very restricted sample and make recruitment difficult and slow. Covariate adjustment is another strategy for dealing with this problem.

The change in the management of stroke has led to a major change in the inclusion of patients in clinical trials. Although most drugs have been tested in animal occlusion models followed by reperfusion, most patients nowadays obtain some form of reperfusion, which is heterogeneous, because of natural reperfusion and treatments. Also, collateral circulation is variable among subjects, leading to different levels of susceptibility to ischemic damage during occlusion. This has to be taken into account in order to select those patients with more similarities to what the drug is intended for. Penumbra imaging (with MR or perfusion CT) could be a clue for detecting those patients to be included in neuroprotection studies.

15.2.3.2 Management Variability among Including Centres

Although both TBI and stroke management approaches have improved due to the widespread use of guidelines and management algorithms, there are still many differences in the management of patients between centres and countries.[3,25] This has been corroborated by many different studies, but especially by the IMPACT investigators regarding TBI.[4,23,26] In these studies, there was a 3.3-fold difference in the odds of having unfavourable outcome at 6 months between very good and very poor centres. In the CRASH studies, also in the TBI area, differences reached 6.6-fold between centres and 15-fold between countries regarding 14 day mortality.[27]

Many clinical trials did not include specific guidelines regarding the management of specific outcome-related parameters. In the TBI world, most of them referred to clinical guidelines when directions regarding the management of important variables such as intracranial pressure were addressed. This has recently been acknowledged in ongoing clinical trials, and treatment harmonization guidelines have been developed for specific trials.

15.2.3.3 Recruitment Problems

Many clinical trials include patients more slowly than expected, making inclusion times longer than those that were planned at the beginning of the trial. This not only makes the trial more expensive, as time is money for sponsors, but it can also cause problems regarding the quality of the performance of the trial. Slow recruitment can be caused by many factors such as the patient pool (how prevalent or frequent is the disease and the inclusion/exclusion criteria), the attractiveness of the trial and the adequacy of the network of the participating clinical sites. When a clinical site is recruiting patients at a much slower pace than expected, it raises doubts regarding the quality of the trial being conducted at that specific site. This could be due to competing clinical trials, which can bias the sample obtained for the different clinical trials, fatigue of investigators or a lack of commitment. It is therefore imperative to analyse the screening logs of clinical trials and to include their analysis as supplementary material of the clinical trial report.[28] Also, centres not following the projected inclusion pace at the beginning of the trial should be thoroughly audited in order to clarify whether inclusion/exclusion criteria were met and representativeness of the sample was achieved.

15.2.3.4 Pitfalls During Data Handling

The consistency and quality of the data reported in multi-centre clinical trials, the frequency of data entry errors, specifically in each participating institution, and the techniques used to clarify data and maintain the database are of critical importance.[4] The use of ambiguous criteria to define adverse events or variables related to outcome can have dramatic effects on the final results of trials. It is imperative that there are strict criteria in the definition and recording of the different clinical definitions and diagnoses regarding investigation variables. In the field of TBI, there has been a major initiative to try to obtain a consensus and to define a basic set of core variables to be collected in all TBI studies, with agreement of appropriate definitions, field names and coding. This international initiative has produced a set of core variables to be recorded, with rigorous definitions on them and how to record them, under the name of common data elements.[26,29] This initiative will certainly improve data management and consistency between centres and countries.

15.2.3.5 Outcome Measures

Although mortality has been used in many clinical trials as a final outcome measure, as it is clear and unambiguous, it is too blunt an outcome measure for clinical trials in stroke and TBI, as survival of brain-injured patients in a severe condition is possible and is sometimes of no benefit to families, society or the patients themselves.[4,17,24,30,31] Most studies have used different clinical scales, describing outcomes in an ordered fashion, from good to poor and death, based on descriptions of patients' needs for their daily living and capabilities in terms of mobility, social life and work. The most frequently used scale is the Glasgow Outcome Scale (GOS). This has been further refined with the description of the Extended GOS (GOS-E), which can be applied by a structured interview. Although efforts have been made to improve GOS and GOS-E evaluation in terms of reproducibility (inter-rater, inter-centre and inter-country reliability) by courses and seminars given to the investigators of clinical trials, there are still concerns regarding the final subjective nature of this scale and the lack of quantitative evaluation of deficits or recovery. The GOS and GOS-E have advantages such as they are inexpensive, they are easy to use, they can be administered by any kind of personnel and need no technology for their administration and the categories described are relevant clinically and to society. However, they: (1) do not describe the quality of life; (2) do not quantitate the deficits patients present with; and (3) have low sensitivity to changes in outcome. To improve the assessment, more detailed evaluation should be obtained. However, the use of extensive neuropsychological testing is much more demanding, as it requires the services of a skilled professional, is time consuming and demanding for patients with disabilities and is hard to interpret, as well as expensive to administer. Therefore, most clinical trials have finally included as a compromise different outcome measures, including the GOS or GOS-E, some quality of life scale and some quantitative description of disability or deficit. This multidimensional approach has been implemented in TBI research only recently.[29,32]

Another issue regarding outcome measures is the time lapse from the brain injury to when outcome is measured. For most clinical trials, this has been performed after 6–12 months of the injury. However, this time lapse might not be enough to differentiate spontaneous recovery from the effect of the drug being tested, and for some patients, especially those suffering from the most serious damage, this might not be time enough to experience recovery of any form, especially when this time is just 6 months. There is a pending debate on how to achieve a compromise between the best timing for outcome analysis and costs of the trial, and also if there should be multiple time testing or just a final one.

15.2.3.6 Outcome Analysis

Not only is the way in which outcome is measured important, but the planned statistical analysis is also critical.[4,33,34] Most studies so far have condensed the results from outcome measures, which have several ordered outcome categories, in a reductionist approach allocating patients into two broad categories of good and poor outcome, referred to as a dichotomized outcome study. This strategy simplifies the interpretation of the findings. However, this approach reduces the statistical power of the trial, as many patients will have very little chance of changing outcome category due to their very poor prognosis owing to their clinical characteristics at baseline. Also, this approach does not show which benefit a patient has experienced when moving to a better outcome inside each category. For instance, moving from moderate disability to good recovery in the good outcome group could be an important achievement for the patient, family and society, which will not be recognized by this strategy.

There are two possible solutions to this matter. One of them is using the sliding dichotomy strategy, in which the definition of good or poor outcome is tailored based on each individual's baseline prognosis on entry into the trial. With this strategy, surviving once a patient has entered the trial with a very poor prognosis could be defined as a good outcome. This strategy relies on information given by prognostic models, which are still developing and are still not completely validated. In the TBI field, the IMPACT prognostic model is now being used in this strategy and has been validated for different countries.[24,35,36] Therefore, it could be a valid strategy. The other strategy is using different tools for analysis that try not to lose information given by scales showing different categories in outcome, such as ordinal regression analysis (using the proportional odds assumption) or multinomial regression analysis. This other strategy, however, presents the drawback of difficulty in interpreting and managing results.

15.3 Conclusions: Recommendations to Improve Translation

We have presented several difficulties that clinical trials on brain injury caused by stroke or TBI present. Many different groups and international consortiums have acknowledged these difficulties, and recommendations to improve the results from experimental and clinical trials have been suggested.

To improve the likelihood of translation of experimental neuroprotective agents, the Stroke Therapy Academic Industry Roundtable (STAIR) made several recommendations (Table 15.1),[37,38] which have demonstrated, when followed, some effect on improving therapeutic efficacy.[39]

In the case of TBI, different groups have suggested similar recommendations, as many of the STAIR recommendations could be applied to TBI. There is increasing evidence that, as in stroke, preclinical testing should include: (1) different ages and sexes; (2) multiple TBI models, severities and species; (3) extensive pharmacokinetic profiles of the drug and brain concentrations associated with treatment efficacy; (4) relevant physiological monitoring during and after TBI surgery and treatment; (5) testing with drug administration closer to the real drug administration in humans; (6) multiple outcome analysis (histological, functional and behavioural); and (7) replication of the results in different laboratories.

In terms of actual clinical trial design, the following recommendations, also based on the recommendations of different authors, can be made:

1. The inclusion criteria should be as broad as is compatible with the mechanism of action of the intervention being evaluated. This will increase recruitment rates and increase generalization of the results.
2. There should be use of standards for data collection. In TBI, the use of common data elements is currently required for all TBI trials funded by the NIH-NINDS and European Commission. This will help data

Table 15.1 Recommendations for preclinical studies of potentially neuroprotective drugs, adapted from the Stroke Therapy Academic Industry Roundtable.[37,38]

Adequate dose–response curve
Minimum effective and maximum tolerated dose should be defined.
There should be a target concentration, a tissue level of effect identified.
It should be documented that the drug in these ranges accesses the target organ.
Define the time window in a well-characterized model
The effectiveness of the drug should be determined at different times of opportunity.
Penumbral imaging may be useful to guide the identification of the therapeutic window in stroke.
Physiologically controlled reproducible studies
Basic physiological parameters such as blood pressure, temperature, blood gases and blood glucose should be routinely monitored.
Cerebral blood flow to document adequate sustained occlusion and monitor reperfusion is advocated in stroke models.
Histological and functional outcomes assessed acutely and long term
Multiple endpoints are fundamental.
Include relevant biomarker endpoints that can also be tested in humans.
Initial rodent studies, then consider gyrencephalic species
It is suggested that treatment efficacy should be established in at least two species, using both histological and behavioural outcome measures.
Reproducibility
Results obtained should be independently replicated in at least two laboratories.
Fundamentals of good scientific enquiry should be applied to preclinical testing
Randomization and sample size calculation description.
Inclusion/exclusion criteria. These criteria should be applied before the allocation to experimental groups.
Allocation concealment.
Reporting of animals excluded from the analysis.
Blinded outcome assessment.
Reporting potential conflicts of interest and study funding.
Include efficacy studies in different sexes and ages and animals with comorbidities (hypertension or diabetes) and drug interactions

collection, data handling, reproducibility and comparability among studies and generalization of the results.

3. Management of patients should be harmonized along the different clinical sites.
4. Screening logs should be used and included in the results of the trial in order to determine the representativeness of the sample and generalizability of the results.
5. Recruitment rates should be monitored and centres and investigators should be encouraged to be committed to the trial.
6. Outcome measures used should include different dimensions, including functional measures and quality of life.
7. Outcome analysis should go beyond dichotomized outcomes in order to maintain statistical power. The use of ordinal analysis is

recommended, in which covariate adjustment could be used in order to reduce the effect of heterogeneity. When prognostic models are available, an analysis based on the sliding dichotomy principle could be used.

References

1. J. A. Langlois, W. Rutland-Brown and M. M. Wald, *J. Head Trauma Rehabil.*, 2006, **21**, 375.
2. P. A. Gomez, A. M. Castano-Leon, J. de-la-Cruz, D. Lora and A. Lagares, *Neurocirugia*, 2014, **25**, 199.
3. A. I. Maas, D. K. Menon, E. W. Steyerberg, G. Citerio, F. Lecky, G. T. Manley, S. Hill, V. Legrand, A. Sorgner, C.-T. Participants and Investigators, *Neurosurgery*, 2015, **76**, 67.
4. A. I. Maas, E. W. Steyerberg, A. Marmarou, G. S. McHugh, H. F. Lingsma, I. Butcher, J. Lu, J. Weir, B. Roozenbeek and G. D. Murray, *Neurotherapeutics*, 2010, **7**, 127.
5. T. L. Briones, *Annu. Rev. Nurs. Res.*, 2015, **33**, 31.
6. M. C. Morganti-Kossmann, E. Yan and N. Bye, *Injury*, 2010, **41**(Suppl. 1), S10.
7. Y. Xiong, A. Mahmood and M. Chopp, *Nat. Rev. Neurosci.*, 2013, **14**, 128.
8. I. M. Macrae, *Br. J. Pharmacol.*, 2011, **164**, 1062.
9. S. Y. Xu and S. Y. Pan, *Med. Sci. Monit. Basic Res.*, 2013, **19**, 37.
10. T. Janowitz and D. K. Menon, *Sci. Transl. Med.*, 2010, **2**, 27rv1.
11. N. Marklund and L. Hillered, *Br. J. Pharm.*, 2011, **164**, 1207.
12. P. W. Ho, D. C. Reutens, T. G. Phan, P. M. Wright, R. Markus, I. Indra, D. Young and G. A. Donnan, *Stroke*, 2005, **36**, 2742.
13. M. Schumacher, C. Denier, J. P. Oudinet, D. Adams and R. Guennoun, *J. Steroid Biochem. Mol. Biol.*, 2016, **160**, 53.
14. D. G. Stein, *Brain Inj.*, 2015, **29**, 1259.
15. E. J. Calabrese, *Crit. Rev. Toxicol.*, 2008, **38**, 557.
16. Editorial, *Lancet*, 2006, **368**, 1548.
17. D. J. Cook and M. Tymianski, *Expert Rev. Cardiovasc. Ther.*, 2011, **9**, 433.
18. P. Jakala and J. Jolkkonen, *Expert Opin. Biol. Ther.*, 2012, **12**, 267.
19. D. J. Loane and A. I. Faden, *Trends Pharm. Sci.*, 2010, **31**, 596.
20. D. K. Menon, *Crit. Care Med.*, 2009, **37**, S129.
21. S. Roth and A. Liesz, *Swiss Med. Wkly.*, 2016, **146**, w14329.
22. R. C. Turner, B. Lucke-Wold, N. Lucke-Wold, A. S. Elliott, A. F. Logsdon, C. L. Rosen and J. D. Huber, *Int. J. Mol. Sci.*, 2013, **14**, 1890.
23. B. Roozenbeek, A. I. Maas, A. Marmarou, I. Butcher, H. F. Lingsma, J. Lu, G. S. McHugh, G. D. Murray and E. W. Steyerberg, *J. Neurotrauma*, 2009, **26**, 1069.
24. H. Lingsma, T. M. Andriessen, I. Haitsema, J. Horn, J. van der Naalt, G. Franschman, A. I. Maas, P. E. Vos and E. W. Steyerberg, *J. Trauma Acute Care Surg.*, 2013, **74**, 639.
25. M. C. Cnossen, S. Polinder, H. F. Lingsma, A. I. Maas, D. Menon, E. W. Steyerberg, C.-T. Investigators and Participants, *PLoS One*, 2016, **11**, e0161367.

26. B. Roozenbeek, A. I. Maas, H. F. Lingsma, I. Butcher, J. Lu, A. Marmarou, G. S. McHugh, J. Weir, G. D. Murray and E. W. Steyerberg, *Crit. Care Med.*, 2009, **37**, 2683.
27. H. F. Lingsma, B. Roozenbeek, B. Li, J. Lu, J. Weir, I. Butcher, A. Marmarou, G. D. Murray, A. I. Maas and E. W. Steyerberg, *Neurosurgery*, 2011, **68**, 601.
28. F. J. Slieker, E. J. Kompanje, G. D. Murray, J. Ohman, N. Stocchetti, S. G. Teasdale and A. I. Maas, *Neurosurgery*, 2008, **62**, 1321.
29. J. K. Yue, M. J. Vassar, H. F. Lingsma, S. R. Cooper, D. O. Okonkwo, A. B. Valadka, W. A. Gordon, A. I. Maas, P. Mukherjee, E. L. Yuh, A. M. Puccio, D. M. Schnyer and G. T. Manley, *J. Neurotrauma*, 2013, **30**, 1831.
30. G. W. Hawryluk and M. R. Bullock, *Neurosurg. Clin. North Am.*, 2016, **27**, 375.
31. M. Sherer, T. Roebuck-Spencer and L. C. Davis, *J. Head Trauma Rehabil.*, 2010, **25**, 92.
32. J. F. Stover, A. Belli, H. Boret, D. Bulters, J. Sahuquillo, E. Schmutzhard, E. Zavala, U. Ungerstedt, R. Schinzel, F. Tegtmeier and N. Investigators, *J. Neurotrauma*, 2014, **31**, 1599.
33. A. I. Maas, G. D. Murray, B. Roozenbeek, H. F. Lingsma, I. Butcher, G. S. McHugh, J. Weir, J. Lu, E. W. Steyerberg and G. International Mission on Prognosis Analysis of Clinical Trials in Traumatic Brain Injury Study, *Lancet Neurol.*, 2013, **12**, 1200.
34. G. S. McHugh, I. Butcher, E. W. Steyerberg, A. Marmarou, J. Lu, H. F. Lingsma, J. Weir, A. I. Maas and G. D. Murray, *Clin. Trials*, 2010, **7**, 44.
35. A. I. Maas, H. F. Lingsma and I. S. Group, *Acta Neurochirgica*, 2008, **101**(Suppl.), 119.
36. A. M. Castano-Leon, D. Lora, P. M. Munarriz, S. Cepeda, I. Paredes, J. de la Cruz, P. A. Gomez Lopez and A. Lagares, *J. Neurotrauma*, 2016, **33**, 1598.
37. R. Stroke, Therapy Academic Industry, *Stroke*, 1999, **30**, 2752.
38. M. Fisher, G. Feuerstein, D. W. Howells, P. D. Hurn, T. A. Kent, S. I. Savitz, E. H. Lo and S. Group, *Stroke*, 2009, **40**, 2244.
39. S. I. Savitz and M. Fisher, *Ann. Neurol.*, 2007, **61**, 396.

CHAPTER 16

Challenges in Using Biomarkers in Central Nervous System Applications

SVETLANA A. DAMBINOVA,*[a] JOSEPH D. WEISSMAN[a] AND JOHN D. MULLINS[b]

[a] DeKalb Medical, Decatur, GA, USA; [b] Piedmont Hospital, Atlanta, GA, USA
*Email: dambinova@aol.com

16.1 Introduction

Molecular biomarkers offer additional information in the differential diagnosis of neurological conditions in general, and the potential rapid response time of these markers has particular utility in acute neurological emergencies.

Molecular biomarkers derive from particular biochemical reactions and must be chosen to reflect key changes in the state of the condition of interest. For neurological trauma, this might include cerebral blood flow and measures of gray and white matter dysfunction.[1] From these specific interactions, clinical research needs to determine correlations with diagnostic accuracy, prognosis, and monitoring treatment.

An important requirement is that biomarkers reflect the severity of dysfunction and correlate to or at least have a defined relationship with acute, subacute, and delayed consequences. In the case of central nervous system (CNS) trauma, this would enable the clinician to intercept the chain of

Drug Discovery Series No. 62
Acute Brain Impairment: Scientific Discoveries and Translational Research
Edited by Philip V. Peplow, Svetlana A. Dambinova, Thomas A. Gennarelli and Bridget Martinez

Published by the Royal Society of Chemistry, www.rsc.org

adverse events, such as raised intracranial pressure, edema formation, or excitotoxicity.[2]

To date, there is no FDA-approved biomarker for traumatic brain injury (TBI), transient ischemic attack (TIA), or stroke available. There are a number of candidates, some commercially available but not FDA approved, and many are in research or undergoing clinical trials (see Chapter 8). However, no current biomarker has been able to predict specific anatomical foci or replace neuroimaging as the gold standard for evaluating neurotrauma and cerebrovascular events.

The present chapter highlights crucial aspects of brain biomarker design, selection, and clinical evaluation. Limitations in sample preparations, storage issues, and testing drawbacks are considered as well.

16.2 Biases and Pitfalls in Choice of Biomarkers

16.2.1 Peculiarities in Primary Biomarker Testing

A majority of current research is focused on molecular genetics biomarkers (DNA, single-nucleotide polymorphisms, and mRNA expression profiles) for various conditions that identify transcriptional patterns with a diagnostic potential.[3,4] Whereas genetic biomarkers are extremely successful for primarily genetic conditions, they may not detect the actual onset and progression of pathological processes, but rather the risk of developing a genetically determined condition.[5] mRNA expression profiles are more closely related to the actual acute pathological processes, but may lag in acute events.

Another approach that differs from genomics—neurodegradomics—refers to the proteolytic breakdown of brain proteins to peptide fragments and focuses primarily on studying the "fate" of these cleaved products[6] and their link to pathological processes. In this case, peptide or protein biomarkers may be a step closer to the pathological process.

It is important to choose biomarkers that reflect the most acute and proximal molecular processes involved in the pathology in question. In the case of neurotrauma, this includes neurotoxicity, oxidative stress, and ischemia at the cellular level. These processes are activated within 24 hours of trauma. Related subacute processes include changes in neuroplasticity and neuroimmune processes.[7] Along these lines of thinking, there are two initial processes involving cell surface molecules involved in neurological trauma.

Use-dependent turnover of cell surface proteins is a post-translational process that releases specific and immunologically active peptide fragments into the extracellular space (through a disrupted blood–brain barrier) and into the bloodstream, generating autoantibodies that can be detected directly.[6] In acute ischemic stroke, neurotoxicity and apoptosis results in the release of *N*-terminal fragments of *N*-methyl-D-aspartate receptors (NMDARs).[8] The C-terminal NMDAR peptide sequence domain is intracellular and is an initial component of the chain of events that results in necrosis after TBI.[9,10]

One of the earliest neurovascular unit consequences in acute concussion injury is the neurotoxic cascade around small subcortical blood vessels (see Chapter 9). This results in a heterogeneous mixture of peptides, including proteolytic products derived from the extracellular N-terminal domain of α-amino-3-hydroxy-5-methyl-4-isoxazolepropionic acid receptors (AMPARs).[11–14] Similarly, the N-terminal domain of kainite receptor biomarkers are associated with impairment of venous function in the vertebral/basilar distribution (see Chapter 11).

AMPAR peptides produced by physical impact to the brain can be detected in plasma within the first hour after a concussion.[11] Multiple concussions not only increase the amounts of AMPAR peptide, but these also circulate in the bloodstream for extended periods (from months to a year). This is in contrast to NMDAR peptides released after TIA/stroke that have a short lifespan in plasma samples.[15,16] The different longevities of glutamate receptor peptides are probably due to the distinct origins of the inciting events. Further research will be required to clarify this situation.

16.2.2 Selection of Indicators for Secondary Injuries

Metabolomics focuses on finding differences in metabolic profiles in serum, cerebral spinal fluid (CSF), or plasma metabolites involved in intermediate metabolism and in the synthesis and turnover of non-protein and non-nucleic acid molecules. This often involves comparison of multidimensional profiles from patients with various conditions (in this case, brain injury) and appropriately matched controls.[17] This approach provides a wide range of non-specific variables that can be correlated with a pathological profile. There are several limitations in metabolomics translational research: the analytical procedures are relatively time consuming and costly for practical application, and they show a low specificity for a given condition at present. Data from standardized measurements of metabolic profiles for disease (stroke, TBI, *etc.*) *vs.* health are collected, analyzed, and biologically interpreted using translational bioinformatics.[18]

Allard *et al.*[19] have studied the molecular genetics of common cerebrovascular diseases and found that the genome has an effect on whether traditional risk factors for hypertension or diabetes are manifested and/or can complicate these conditions. Matarin *et al.*[20] predicted that genome-wide studies will result in mapping genes with major and minor effects on stroke susceptibility, allowing a definitive genetic map to be drawn. Understanding the molecular pathogenesis of stroke and TBI will present an opportunity for a third generation of biomarkers that will allow more accurate diagnosis and triaging for treatment.[5,21,22]

Proteomics has defined numerous biomarkers of neuroplasticity, implementing pivotal glial, neuronal, and immune processes.[12,23] Most of these biomarkers are associated with secondary or delayed (beyond 24 hours after injury or onset) events and may reflect *de novo* synthesis, implying a time delay. Other proteomic biomarkers like S100B, neuron-specific enolase (NSE), matrix metallopeptidase 9, D-dimer, and tau protein are present in biological fluids due to cell matrix degradation, delayed ischemia, and neuroinflammation. A single proteomic biomarker may be inadequately sensitive to the detection of a given pathological process (see Chapter 8) due to the heterogeneous nature of pathology.

A panel of multiple biomarkers may be more sensitive to a heterogeneous pathophysiology. Integrating results from metabolomic, proteomic, and genomic testing, coupled with a patient's history and physical examination, could offer the most sensitive, specific, and clinically relevant test for brain damage and prognosis of secondary neurological deterioration.

16.2.3 Biomarkers of Autoimmune Response

Brain-derived peptides and small proteins released in the neurotoxicity cascade after TBI and stroke might carry immune active epitopes that activate neuroimmune cells (microglia) and ultimately generate circulating antibodies. Antibodies are the normal reaction of the immune system to "foreign" or previously naive peptides with active epitopes that are usually not present in the bloodstream.[24] The latter should not be confused with pathological immune reactions to "self-antigens" that produce pathogenic autoantibodies[25] that form circulating immune complexes and destroy healthy normal cells, including neurons, depending on the specificity of the autoantibody.

The concept concerning the appearance of antibodies to brain proteins in epilepsy has been presented.[26] Translational studies have shown that ionotropic glutamate receptors (GluRs) are activated in neurotoxicity and abnormal epileptic spiking activity.[27] Antibodies against specific recombinant subunits of NMDAR peptides correlate with cerebral ischemic processes,[28,29] while AMPAR peptides are more specific to TBI.[30–33] Recently, antibodies to recombinant kainite receptor subunits detected in CSF have been linked to spinal cord compression and incomplete injury (see Chapter 11).

16.3 Possible Omissions in Brain Biomarker Testing

16.3.1 Single *vs.* Multiple Biomarker Testing

In practice, it is a very attractive approach to deal with a single biomarker defining a pathological condition of interest. Examples would be troponin for cardiovascular disease or brain natriuretic peptide for progression of ischemic heart infarction when studying the assay value, particularly as an emergency tool.[34] Due to the heterogeneous nature of events that progress in various CNS locations following physical impact, a panel of multiple biomarkers might be helpful.

However, if the approach from a single pathological process is considered, such as the ischemic neurotoxicity cascade, then separate biomarkers associated with lesions in cortical, subcortical, and vertebral/basilar areas could be identified.[13] Detected together in multiple panel assay, such biomarkers could supply information that would be useful in the development of a brain injury severity scoring system to triage individuals for personalized treatment.

There is a trend toward the development of panels of multiple biomarker assays that might satisfy the accuracy of a standalone or binary clinical indication (ischemic stroke *vs.* hemorrhage, or stroke *vs.* mimics) and possibly reflect the complexity of brain injury.

Results with biomarkers in brain injury have been mixed. It was recently demonstrated that a multiple blood biomarker panel contributed low value to the assessment of acute magnetic resonance imaging (MRI)-proven stroke.[35] NSE levels peaked in serum within the first 96 hours after cerebral infarction,[36] as well as in persons with TBI showing a higher NSE value corresponding to higher mortality and a more severe score on the Glasgow Coma Scale.[37] Similar results were obtained for S100B, glial fibrillary acidic protein, and others.[36,38] However, a panel of biomarkers performed better in predicting mortality rate[39] and showed a good correlation with clinical severity of stroke.[40]

Recent research in disease biomarkers has focused on human pathological specimen banks. These present investigators with tissue from pathologically confirmed situations and avoid the issue of differences between human conditions and animal models.[41,42]

In future, a multiple biomarker panel built on a combination of metabolomic, proteomic, degradomic, and genomic approaches could provide emergent "fingerprints" of systemic conditions.

16.3.2 Translational Study Design Considerations

Several major aspects should be considered while planning translational research and before conducting extremely costly clinical trials of biomarkers. First, the biomarker or biomarkers must be selected. An early consideration is whether to pair the testing of a biomarker with a therapeutic trial (*i.e.*, a "companion diagnostic test"). This has obvious advantages from scientific and feasibility points of view,[43] as the benefits and risks of using the biomarkers or tests are going to be well defined. The FDA regulatory approach is determined by the degree of risk involved in the use of diagnostic tests.[44] It is necessary to define precisely the clinical condition, inclusion and exclusion criteria, and the setting (community, emergency room, physician office, hospital, *etc.*), and neither to be too narrow as to limit clinical utility nor too broad as to obscure observed differences. Surrogate endpoints must ultimately satisfy medical practitioners and correspond to regulatory requirements.[45] It is important to determine what constitutes a suitable control population, as the range of biomarker values may be greater in older individuals with multiple risk factors than in young, healthy controls.

In the case of an emergency room and emergent pre-hospital setting, then rapid differential diagnosis (*i.e.*, "stroke *vs.* non-stroke" or "hemorrhagic *vs.* non-hemorrhagic stroke") assumes greater importance, and on-site diagnostic biomarker testing may complement off-site teleneurology and delays in getting MRI scans. In these cases, a rapid blood biomarker assay would help to make an accurate diagnosis and guide initial urgent therapy (*i.e.*, intravenous recombinant tissue plasminogen activator). In this setting, the turnaround time is very important and a point-of-care assay format would be the best choice (see Chapter 17).

In the non-urgent outpatient office setting, biomarkers are more likely to be useful for risk assessment (*i.e.*, optimization of antiplatelet therapy or detection of future stroke risk). The non-neurologist general practitioner could promptly adjust secondary preventative stroke therapy or refer patients for stroke specialist evaluation. Turnaround time is less important here and cost is perhaps more important.

It is important to control for differences in risk factors by controlling for ethnic and comorbid conditions in patient selection. If significant numbers are present, it may be possible to define different cut-offs based on ethnicity, age, and gender. In the absence of this, control and patient groups should be made comparable by careful stratifying of patient enrollment.

Control groups should include not only healthy individuals, but also those with typical comorbid conditions and particularly those with other neurological "mimic" conditions. This would include, in the case of strokes, cranial nerve palsies, migraine, brain tumors, seizures, and functional neurological disorders.

16.3.3 Determining Study Power

The essence of diagnostic biomarker research is to detect a difference between two groups. The ability to do this is formulated in the statistical power of a given test or combination of biomarkers. The power may be calculated by determining the relative difference of a biomarker result between two groups in relation to the distribution of biomarker results within each group. To date, most biomarker assay studies have had relatively low statistical power and have not been validated in a high-quality testing,[46] mostly due to small sample sizes and comparisons with healthy controls only. Heterogeneous populations require a larger study population.

The accuracy of individual biomarker or panels of biomarkers can be assessed by the receiver operating characteristic (ROC) analysis. The ROC curve is plotted with the *y*-axis as the true-positive rate (sensitivity, percentage of patients with disorder who tested positive for biomarkers) and the *x*-axis as the false-positive rate (percentage of control subjects classified as having disease) as the cut-off or threshold is varied. In this method, the area under the ROC curve is used as an indicator of biomarker performance regarding the ability to distinguish between control and patient subjects. The false-positive and false-negative rates should be taken in account to establish an optimal threshold at a desired specificity and sensitivity level. This threshold is used to calculate the power of the study, usually by determining the necessary number of subjects.

For instance, the specificity is set between 0.70 and 0.95 as a global measure of the diagnostic effectiveness for clinical interest. The sample size calculation is based on the following assumptions: (1) the prevalence rate of disease in the recruited cohort is 70%; and (2) a transformation of the biomarker measure is normally distributed for feasibility study participants with and without disease. Then a sample size $n = 210$ per diseased group and $n = 90$ per control group will be adequate for study with a power of no less than 83% for the setting.

In practice, it would be recommended to initiate enrollment for twice the number of patients and controls due to multiple omissions.[47] Up to 30–50% could have no endorsed study consent form, about 10% of collected samples will have no (or lost or misplaced) requested information (did not insert information in the database on enrollment), and up to 10–15% of samples will be misplaced or of too poor a quality to assay (hemolysis, incorrect storage and transportation, *etc.*).

For biomarker studies detecting for screening capacity, besides a high sensitivity, it is more important to achieve a low false-positive level. To improve the outcome, false-positive and false-negative cases have to always be considered as a benefit-to-harm ratio. The calculation of a disease's positive and negative predictive values provides information about the probability of disease occurrence. These values are dependent on disease prevalence and can be assessed only in prospective studies.

16.4 Biological Fluid Testing Drawbacks

In a healthy brain, low concentrations of proteolytic breakdown proteins might be detected in the biological fluids of subjects as a part of normal metabolism. The natural peptide fragments with weak epitopes usually have a muted immune response and produce natural antibodies in meager quantities.[48]

Careful choices of biospecimen, sample collection and preparation, handling and storage, sample comparison, methods of analysis, and number of replicates should be made before trials to avoid omissions in study executions.

16.4.1 Biological Fluid Selection

The choice of biospecimens for biomarker assays should be based on the following considerations: (i) the quantities of the biomarker should be in the measurable range (nanograms to micrograms); (ii) the sample withdrawal procedure should be as minimally invasive as possible; (iii) the biomarker should be relatively stable in the sample of choice at room temperature during the processing; (iv) techniques should be implemented to protect sample integrity when shipping, analyzing, and for prolonged storage at low temperatures. These issues are of great importance due to potential incorrect diagnosis and treatment, particularly for biomarker assay(s) seeking diagnostic application.

Plasma samples would be a good choice for peptide and protein biomarker detection and are better drawn in anti-peptidase-treated tubes to protect more liable peptides. Various peptide biomarkers have different sensitivities to coagulants (heparin, EDTA, and citric acid) and assays should be tested at the discovery stage. Compared to plasma, serum samples contain about 20–30% more antibodies and should be selected for antibody biomarker measurements. CSF is rarely used due to the invasive nature of withdrawal and consequent dilution; however, it is utilized to prove the origin of brain biomarkers.

GluR biomarkers for the assessment of stroke and TBI are different specific peptide fragments of vascular origin that might be responsible for the progression of white or grey matter impairments. For instance, ischemic events in stroke are primary impairments with NR2 peptide as a suitable biomarker,[7] while the same events will be delayed for TBI, and NR2 antibodies will describe it better.[12] Parameters of these biomarkers should be detected using conventional immunoassays (see Chapter 17) in plasma for peptides and in serum samples for antibodies. Due to the invasive nature of CSF withdrawal, the use of this specimen is rare and only employed in cases of uncertain diagnosis or spinal cord injuries.

The major initial point is the patient consenting followed by sample drawing. The blood should be collected in a controlled environment (a hospital or other medical facility) and it is important that the sampling procedure should be kept as constant as possible. Here, the cycle starts and includes the type of collection tubes, processing temperature and time, storage issues before and after assay, labeling, and secure storage of data.

16.4.2 Sample Preparation and Storage

Necessary information guides for blood sample processing, testing, and storage is depicted in Tables 16.1 and 16.2. Usually, two 5 mL samples of blood are drawn for peptides in plasma (collection tubes containing EDTA or, even better, anti-protease-treated tubes containing EDTA) and antibodies in serum (gel separation tubes) detection. Plasma tubes should be immediately placed on an ice pack while serum tubes should be stored at room temperature to allow for serum separation.

Tubes should be centrifuged in refrigerated conditions, aliquoted into three or four labeled 1.5 mL Eppendorf tubes (specimen per patient per draw, about 0.5 mL volume), and stored depending on time of assay. For better handling in future, it is recommended to use clear (serum) and colored Eppendorf (plasma) tubes to help easily distinguish serum from plasma samples. Individual specimens should be identified with both an alphanumeric study number and patient's abbreviated name.

Table 16.1 Avoiding pitfalls in sample quality for peptide assays.

Description	Requests	Notes
Blood sample	Plasma only	For analysis of peptides, the use of platelet-depleted EDTA or citrate plasma is recommended
Plasma quality	Clear, not hemolyzed specimen	Interference with assays
Blood withdrawal	Should be withdrawn before thrombolytic administration	Immediate placement on ice and then separation by centrifugation (3000 *g*) is recommended
Plasma preparation	Aliquot sample after centrifugation in three to five Eppendorf tubes (0.5–1.0 mL)	Allows separate storage of the material for different assays
Perform assay	Fresh sample of plasma is preferred	Accurate amount of peptides will be detected
Plasma storage	−80 °C: up to 1 year −15 to −20 °C: up to 2 months	Longer storage reduces peptides (20–30%)
Frozen plasma	One thaw only	Thawing activates serine proteases: one thawing reduces peptide amounts by 30–50%

Table 16.2 Avoiding pitfalls in sample quality for antibody assays.

Description	Requests	Notes
Serologic test	Serum only	Serum contains 20–25% more antibodies than plasma
Serum quality	Clear, not hemolyzed specimen without gelatinous substance	If hemolyzed, interference with assays will give a wrong results
Blood withdrawal	Best results if gel separation tubes used	Keep at room temperature for no longer than 30 minutes for serum separation then centrifuge
Sample preparation	Aliquot sample after centrifugation in three to five Eppendorf tubes (0.5–1.0 mL)	Allows separate storage of the material for different assays
Sample storage	−80 °C: up to 1.5 years −15 to −20 °C: up to 3 months	Longer storage reduces antibodies (20–50%)
Sample thawing	One thaw preferable	First thawing: IgG down by 10% Second thawing: IgG down by 25–30% (relative to first thaw)

If analyzed the same day, then a couple of specimens per patient (plasma and serum) are stored at 2–8 °C. The rest of the correctly labeled aliquots are placed in a storage box containing 9×9 cells (holding 81 samples) and should be moved to a freezer at or below −20 °C. For biobanking, aliquots should be frozen at −80 °C (Tables 16.1 and 16.2).

In case of needing to transport samples, storage boxes should be arranged in a Styrofoam box having enough space for dry ice (about 7–10 pounds per box) and shipped to the central laboratory *via* overnight courier to maintain frozen status until analyzed.

It is necessary to note that all samples should be coded to maintain blinding prior to shipping and should be identified only with the unique study number assigned to that participant by the clinical investigator at each clinical site, ensuring that the central laboratory staff will be blinded to all demographic and clinical data.

16.5 Conclusion

Clinical studies for biomarker testing should be a collaborative effort of professionals (neurologists, radiologists, clinical laboratorians, and biostatisticians) in order to better design and perform the trials. Several major requests should be satisfied in order not to fail in the translational research of promising biomarker(s) for urgent care.

To overcome obstacles and avoid biases, the study design and surrogate endpoints should be clearly stated. The testing of multiple panel biomarkers should be considered, as a single biomarker cannot reflect the complexity of brain structures involvement and the rate of disease progression for stroke or TBI. Then, biomarkers in the panel should be built up according to the pathological process, the surrogate endpoints, and the clinical setting (emergency, surgery, stroke neurology, or primary physician office).

The type of specimen and methods of withdrawal, storage, and transportation should be standardized within all participating clinical sites approving the institutional review board protocol. The power of study calculations must be considered not only for "disease" *vs.* "control" groups, but also when expanding the population to "other or mimicking diseases". The principal investigator should be aware of enrollment omissions and pitfalls due to human factors (unsigned consent forms, loss of patient information, and poor quality of samples).

Consideration of the challenges and biases in biomarker testing might help in study planning to avoid negative results and to save time and funding. While moving forward from initial discovery of candidate biomarkers to translational medicine is a laborious process, it could be beneficial to patients and healthcare providers.

References

1. J. J. Chen, H. D. Rosas and D. H. Salat, *PLoS One*, 2013, **8**, e56733.
2. J. A. Stokum, V. Gerzanich and J. M. Simard, *J. Cereb. Blood Flow Metab.*, 2016, **36**, 513.
3. G. C. O'Connell, A. B. Petrone, M. B. Treadway, C. S. Tennant, N. Lucke-Wold, P. D. Chantler and T. L. Barr, *Genomic Med.*, 2016, **1**, 16038.

4. G. C. O'Connell, P. D. Chantler and T. L. Barr, *bioRxiv*, 2017, DOI: 10.1101/109777.
5. G. C. Jickling and F. R. Sharp, *Neurotherapeutics*, 2011, **8**, 349.
6. S. A. Dambinova, in *Biomarkers for TBI*, ed. S. A. Dambinova, R. L. Hayes and K. K. W. Wang, Royal Society of Chemistry, London, 2012, ch. 4, p. 66.
7. S. A. Dambinova, K. Bettermann, T. Glynn, M. Tews, D. Olson, J. D. Weissman and R. L. Sowell, *PLoS One*, 2012, **7**, e42362.
8. M. U. Gappoeva, G. A. Izykenova, O. K. Granstrem and S. A. Dambinova, *Biokhimia*, 2003, **68**, 696.
9. S. Shu, L. Pei and Y. Lu, *Regen. Med. Res.*, 2014, **2**, 8.
10. F. J. Carvajal, H. A. Mattison and W. Cerpa, *Neural. Plast.*, 2016, **2016**, 2701526.
11. S. A. Dambinova, A. V. Shikuev, J. D. Weissman and J. D. Mullins, *Mil. Med.*, 2013, **178**, 285.
12. S. A. Dambinova, R. L. Sowell and J. C. Maroon, *J. Mol. Biomark Diagn.*, 2013, **S3**, DOI: 10.4172/2155-9929.S3-003.
13. S. A. Dambinova, J. C. Maroon, A. M. Sufrinko, J. D. Mullins, E. V. Alexandrova and A. A. Potapov, *Front. Neurol.*, 2016, **7**, 172.
14. S. A. Dambinova, J. D. Mullins, T. Gennarelli, in *Neurotrauma: A Comprehensive Book On Brain Injury and Spinal Cord Injury*, ed. K. K. Wang, Oxford University Press, New York, 2017, ch. XI, 4, in press.
15. J. D. Mullins, A. V. Shikuev, U. I. Danilenko and S. A. Dambinova, presented in part at Military Health System Research Symposium USAMRMC, Ft. Lauderdale, August 2012.
16. J. D. Mullins, presented in part at AMSUS 118th Annual Continuing Education Meeting, Seattle, November 2013. http://amsusce.org/wp-content/uploads/2015/05/Abstract-Summaries-10.22.13.2.pdf.
17. I. Aretz and D. Meierhofer, *Int. J. Mol. Sci.*, 2016, **17**, pii: E632, DOI: 10.3390/ijms17050632.
18. J. D. Tenenbaum, *Genomics, Proteomics Bioinf.*, 2016, **14**, 31.
19. L. Allard, D. F. Hochstrasser and J.-C. Sanchez, in *Biomedical Applications of Proteomics*, ed. J.-C. Sanchez, G. L. Corthals, D. F. Hochstrasser, Wiley-VCH, Verlag, Weinhein, 2004, ch. 4, p. 57.
20. M. Matarin, J. Simon-Sanchez, H. C. Fung, S. Scholz, J. R. Gibbs, D. G. Hernandez, C. Crews, A. Britton, F. W. De Vrieze, T. G. Brott, R. D. Brown Jr., B. B. Worrall, S. Silliman, L. D. Case, J. A. Hardy, S. S. Rich, J. F. Meschia and A. B. Singleton, *Neurogenetics*, 2008, **9**, 101.
21. Y. Wang, H. Liu, Y. Lin, G. Liu, H. Chu, P. Zhao, X. Yang, T. Zheng, M. Fan, X. Zhou, J. Meng and C. Sun, *Sci. Rep.*, 2017, **7**, 40137.
22. E. R. Bennett, K. Reuter-Rice and D. T. Laskowitz, in *Translational Research in Traumatic Brain Injury*, ed. D. Laskowitz, G. Grant, C. R. C. Press/Taylor and Francis Group, Boca Raton, 2016, ch. 9.
23. G. A. Sarkis, M. D. Mangaonkar, A. Moghieb, B. Lelling, M. Guertin, H. Yadikar, Z. Yang, F. Kobeissy and K. K. Wang, *Curr. Neurol. Neurosci. Rep.*, 2017, **17**, 23.

24. S. A. Dambinova, in *Biomarkers for TBI*, ed. S. A. Dambinova, R. L. Hayes, K. K. W. Wang, Royal Society of Chemistry, London, 2012, ch. 4, p. 66.
25. S. Amor and R. Huizinga, *Neurological Disorders (Neurology-Laboratory and Clinical Research Developments)*, Novinka Books, Bel Air, 2011.
26. S. A. Dambinova, *Glutamate Neuroreceptors*, Nauka, Leningrad, 1989.
27. S. A. Dambinova, G. A. Izykenova, S. V. Burov, E. V. Grigorenko and S. A. Gromov, *J. Neurol. Sci.*, 1997, **152**, 93.
28. M. U. Gappoeva, G. A. Izykenova, O. K. Granstrem and S. A. Dambinova, *Biokhimia*, 2003, **68**, 696.
29. E. I. Gusev and V. I. Skvortsova, in *Brain Ischemia*, Kluwer Academic Publishers, Berlin, 2003, p. 151.
30. M. M. Odinak, D. E. Dyskin, I. A. Toropov, A. Y. Emelin, A. A. Cherepanov and S. A. Dambinova, *Zh. Nevrol. Psikhiatr. Im. S.S. Korsakova*, 1996, **96**, 45.
31. S. A. Gromov, S. K. Khorshev, Yu. I. Poliakov, L. G. Gromova and S. A. Dambinova, *Zh. Nevropatol. Psikhiatr. Im. S.S. Korsakova*, 1997, **97**, 46.
32. A. V. Goryunova, N. A. Bazarnaya, E. G. Sorokina, N. Y. Semenova, O. V. Globa, Z. B. Semenova, V. G. Pinelis, L. M. Roshal and O. I. Maslova, *Neurosci. Behav. Physiol.*, 2007, **37**, 761.
33. E. G. Sorokina, Z. B. Semenova, N. A. Bazarnaya, S. V. Meshcheryakov, V. P. Reutov, A. V. Goryunova, V. G. Pinelis, O. K. Granstrem and L. M. Roshal, *Behav. Physiol.*, 2009, **39**, 329.
34. E. Ter Avest, A. Visser, B. Reitsma, R. Breedveld and A. Wolthuis, *Eur. J. Emerg. Med.*, 2016, **23**, 95.
35. C. Knauer, K. Knauer, S. Müller, A. C. Ludolph, D. Bengel, H. P. Müller and R. Huber, *BMC Neurol.*, 2012, **12**, 14.
36. D. M. Stanca, I. C. Mărginean, O. Sorițău and D. F. Mureşanu, *J. Med. Life*, 2015, **8**, 21.
37. H. Zetterberg, D. H. Smith and K. Blennow, *Nat. Rev. Neurol.*, 2013, **9**, 201.
38. J. Zhang, V. Puvenna, and D. Janigro, in *Translational Research in Traumatic Brain Injury*, ed. D. Laskowitz, G. Grant, CRC Press/Taylor and Francis Group, Boca Raton, FL, 2016, ch. 12.
39. F. Iemolo, E. Sanzaro, G. Duro, A. Giordano and M. Paciaroni, *Immun. Ageing*, 2016, **13**, 19.
40. A. Barbieri, E. Giuliani, C. Carone, F. Pederzoli, G. Mascheroni, G. Greco, C. Stucchi and S. Genedani, *Minerva Anestesiol.*, 2013, **79**, 750.
41. T. D. Kwiecien, C. Sy and Y. Ding, *Neurol. Res.*, 2014, **36**, 417.
42. Y. Xiong, A. Mahmood and M. Chopp, *Nat. Rev. Neurosci.*, 2013, **14**, 128.
43. FDA, 2005 Concept paper, Online, https://www.fda.gov/downloads/drugs/science research/researchareas/pharmacogenetics/ucm116689.pdf (assessed on April 10, 2017).
44. J. Shuren, Statement to the Subcommittee on Health, Committee on Energy and Commerce, U.S. House of Representatives, Online, https://www.fda.gov/NewsEvents/Testimony/ucm473922.htm (assessed on April 10, 2017).

45. S. A. Dambinova and R. L. Hayes, in *Biomarkers for TBI*, ed. S. A. Dambinova, R. L. Hayes, K. K. W. Wang, Royal Society of Chemistry, London, UK, 2012, ch. 13, p. 200.
46. E. Drucker and K. Krapfenbauer, *EPMA J.*, 2013, **4**, 7.
47. J. D. Weissman, G. A. Khunteev, R. Heath and S. A. Dambinova, *J. Neurol. Sci.*, 2011, **300**, 97.
48. S. Panda and J. L. Ding, *J. Immunol.*, 2015, **194**, 13.

CHAPTER 17

Resolving Difficult Case Scenarios by Incorporating Stroke Biomarkers in Clinical Decision-making

GEELYN J. L. NG,[a,b] EI Z. THE,[b] MEI YEN NG,[a,b] AMY M. L. QUEK[a,b] AND RAYMOND C. S. SEET*[a,b]

[a] Department of Medicine, Yong Loo Lin School of Medicine, National University of Singapore, Singapore; [b] Division of Neurology, Department of Medicine, National University Health System, NUHS Tower Block Level 10, 1E Kent Ridge Road, Singapore 119228, Singapore
*Email: raymond_seet@nuhs.edu.sg

17.1 Introduction

The world faces an impending epidemic of stroke. These epidemiologic concerns are fueled by a rapidly aging society where population estimates indicate that close to 1.5 billion people in the world will be 65 years and older by 2050.[1] Presently, stroke stands as the most prominent cause of permanent disability and is the second-leading cause of death for people older than 60 years. Each year, approximately 3–7% of total healthcare resources are spent on acute stroke treatment and rehabilitation in high-income countries.[1]

Drug Discovery Series No. 62
Acute Brain Impairment: Scientific Discoveries and Translational Research
Edited by Philip V. Peplow, Svetlana A. Dambinova, Thomas A. Gennarelli and Bridget Martinez

Published by the Royal Society of Chemistry, www.rsc.org

Cerebrovascular illnesses are often the aftermath of compounding risk factors that drive atherosclerosis in the small and large arteries of the brain, manifesting initially as a subtle subclinical disease before evolving into an overt ischemic stroke.[2,3] Ischemic stroke occurs as a result of a complete cessation or significant reduction of cerebral blood flow to the brain or from an embolic phenomenon of blood thrombus from the heart or aorta.[4] Depending on the extent of hypoperfusion, an area with complete interruption of blood flow can evolve into an infarct core (where neuronal death occurs within a few minutes) and, lying in its periphery, the penumbra, which suffers from limited blood flow and houses functionally impaired but semi-viable brain tissues.[5] In the ischemic core region, hemostatic activation (due to an overproduction of thrombin) hastens necrotic cell death, giving rise to an area that is electrically, metabolically and functionally inactive. By contrast, cell viability in the ischemic penumbra is still preserved, where neurons are thought to be metabolically active, but electrically and functionally compromised. If blood flow is not restored within a relatively short time, the penumbra acquires the same destiny as the infarct core region.[4]

Arterial recanalization and reperfusion *via* intravenous recombinant tissue plasminogen activator (TPA)[6] and endovascular treatment through device-driven retrieval or aspiration of blood thrombus[7] are presently the only two means of acute stroke treatment adopted clinically. Early reperfusion has the potential of salvaging ischemic brain tissues and is associated with a five-fold improvement in functional stroke recovery.[7] In practice, the benefits of arterial recanalization and reperfusion are often balanced against the dreaded risk of intracranial hemorrhage that is associated with early neurologic deterioration, malignant cerebral infarction and high mortality.[8]

The ability to identify high-risk stroke patients is desirable for clinicians to accurately triage patients to specialized stroke units for closer monitoring, to individualize treatment in anticipation of stroke-related complications and to accurately inform long-term stroke prognosis. Existing methods to identify high-risk individuals depend largely on a physician's subjective clinical assessment of neurologic deficits and their evolution and neuroimaging findings. These methods, however, are limited by the lack of round-the-clock availability of clinical expertise, the use of sedating medications that might affect accurate assessment of neurologic deficits and the safety of frequent computed tomographic (CT) scans and contrast agent use. Biological signatures of cerebral ischemia could reveal a unique "window" to the complex pathological consequences following cerebral ischemia, thereby providing clinicians with an objective measure of risks and benefits when weighing stroke treatments.

Biomarkers are biological hallmarks of normal and pathologic processes that are objectively measured and can serve a diverse range of purposes such as risk stratification, therapeutic assessment strategies, clinical trial design and drug development.[10] Generally, a biomarker is clinically well accepted if the proposed biomarker is accurate, is acceptable to the patient, is easy to interpret by clinicians, has a high sensitivity and specificity for the outcome it is expected to identify and explains a reasonable proportion of the outcome independent of established predictors consistently across multiple studies. Biomarkers have also become a means for researchers to gain a mechanistic understanding of the differences in pharmacological responses to drugs in preclinical and human models and to improve the design of clinical trials by strengthening the selection criteria in order to better target the application of a particular compound of interest.[11]

As clinical events take up to years to develop, the use of biomarkers as surrogate endpoints could potentially reduce the duration of a study and lower the overall cost of screening the efficacy of a compound of interest, making biomarkers an appealing option for clinical trialists and epidemiologists. A good surrogate biomarker has features such as the ability to track closely with changes in the outcome of interest and being validated in independent cohorts.[12,13] As stroke is a heterogeneous disease, the use of multiple biomarkers targeting various pathways may be imperative in order to reflect the different facades of this complex disease and to yield surrogate endpoints that offer a more representative effect of the intended treatment.

In stroke patients, the inaccessible nature of human brain tissues makes it impracticable to measure a biomarker within the brain; and the chances of routinely obtaining cerebrospinal fluid from stroke patients who are receiving either antiplatelet or anticoagulation treatment are also low, as performing a lumbar puncture in these patients puts them at an increased risk of bleeding. For patients suffering from conditions such as intracerebral hemorrhage and tumor, performing a lumbar puncture may result in a shift in intracranial pressure, potentially triggering cerebral herniation. Lumbar puncture can also be technically difficult to perform, especially among the elderly, who are plagued with degenerative disc disease. In normal individuals, the presence of an intact blood–brain barrier (BBB) makes it difficult to detect large amounts of brain-derived biomarkers in the peripheral circulation. Stroke, however, results in a breakdown in the BBB, which allows for the release of brain-derived biomarkers into the blood circulation.

We have previously examined notable blood-based stroke biomarkers that may assist in the clinical decision-making process.[9] This chapter serves to follow up on this earlier review and to highlight unique clinical scenarios where there is a need to improve clinical decision-making processes at crucial time-points in stroke management. Biomarkers that are measurable in blood are focused on in this chapter, as blood is widely accessible in human stroke. We searched medical databases such as MEDLINE, PubMed and Ovid for publications that highlight the use of blood-based biomarkers in clinical scenarios where comparisons between biomarkers are made with outcomes such as hemorrhagic transformation (HT), early neurologic deterioration (END) and malignant infarction.

17.2 Blood-based Stroke Biomarkers in Clinical Scenarios

In ischemic stroke, studies have evaluated biomarkers to distinguish ischemic stroke from its mimics, to determine stroke etiology, to predict stroke severity and outcomes (*e.g.* END and hemorrhagic complications) and to identify patients who may benefit from decompressive hemicraniectomy (Table 17.1). Several studies have also examined the use of biomarkers to prognosticate patients for outcomes such as functional recovery and vascular events (*e.g.* recurrent stroke, myocardial infarction and death).[10] Several researchers have investigated biomarkers that estimate infarct volume and have found significant associations between infarct volume and biomarkers such as S100β, matrix metalloproteinase-9 (MMP-9), interleukin (IL)-6, tumor necrosis factor-α (TNF-α), intercellular adhesion molecule-1 (ICAM-1) and glutamate.[11–13]

Such findings, however, have limited clinical application, as stroke severity is determined by physicians mostly through a detailed neurological examination and interpretation of neuroimaging findings.[14] Furthermore, stroke volume itself may not be a reliable indicator of stroke severity, as a small stroke may not be any less dangerous, especially when critical areas of the brain such as the brainstem or language areas are affected. By contrast, large areas of injury may present with subtle clinical features (*e.g.* strokes affecting the non-dominant temporal lobe). Detection of blood-based biomarkers immediately after stroke onset could triage individuals for stroke treatment and provide critical information that is otherwise elusive to clinical and neuroimaging modalities. A quicker clinical assessment could then be made, having taken into account the likelihood of cerebral hemorrhage, END and malignant cerebral edema.

17.2.1 Hemorrhagic Transformation

Despite a precedent ischemic insult, hemorrhage can secondarily develop within the infarcted brain tissues (*i.e.* HT). Although the outcome of HT is generally favorable, HT can result in catastrophic outcomes when the hemorrhage: (1) extends into the ventricles or subarachnoid spaces; (2) results in cerebral edema; and (3) is associated with concomitant use of antifibrinolytic, antiplatelet and anticoagulant agents.[15,16] Figure 17.1 highlights neuroimaging features of ischemic patients with an uncomplicated HT and HT that is associated with intravenous thrombolysis. HT occurrence may also be influenced by stroke etiology; ischemic stroke that is associated with cardioembolism is more prone to HT.

HT can be categorized into two main subtypes, namely hemorrhagic infarction (HI) and parenchymatous hematoma (PH). HI refers to the pathological condition in which petechial or more confluent hemorrhages occupy a portion of an area of ischemic infarction. By contrast, PH in an area of

Table 17.1 Summary of stroke biomarkers in different clinical situations.

Type	Biomarker	Ref.
Hemorrhagic transformation: the spontaneous secondary bleeding of cerebral infarction	Matrix metalloproteinases	11, 19–23, 28, 51, 54–56, 82
	Cellular fibronectin	10, 23, 28
	Plasminogen activator inhibitor type 1	24, 25
	Thrombin-activated fibrolysis inhibitor	24, 25
Malignant infarction: development of space-occupying edema and fatalistic cerebral herniation following stroke	S100β	10, 26, 30–35, 54–56
	Cellular fibronectin	10, 23, 28
	Matrix metalloproteinases	11, 19–23, 28, 51, 54–56, 82
Early neurologic deterioration: worsening of neurological status from stroke onset to 48–72 hours after admission	Glutamate	4, 10, 13, 28, 42, 49, 50, 110
	γ-Aminobutyric acid	28, 41
	Ferritin	10, 42
	Tumor necrosis factor-α	14, 19, 28, 82, 113
	Intercellular adhesion molecule 1	12, 14, 113
	Matrix metalloproteinases	11, 19–23, 28, 51, 54–56, 82
	C-reactive protein	10, 11, 44, 46, 55, 58, 60, 65, 82, 88–90, 99, 100, 106, 107, 112, 113, 118
Stroke diagnosis: differentiating stroke conditions from stroke mimics	Glutamate	4, 10, 13, 28, 42, 49, 50, 110
	S100β	10, 26, 30–35, 54–56
	von Willebrand factor	51–55, 60, 61
	Matrix metalloproteinases	11, 19–23, 28, 51, 54–56, 82
	Monocyte chemotactic protein-1	51, 64, 113
	Vascular cell adhesion molecule 1	54
	N-terminal pro-brain natriuretic peptide	10, 55, 56, 80, 81, 83–87, 98, 104, 118
	C-reactive protein	10, 11, 44, 46, 55, 58, 60, 65, 82, 88–90, 99, 100, 106, 107, 112, 113, 118

Table 17.1 (*Continued*)

Type	Biomarker	Ref.
Thrombus formation: atypical platelet adherence and aggregation to the vessel wall	D-dimer	10, 49, 55, 56, 58–60, 78–80
	Uric acid	66–74
Stroke etiology: underlying cause of stroke	N-terminal pro-brain natriuretic peptide	10, 55, 56, 80, 81, 83–87, 98, 104, 118
	D-dimer	10, 49, 55, 56, 58–60, 78–80
Stroke risk: risk of recurrent stroke event	Lipoprotein-associated phospholipase A_2	10, 88–90, 107, 112
	Lipoproteins	91–95
	C-reactive protein	10, 11, 44, 46, 55, 58, 60, 65, 82, 88–90, 99, 100, 106, 107, 112, 113, 118
Atrial fibrillation: erratic firing of electrical activities that causes pooling of blood within the heart	N-terminal pro-brain natriuretic peptide	10, 55, 56, 80, 81, 83–87, 98, 104, 118
	High-sensitivity troponin T	10, 14, 51, 54, 56, 98, 115–118
	F_2-isoprostanes	11, 100, 101, 105–114

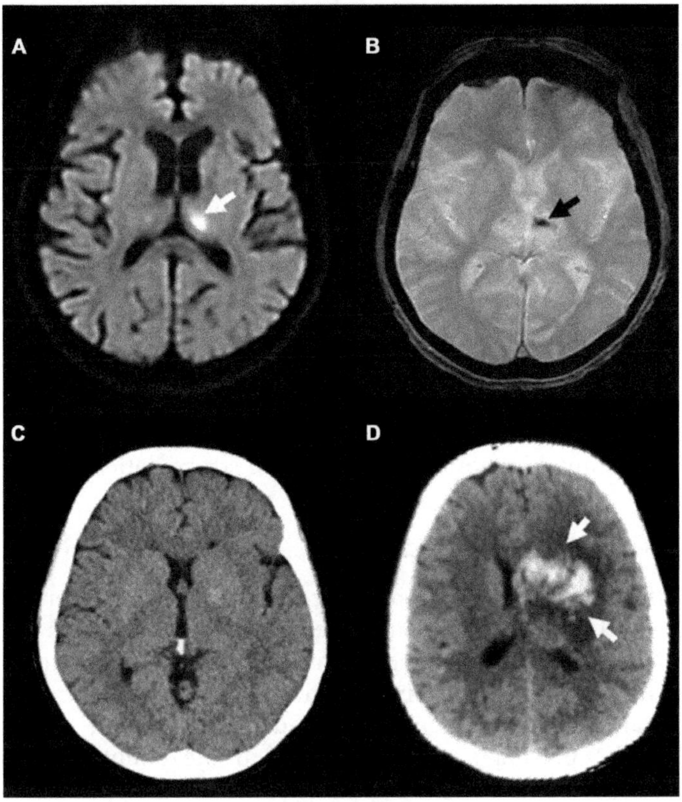

Figure 17.1 Hemorrhagic transformation. (A) Magnetic resonance imaging of the brain showing an area of restricted diffusion that is consistent with an acute ischemic stroke in the left thalamus (white arrow). (B) An area of hemorrhage (black arrow) is observed within the infarcted brain tissue that is consistent with hemorrhagic transformation. (C) Computed tomography (CT) scan of a patient who presented with right-sided weakness within 3 hours of symptom onset. As the CT scan had excluded cerebral hemorrhage as the cause of his stroke, he was treated with intravenous recombinant tissue plasminogen (Alteplase) 0.9 $mg\,kg^{-1}$. (D) CT scan, taken 12 hours later after he developed neurologic deterioration, showed hemorrhagic transformation (white arrows) within the territory of the left middle cerebral artery.

infarction could cause mass effect, potentially displacing and destroying brain tissues.[17] HI has been more closely associated with cerebral embolism. When assessing stroke patients for reperfusion treatment, apart from the beneficial effects of achieving arterial recanalization *via* TPA administration, the risk of symptomatic intracranial hemorrhage (SICH) often needs to be taken into consideration as well.[8,18] Early recognition of patients at high risk of SICH makes it possible to determine their suitability for TPA treatment while simultaneously facilitating the process of triaging these patients into specialized stroke units for closer monitoring and tighter control of risk factors (especially blood pressure). As disruptions in the BBB have been implicated in HT, current research has focused on biomarkers that indicate early BBB as predictors for HT.

17.2.1.1 Matrix Metalloproteinase-9

To date, the roles of biomarkers have been explored through the measurement of MMP-9, which *in vivo* is involved in the destruction of microvascular integrity by degradation of the basal lamina and extracellular matrix.[19] MMP-9 belongs to a family of zinc-binding enzymes that plays a part in the breakdown of the extracellular matrix, including the components surrounding cerebral blood vessels. During TPA treatment, MMP-9 becomes activated and is capable of disrupting the BBB, resulting in HT.[20,21] Before TPA treatment, serum MMP-9 levels >140 ng mL^{-1} predicted the development of intracranial hemorrhage in ischemic stroke patients.[22]

17.2.1.2 Cellular Fibronectin

Cellular fibronectin (c-Fn) is another factor that has been associated with increased HT and malignant cerebral infarction. c-Fn is synthesized by endothelial cells and is largely involved in the interaction between endothelial cells and its matrix. The presence of c-Fn in vascular endothelium makes it a good marker of endothelial health; elevated levels of c-Fn in the plasma suggests damage to the endothelium. A study showed c-Fn levels >3.6 μg mL^{-1} predicted the development of HT following TPA use.[23]

17.2.1.3 Plasminogen Activator Inhibitor type 1 and Thrombin-activated Fibrinolysis Inhibitor

Other biomarkers that have been evaluated to predict the development of intracranial hemorrhage include plasminogen activator inhibitor type 1 (PAI-1) and thrombin-activated fibrinolysis inhibitor (TAFI). PAI-1 suppresses fibrinolysis by inhibiting the effects of TPA, thereby causing thrombosis.[24] TAFI removes C-terminal lysine residues from partially degraded fibrin, reducing plasminogen binding to fibrin and subsequent activation by TPA. A study of 77 ischemic stroke patients identified lower levels of PAI-1 and TAFI to be associated with HT.[25] In combination, PAI-1 levels >180% and TAFI levels <21.4 ng mL^{-1} predicted SICH after TPA treatment.[26]

17.2.2 Malignant Cerebral Infarction

Malignant cerebral infarction refers to the development of space-occupying edema and cerebral herniation following an ischemic stroke occlusion affecting the anterior circulation, such as the internal carotid and middle cerebral artery[27] (Figure 17.2). In one study, c-Fn and MMP-9 levels were found to be higher in patients with malignant cerebral infarctions. c-Fn >16.6 $\mu g\,mL^{-1}$ had a sensitivity of 90% and a specificity of 100% for the prediction of a malignant course of cerebral infarction.[28] A randomized multicenter study suggested early decompressive surgery to be beneficial for patients with "malignant cerebral infarctions", where the overarching aim is to divert raised intracranial pressure away from the brainstem.[29] In this study, plasma S100β level >0.35 $g\,L^{-1}$ predicted the development of malignant cerebral infarction at 12 hours with 75% sensitivity and 80% specificity, and at 24 hours with 94% sensitivity and 83% specificity. Trial data on the use of these biomarkers to guide clinical decisions leading to early decompressive surgery are currently lacking.

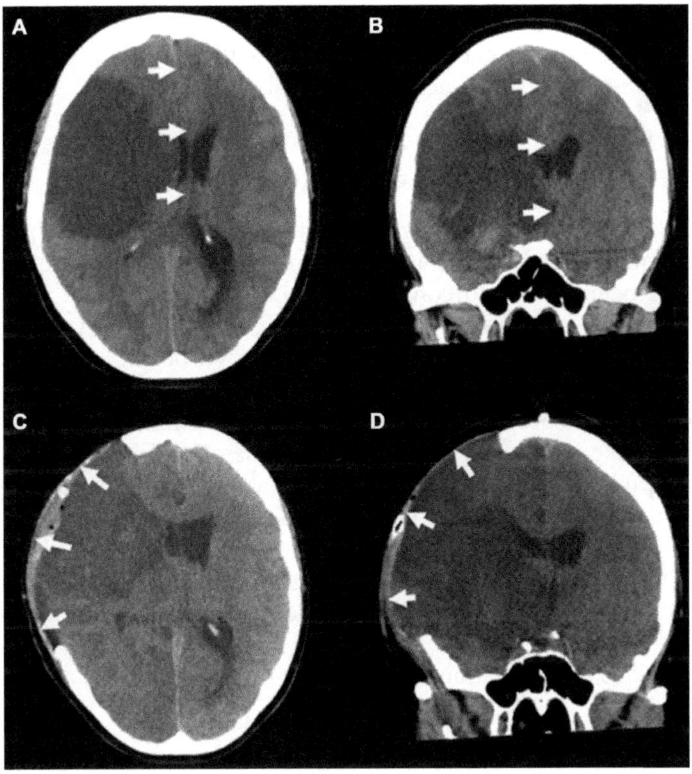

Figure 17.2 Malignant cerebral infarction. (A) Axial and (B) coronal computed tomography (CT) scans showing a right middle cerebral infarction with cerebral edema causing a mass effect and midline shift (white arrows). The ventricles are asymmetrically enlarged due to raised intracranial pressure. (C) Axial and (D) coronal CT scans of the same patient showing diversion of intracranial pressure outwards (white arrows) following a right decompressive hemicraniectomy.

17.2.2.1 *S100β*

Belonging to the S100 family is the S100β homodimeric glial protein that modulates cellular activity through binding with calcium. S100β comprises two β-subunits that are mostly expressed by astrocytes and, in minute concentrations, by neurons, microglia and oligodendrocytes.[30,31] Through the monitoring of secondary messenger calcium signaling, S100β regulates cell growth and participates in cellular differentiation and motility.[32] Even though S100β has been shown to be protective and trophic at low concentrations, increased levels of S100β indicate astrocytic damage and can be neurotoxic.[31] When structural damage such as infarction occurs, S100β is released into the cerebrospinal fluid and, coupled with the breakdown of the BBB, is circulated into the blood. Serum S100β values were shown to correlate with infarct size; the larger infarcts had higher concentrations of S100β as compared to lacunar infarcts.[33] Previous studies by Aurell *et al.* and Buttner *et al.* have demonstrated glial protein S100β to be increased in ischemic stroke patients who present 1–7 days from symptom onset.[34,35] An elevated S100β serum level before thrombolytic therapy has also been shown to be a risk factor for HT in acute stroke patients.[36]

17.2.3 Early Neurologic Deterioration

END, defined as worsening of neurological status within 48–72 hours of hospitalization, affects up to 40% of stroke patients.[37,38] Risk factors for stroke have been shown to contribute to END as well. These include age, diabetes mellitus, hypertension, high stroke severity, cardiac failure and high fibrinogen concentration[38] (Table 17.2). END occurs due to the development of complications following stroke such as failure of collateral circulation, thrombus progression, brain edema or herniation and HT. Often, a recurrent stroke event can be mistakenly diagnosed as an END due to overlapping clinical features.[39] Although existing management of END targets the prevention of these complications, early recognition of END and admission of stroke patients into specialized stroke units have been shown to reduce the rate of poor outcome.[40] With limited stroke care facilities, a quick and conclusive identification of patients is pivotal to the cost-effective

Table 17.2 Potential reasons for early neurologic deterioration.[119,120]

Brain factors	Systemic factors
Intracerebral hemorrhage	Atrial fibrillation
Severe stroke	High serum urea and leukocyte count
Space-occupying edema	Diabetes mellitus
Cerebral herniation	Inflammation
Poor collateral circulation	Hyperthermia
Thrombus progression	Hyperglycemia
	Hypoxia
	Blood pressure

allocation of these scarce resources. Biological signatures have the potential to identify these high-risk patients, especially since biomarkers such as ferritin, glutamate, γ-aminobutyric acid (GABA), TNF-α, ICAM-1 and MMPs have been associated with END.[28,41,42] For example, plasma glutamate levels of >200 μmol L^{-1} predicted END in patients with hemispheric stroke[42] and correlated with infarct growth determined using diffusion-weighted imaging.[26] More recently, inflammation and its surrogate markers such as IL-6 and C-reactive protein (CRP) have also been suspected to be driving mechanisms underlying END.[43] High-sensitivity CRP has been found to be associated with END alongside high-density lipoprotein (HDL) and the presence of internal carotid artery occlusion.[44]

17.2.3.1 C-reactive Protein

CRP is a type of pattern recognition receptor and is also a calcium-dependent ligand-binding plasma protein. CRP is promptly generated as an acute-phase response to inflammation and is often lauded for its sensitivity as a systemic marker of inflammation. In healthy individuals, its median concentration is 0.8 mg L^{-1}. However, during the acute phase of inflammation, CRP values may rise considerably from less than 50 μg L^{-1} to more than 500 mg L^{-1}. The production of plasma CRP in the liver is predominantly controlled by IL-6 during the transcription stage.[45] CRP is a well-studied biomarker of ischemic stroke. One study noted that CRP is the strongest independent marker of adverse outcome at hospital discharge, as ischemic stroke patients with CRP levels higher than 1.5 mg dL^{-1} at discharge develop significantly worse outcome.[46]

17.2.4 Stroke Diagnosis

A sudden-onset, focal neurological deficit experienced by a patient may occur as a result of stroke or as a consequence of a "stroke mimic". Examples of conditions that could mimic a stroke presentation include the aura of migraine, postictal deficits following a focal seizure, metabolic derangements (especially hypoglycemia), recrudescence of prior deficits in the setting of an infection or metabolic derangement, mass lesions or a functional psychogenic spell (Table 17.3). In a prospective series, a third of patients presenting with stroke-like symptoms may actually harbor a "stroke mimic".[47] While there exist clinical means to aid in differentiating between these possibilities (*e.g.* typical migraine preceding neurologic symptoms, jerking movements and sphincteric disturbances following a seizure), these features are not always present and often occur unwitnessed. As a significant proportion of such patients have altered consciousness and limited speech abilities, this information may not be conveyed to clinicians when time-sensitive decisions relating to acute reperfusion treatment are being considered.

While many cutting-edge neuroimaging methods such as magnetic resonance imaging (MRI) and vascular and perfusion scans exist to help

Table 17.3 Features of clinical scenarios mimicking stroke.[47,121,122]

Stroke mimics	Features
Migraine with aura	Preceding aura, headache
Epileptic seizure	Witnessed seizure activity, postictal period
Central nervous system abscess	Drug abuse, endocarditis
Central nervous system tumor	Gradual progression of symptoms, seizure at onset
Hypoglycemia	Diabetes, low serum glucose, reduced level of consciousness
Psychogenic	Neurological findings in a non-vascular distribution
Dementia	Memory loss, headache
Hypertensive encephalopathy	Headache, delirium, significant hypertension, cerebral edema
Wernicke's encephalopathy	Alcohol abuse, ataxia, confusion, ophthalmoplegia
Syncope	Insufficient blood flow to brain, fainting
Anxiety disorder	Hyperventilation, palpitation
Vestibular dysfunction	Chronic dizziness, imbalance

distinguish stroke from its mimics,[48] these options are not often available, or the investigations may be too time consuming to organize, even in established stroke centers. With a single or panel of biomarkers to complement basic CT scan findings, the presence of brain-derived biomarkers in the circulation could be confirmed, allowing stroke diagnosis to be made more quickly and accurately. In the setting of pre-hospital stroke care, these diagnostic stroke biomarkers could be used to promptly triage patients for neuroprotection agents to either halt or delay the detrimental cascade of events triggered by the ischemic insult before a definitive reperfusion treatment is administered. Studies have reported the ability of single and multiple-panel biomarkers to distinguish stroke from controls with high sensitivity and specificity by targeting antibodies to the glutamate NMDA-R (NR2A/NR2B subunits)[13,49] and an NMDA-R peptide.[50] NMDA-R, being a key regulator of nerve cell membrane functions, is cleaved by thrombin-activated serine proteases and triggers immune response and inflammation.[50] Reynolds and colleagues had previously screened plasma samples from 223 stroke patients (including ischemic stroke, intracerebral hemorrhage and subarachnoid hemorrhage) and from 214 healthy individuals for more than 50 serum biomarkers.[51] Univariate analysis revealed astroglial protein S100β, B-type neurotrophic growth factor, von Willebrand factor (vWF), MMP-9 and monocyte chemotactic protein-1 (MCP-1) to be associated with stroke.

17.2.4.1 von Willebrand Factor and Monocyte Chemotactic Protein-1

vWF is a plasma glycoprotein that spontaneously binds with platelet glycoprotein, thereby activating platelets. Its participation in platelet aggregation and thrombus formation escalates stroke development.[52] MCP-1, on the other hand, recruits macrophages and neutrophils, contributing to inflammatory injury following stroke onset.[53] The same group published data on 65 patients with suspected ischemic stroke admitted within 24 hours of symptom onset and 157 healthy controls.[54] Twenty-six blood-based markers with relation to the ischemic cascade were analyzed. Protein S100β, MMP-9, vascular cell adhesion molecule-1 and vWF were identified as having correlations with ischemic stroke. Subsequently, Laskowitz and colleagues obtained data on 130 patients admitted with acute focal neurologic deficits within 6 hours of symptom onset.[55] Forty-one patients were later diagnosed with ischemic stroke. The predictive model included brain natriuretic peptide (BNP), CRP, MMP-9, D-dimer and S100β.

Driven by these encouraging results, the diagnostic accuracy of a biomarker panel including BNP, MMP-9, D-dimer and S100β was evaluated in a prospective multicenter trial.[56] Within a 3 year period, more than 1100 patients presenting with symptoms suspicious for stroke were enrolled within 24 hours of symptom onset. The multivariate model was capable of only moderately differentiating between patients with and without stroke. Setting the threshold of the model to the 25th percentile revealed a sensitivity of 86% and a specificity of 37% for discriminating stroke from its mimics, which is of limited use to guide decision-making relating to thrombolysis. There is still much to explore for the use of a single or a panel of biomarkers to aid in the exclusion of stroke mimics in order to triage patients for neuroimaging confirmation of an ischemic stroke.

17.2.5 Thrombus Formation

Thrombus formation, or thrombogenesis, results from atypical platelet adherence and aggregation to the vessel wall, leading to disruption of cerebral blood flow into brain tissues. In thromboembolic stroke, blood thrombus that is formed from the large arteries of the aorta or the heart can be dislodged into the cerebral circulation. Antiplatelet drugs are used to prevent the formation of such a thrombus. Notable antiplatelet agents (*e.g.* aspirin and clopidogrel), used either singly or in combination, have been shown to confer significant benefit to stroke patients by lowering the risk of recurrent vascular events and death[57] (Figure 17.3). Yet, resistance to the action of aspirin can occur, and this might decrease its benefit in mitigating against stroke risks. The definitions of aspirin resistance vary considerably across studies depending on the type of assays used. Despite the importance of aspirin resistance in clinical practice, the heterogeneity in the definitions of aspirin resistance has led to a lack of consensus on the roles of platelet function testing in stroke management.

17.2.5.1 D-dimers

Plasma D-dimer is a product formed from the degradation of fibrinogen and reflects thrombin production and fibrinolysis. D-dimer levels usually vary in

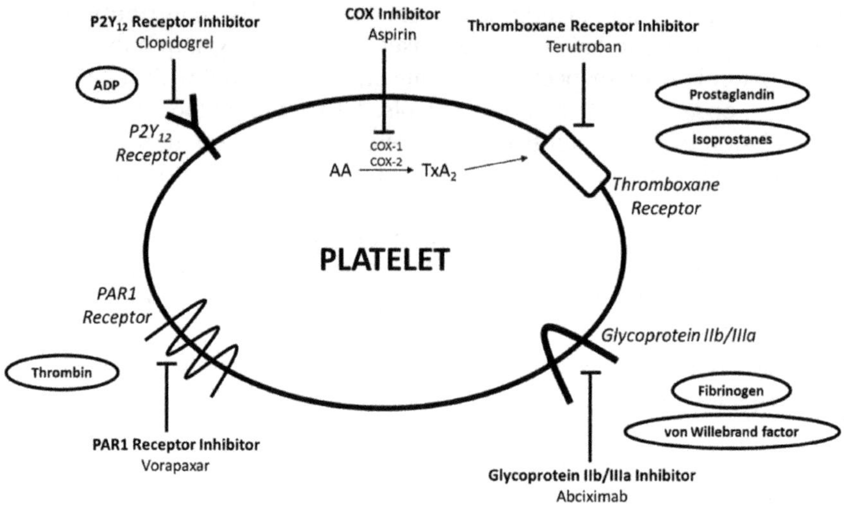

Figure 17.3 Inhibitors and their respective target sites on a platelet. Clopidogrel prevents platelet activation by competing with ADP for receptor $P2Y_{12}$ and binding irreversibly to the target site. Aspirin is a COX inhibitor that suppresses the conversion of arachidonic acid (AA) to thromboxane A_2 (TxA_2), which induces platelet aggregation through its cell surface receptor, thromboxane receptor. The inhibition of thromboxane receptor by terutroban prevents agonists such as prostaglandins and isoprostanes from activating the platelet. The glycoproteins von Willebrand factor and fibrinogen bind to the cell surface receptor glycoprotein IIb/IIIa to initiate platelet aggregation, and this process can be inhibited by abciximab. Protease-activated receptor 1 (PAR1) is another receptor that is present on human platelets and is activated primarily by thrombin, leading to thrombosis. This can be suppressed by vorapaxar, a PAR1 receptor inhibitor.

healthy individuals, but tend to correlate positively with other cardiovascular risk factors,[58] where the presence of D-dimer indicates the generation of thrombin and plasmin.[59] As cross-linked fibrin assembled during thrombus formation is thought to give rise to D-dimer, an elevation in fibrin D-dimer levels usually confirms thrombus formation. D-dimer has been demonstrated to be more stable as compared to the other thrombotic markers following ischemic stroke.[60] D-dimer has also been used in various studies and has proven to be a good predictive biomarker of stroke.[55,56]

17.2.5.2 Platelet Reactivity

Under normal physiological conditions, platelets circulate in close proximity to the surface of the vascular walls, but are protected from premature activation by the healthy endothelial cells, which serve as a natural barrier to prevent thrombosis (Figure 17.4). Activation of platelets occurs when the endothelial layer becomes damaged, exposing the underlying subendothelial

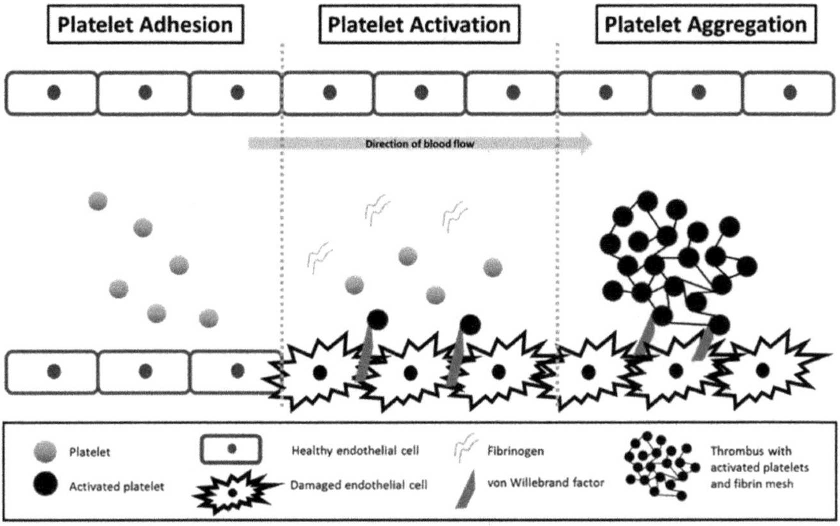

Figure 17.4 Thrombus formation. In their inactivated state, platelets circulate in close proximity to the healthy endothelial cells on the surface of the vascular walls. When the endothelial cells become damaged, the underlying subendothelial matrix becomes exposed, revealing blood glycoproteins such as von Willebrand factor, among others. These proteins activate the platelets by capturing them and binding with the platelet receptors. Glycoprotein fibrinogen becomes activated by thrombin, converting into fibrin strands. Fibrin strands crosslink with one another to form a fibrin mesh that binds the aggregating platelets together to develop into a thrombus.

matrix. vWF, among others, catches the drifting platelets, binds with platelet receptors and induces activation signals.[61] Thrombin-activated fibrin strands cross-link with one another to form a fibrin mesh. Together with the fibrin mesh, the aggregating platelets develop into a thrombus. As thrombogenesis is the consequence of occlusive platelet aggregation, inhibiting the latter is important, not only in the chronic phase, but also in the acute phase of brain ischemia, to prevent platelet aggregation and early vascular recurrence.[62]

The ability to identify patients who are resistant to antiplatelet drugs such as aspirin or clopidogrel early during their course of stroke is advantageous, as a suitable antiplatelet agent can be used for secondary prevention. The concept of antiplatelet resistance is borne out of observations that certain patients do not derive protection from antiplatelet treatment. For example, aspirin resistance can be identified by measuring thromboxane A_2 production and platelet aggregation,[57] or by measuring urinary levels of 11-dehydrxythrombxane B_2.[63] The value of platelet function testing is not clear in stroke patients given the relative scarcity of data in this area. It remains unclear whether changing a therapy in patients identified to be aspirin resistant results in improved outcomes. One study suggested that in stroke patients with a level of MCP-1 >217 $pg\,mL^{-1}$, outcomes were better at 90 days in those treated with aspirin plus extended-release dipyridamole (Aggrenox) compared with aspirin alone.[64] High uric acid levels have been shown to predict aspirin resistance where a cut-off value of 6.45 $mg\,dL^{-1}$ of uric acid predicted aspirin resistance with 79% sensitivity and 65% specificity.[65] These findings suggest a potential application of platelet biomarkers in personalizing the choice of antiplatelet treatment for cardiovascular disease prevention.

17.2.5.3 Uric Acid

Uric acid, which exists almost entirely in its ionized form urate at physiological pH, is derived from adenine- and guanine-based purine compounds.[66] As uric acid possesses the ability to scavenge peroxynitrite, hydroxyl radicals and singlet oxygen, as well as being able to chelate transition metals, urate is widely considered as a potent antioxidant in humans.[66] Uric acid is degraded to allantoin in most mammals by the hepatic enzyme urate oxidase, or uricase. In humans, however, mutations in early hominoids that rendered the uricase gene non-functional have resulted in urate levels in human plasma that are approximately ten-times those of most other mammals.[67] *In vivo*, uric acid is a potent water-soluble antioxidant that targets free radicals caused by oxidative damage, including hydroxyl radicals and superoxide,[68,69] but at higher levels could exhibit paradoxical pro-oxidant effects.[70] Data supporting the neuroprotective functions of uric acid in ischemic stroke are largely borne out of observational and interventional trials. In a prospective study involving 881 consecutive patients, uric acid levels were inversely associated with the extent of neurological deficits on admission and the final infarct volume on CT/MRI scans.[71] In patients receiving TPA treatment, uric acid levels correlated with better neurological outcomes, whereas patients with malignant middle cerebral artery infarction and SICH have significantly lower uric acid levels.[72] The value of replenishing uric acid in the latter group of patients has been investigated in trials that examined the dual administration of uric acid and TPA during acute ischemic stroke.[73] Despite early promising findings, in the definitive URICO-ICTUS Phase IIb/III trial that recruited 411 patients with acute ischemic stroke, the addition of uric acid to thrombolytic therapy did not increase the proportion of patients who achieved excellent outcome after stroke compared with placebo.[74]

17.2.6 Stroke Etiology

Various etiologies of stroke exist, namely large artery disease, cardioembolism and lacunar stroke (Figure 17.5). Determining the underlying cause of stroke is paramount to preventing a recurrent stroke. For example, patients with a cardioembolic stroke benefit significantly from anticoagulation treatment, whereas antiplatelet treatment and strict risk factor control are

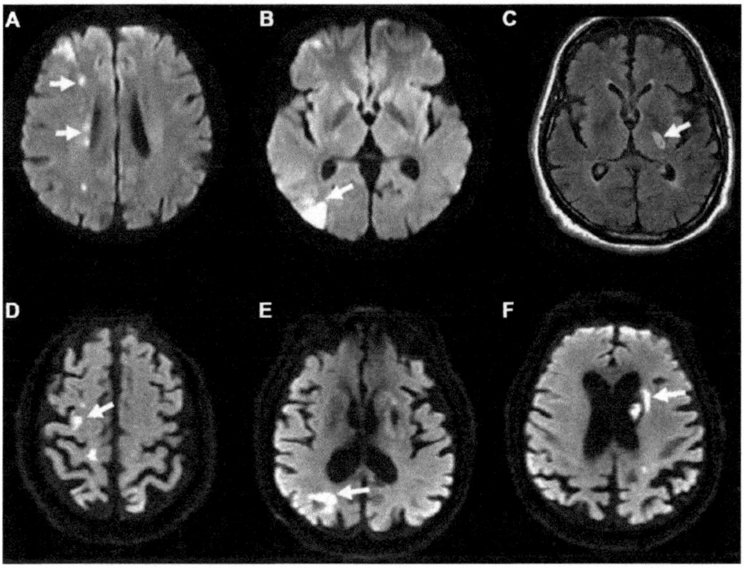

Figure 17.5 Stroke mechanisms. (A, B) Large artery disease. Magnetic resonance imaging (MRI) of the brain of a patient with a severe stenosis of the right middle cerebral artery causing a hypoperfusion cerebral infarction. (C) Small vessel disease. MRI of the brain of a patient showing a lacunar infarction in the posterior limb of the left internal capsule. (D–F) Cardioembolism. MRI of the brain of a patient showing simultaneous cerebral infarction involving multiple vascular territories. White arrows highlight areas of cerebral infarction.

noted to work more effectively for patients with a large artery disease and lacunar stroke.[75] Additionally, patients with a symptomatic large artery disease due to carotid stenosis could benefit from either endarterectomy or stenting. Despite an extensive determination of stroke etiology, the mechanism of stroke is unknown in up to 35% of patients.[76,77] With a biomarker-based platform to identify the likely cause of stroke, recognizing the stroke etiology of a patient could be achieved more quickly, thus rationalizing the need for expensive investigations (*e.g.* angiographic studies, echocardiograms and prolonged cardiac rhythm monitoring). Two such useful biomarkers are pro-BNP and D-dimer, which have been shown to reliably distinguish cardioembolic from non-cardioembolic strokes.[78–80] Pooled data from 23 studies that included clinical information from 2834 patients showed significant elevation in pro-BNP levels in cardioembolic stroke within 72 hours of symptom onset.[81] Conversely, RNA expression in blood has been shown to distinguish cardioembolic stroke from large artery disease. In a study of 194 samples from 76 patients with acute ischemic stroke, a 40 gene panel was able to distinguish strokes of a cardioembolic from large artery disease etiology with >95% sensitivity and specificity at 3, 5 and 24 hours after stroke onset.[82]

17.2.6.1 *N-terminal Pro-brain Natriuretic Peptide*

BNP is a neurohormone that is released by the cardiac ventricles and the brain and circulates in the blood.[83,84] BNP acts as a vasodilator that counters vasoconstrictor effects, stimulating both natriuresis and diuresis in the body.[85,86] Apart from the biologically active BNP, the cleavage of a propeptide also gives rise to N-terminal proBNP (NT-proBNP), which does not have biological activity. Both of these biomarkers are present in equal amounts in normal individuals.[87] NT-proBNP fragment levels have been proven to be useful in identifying patients with cardioembolic stroke by a recent pooled meta-analysis,[81] while BNP continues to be widely studied in ischemic stroke patients.

17.2.7 Stroke Risk

Stroke patients are at increased risk of developing a recurrent stroke. Several biomarkers have been shown to predict this risk.

17.2.7.1 *Lipoprotein-associated Phospholipase A_2*

An enzyme derived from leukocytes, lipoprotein-associated phospholipase A_2 (Lp-PLA_2) is involved in the metabolism of low-density lipoprotein (LDL), releasing pro-inflammatory molecules. Associations have been observed between Lp-PLA_2 and risk of recurrent stroke event, resulting in a two-fold increase in stroke occurrence.[88–90]

17.2.7.2 *Lipoproteins*

While different forms of lipoproteins exist, oxidized LDL (oxLDL) and HDL are associated with the risk of stroke events, though in opposite ways. oxLDL is a biomarker closely linked with atherosclerosis. High levels of oxLDL have recently been revealed to have correlations with 1 year mortality and poor functional outcome, especially among patients with large artery and lacunar stroke mechanisms.[91] In contrast, high levels of HDLs may have beneficial effects in stroke patients, as HDLs have been noted to exert anti-atherosclerotic,[92] antithrombotic,[93] anti-inflammatory[94] and antioxidant[95] properties.

17.2.8 Atrial Fibrillation

Atrial fibrillation (AF) is an erratic firing of electrical activities that causes pooling of blood within the heart. Together with thrombogenic factors, this condition increases the chances of thrombus formation. AF is the prevailing cardiac rhythm disorder that has led to increased mortality and morbidity from stroke and thromboembolism.[96,97] Using data from the ARISTOTLE and STABILITY trials, Hijazi and colleagues developed a risk assessment tool that is based on multiple biomarkers including NT-proBNP (indicating myocyte stress) and cardiac high-sensitivity troponin T (hsTnT; indicating myocardial injury) to assign AF-associated stroke risk. The ABC stroke (age, biomarkers and clinical history of stroke or transient ischemic attack) yielded higher C-indices than the widely used CHA_2DS_2-VASc scores to guide selection of high-risk patients for long-term anticoagulation.[98] In a study of 880 patients with non-valvular AF, CRP levels positively correlated with the $CHADS_2$ stroke risk assessment score and were higher in patients with all-cause mortality and vascular events. However, no significant correlation was found with stroke risk.[99] Among individuals with high cardiovascular risk, our group has shown significant elevations of multiple biomarkers of oxidative damage such as F_2-isoprostanes, neuroprostanes, 7α-hydroxycholesterol and γ-glutamyltransferase.[100] In anticoagulated non-valvular AF patients, urinary F_2-isoprostanes predicted cardiovascular events and all-cause mortality, and the addition of tertiles of urinary F_2-isoprostanes to CHA_2DS_2-VASc scores improved the performance of each outcome.[101]

In patients with cryptogenic stroke, sieving out patients at risk of the occurrence of AF is pivotal to determining the necessity for and length of prolonged cardiac monitoring.[102] Current methods of patient selection are largely based on loose clinical criteria such as age, the pattern of stroke on neuroimaging studies and the number of cardiovascular risk factors.[102,103] A study by Rodriguez-Yanez and colleagues demonstrated the value of measuring proBNP where proBNP ≥360 pg mL^{-1} was the only variable that was independently associated with the risk of developing AF.[104]

17.2.8.1 *F_2-isoprostanes*

F_2-isoprostanes are a family of prostaglandin isomers formed by free radicals that induce the peroxidation of arachidonic acid.[105–110] F_2-isoprostanes can be precisely quantified through various analytical platforms such as high-performance liquid chromatography, gas chromatography–mass spectrometry and liquid chromatography–mass spectrometry using the plasma, serum, urine, saliva or cerebrospinal fluid of a patient.[111] They are widely regarded as accurate and reliable biomarkers of oxidative damage. F_2-isoprostanes have been studied among individuals with various neurological conditions such as Parkinson disease[107] and among individuals with vascular risk factors such diabetes mellitus,[112] hypertension[100] and cigarette smoking.[108]

Despite mounting evidence implicating F_2-isoprostanes in human diseases, few studies have examined F_2-isoprostanes in human stroke. Sanchez-Moreno and colleagues first reported higher F_2-isoprostane levels within 5 days following stroke.[113] Ward and colleagues supported these findings and similarly showed an increase in F_2-isoprostane levels during the early course of ischemic stroke.[114] Our group subsequently showed that the rise in F_2-isoprostanes occurred as early as 3 hours after ischemic stroke onset and remained elevated for several days.[106] In the Biomarker Evaluation for Antioxidant Therapies in Stroke (BEAT-Stroke) study, Kelly and colleagues observed that early F_2-isoprostane levels within 8 hours following the onset of ischemic stroke correlated with MMP-9 levels.[11] Taken together, these data suggest that F_2-isoprostanes are sensitive markers of oxidative damage in human stroke that are elevated early and remain elevated for several days.

As the brain consumes large amounts of oxygen each day, it is largely dependent on the heart to effectively pump and deliver sufficient blood to the cerebrovascular circulation. Disruptions in blood flow to the heart–brain axis could throw the brain into a state of cerebral hypoperfusion and promote the development of blood thrombus within the cardiac atrium and left atrial appendage. Besides BNP, another cardiac biomarker that has been extensively studied is the hsTnT.

17.2.8.2 High-sensitivity Troponin T

Being a highly sensitive and specific marker of myocardial necrosis, troponin has the potential to be used in the diagnosis of acute myocardial infarction and to stratify patients according to the severity of their acute coronary symptoms.[115] Some studies have drawn a link between elevated levels of cardiac troponin T and poor outcomes in acute stroke.[116–118] In a meta-analysis, an elevation in cardiac troponin levels in stroke patients was associated with an overall increased risk of death and disability.[115]

17.3 Conclusions

Biomarkers provide critical insights into the pathological events following stroke onset and could add to the armamentarium of clinical tools available to a clinician in stroke management. Although the ischemic cascade has been extensively studied in different animal stroke models, few studies have examined the prognostic significance and release of transient biomarkers that examine the different states of damage and dysfunction in human stroke. More efforts should focus on the standardization of study cohorts and analytical methods, the incremental yield of different markers and the cost–effectiveness of incorporating biomarker measurement in order to guide clinical decision-making.

Acknowledgements

We would like to thank the National Medical Research Council, Singapore (NMRC/CSA-SI/0003/2015, NMRC/CNIG/1115/2014 and NMRC/MOHIAF-Cat1/0015/2014) for their generous support.

References

1. V. L. Feigin, M. H. Forouzanfar, R. Krishnamurthi, G. A. Mensah, M. Connor, D. A. Bennett, A. E. Moran, R. L. Sacco, L. Anderson, T. Truelsen, M. O'Donnell, N. Venketasubramanian, S. Barker-Collo,

C. M. Lawes, W. Wang, Y. Shinohara, E. Witt, M. Ezzati, M. Naghavi, C. Murray, Global Burden of Diseases, Injuries, Risk Factors study and G. B. D. Stroke Experts Group, *Lancet*, 2014, **383**, 245–254.
2. L. H. Kuller, L. Shemanski, B. M. Psaty, N. O. Borhani, J. Gardin, M. N. Haan, D. H. O'Leary, P. J. Savage, G. S. Tell and R. Tracy, *Circulation*, 1995, **92**, 720–726.
3. B. M. Psaty, C. D. Furberg, L. H. Kuller, D. E. Bild, P. M. Rautaharju, J. F. Polak, E. Bovill and J. S. Gottdiener, *Arch. Intern. Med.*, 1999, **159**, 1339–1347.
4. E. H. Lo, M. A. Moskowitz and T. P. Jacobs, *Stroke*, 2005, **36**, 189–192.
5. J. Astrup, B. K. Siesjo and L. Symon, *Stroke*, 1981, **12**, 723–725.
6. N. R.-P. S. S. Group, *N. Engl. J. Med.*, 1995, **333**, 1581–1587.
7. W. J. Powers, C. P. Derdeyn, J. Biller, C. S. Coffey, B. L. Hoh, E. C. Jauch, K. C. Johnston, S. C. Johnston, A. A. Khalessi, C. S. Kidwell, J. F. Meschia, B. Ovbiagele, D. R. Yavagal and C. American Heart Association Stroke, *Stroke*, 2015, **46**, 3020–3035.
8. R. C. Seet and A. A. Rabinstein, *Cerebrovasc. Dis.*, 2012, **34**, 106–114.
9. G. J. L. Ng, A. M. L. Quek, C. Cheung, T. V. Arumugam and R. C. S. Seet, *Neurochem. Int.*, 2017, **107**, 11–22.
10. W. Whiteley, W. L. Chong, A. Sengupta and P. Sandercock, *Stroke*, 2009, **40**, e380–e389.
11. P. J. Kelly, J. D. Morrow, M. Ning, W. Koroshetz, E. H. Lo, E. Terry, G. L. Milne, J. Hubbard, H. Lee, E. Stevenson, M. Lederer and K. L. Furie, *Stroke*, 2008, **39**, 100–104.
12. D. Orion, Y. Schwammenthal, T. Reshef, R. Schwartz, R. Tsabari, O. Merzeliak, J. Chapman, Y. A. Mekori and D. Tanne, *Eur. J. Neurol.*, 2008, **15**, 323–328.
13. S. A. Dambinova, G. A. Khounteev and A. A. Skoromets, *Stroke*, 2002, **33**, 1181–1182.
14. D. T. Laskowitz, H. Grocott, A. Hsia and K. R. Copeland, *J. Stroke Cerebrovasc. Dis.*, 1998, **7**, 234–241.
15. R. C. Seet, Y. Zhang, S. A. Moore, E. F. Wijdicks and A. A. Rabinstein, *Stroke*, 2011, **42**, 2333–2335.
16. R. C. Seet, A. A. Rabinstein, T. J. Christianson, G. W. Petty and R. D. Brown Jr., *J. Stroke Cerebrovasc. Dis.*, 2013, **22**, 561–569.
17. M. S. Pessin, G. J. Del Zoppo and C. J. Estol, *Clin. Neuropharmacol.*, 1990, **13**, 271–289.
18. R. C. Seet, E. F. Wijdicks and A. A. Rabinstein, *Arch. Neurol.*, 2012, **69**, 1615–1620.
19. S. Mun-Bryce and G. A. Rosenberg, *J. Cereb. Blood Flow Metab.*, 1998, **18**, 1163–1172.
20. T. L. Barr, L. L. Latour, K. Y. Lee, T. J. Schaewe, M. Luby, G. S. Chang, Z. El-Zammar, S. Alam, J. M. Hallenbeck, C. S. Kidwell and S. Warach, *Stroke*, 2010, **41**, e123–e128.

21. M. Ning, K. L. Furie, W. J. Koroshetz, H. Lee, M. Barron, M. Lederer, X. Wang, M. Zhu, A. G. Sorensen, E. H. Lo and P. J. Kelly, *Neurology*, 2006, **66**, 1550–1555.
22. M. Castellanos, R. Leira, J. Serena, J. M. Pumar, I. Lizasoain, J. Castillo and A. Davalos, *Stroke*, 2003, **34**, 40–46.
23. M. Castellanos, R. Leira, J. Serena, M. Blanco, S. Pedraza, J. Castillo and A. Davalos, *Stroke*, 2004, **35**, 1671–1676.
24. T. Wyseure and P. J. Declerck, *Drug Discovery Today*, 2014, **19**, 1476–1482.
25. M. Ribo, J. Montaner, C. A. Molina, J. F. Arenillas, E. Santamarina, M. Quintana and J. Alvarez-Sabin, *Stroke*, 2004, **35**, 2123–2127.
26. C. Foerch, M. T. Wunderlich, F. Dvorak, M. Humpich, T. Kahles, M. Goertler, J. Alvarez-Sabin, C. W. Wallesch, C. A. Molina, H. Steinmetz, M. Sitzer and J. Montaner, *Stroke*, 2007, **38**, 2491–2495.
27. W. Hacke, S. Schwab, M. Horn, M. Spranger, M. De Georgia and R. von Kummer, *Arch. Neurol.*, 1996, **53**, 309–315.
28. J. Serena, M. Blanco, M. Castellanos, Y. Silva, J. Vivancos, M. A. Moro, R. Leira, I. Lizasoain, J. Castillo and A. Davalos, *Stroke*, 2005, **36**, 1921–1926.
29. K. Vahedi, J. Hofmeijer, E. Juettler, E. Vicaut, B. George, A. Algra, G. J. Amelink, P. Schmiedeck, S. Schwab, P. M. Rothwell, M. G. Bousser, H. B. van der Worp, W. Hacke, D. Decimal and H. investigators, *Lancet Neurol.*, 2007, **6**, 215–222.
30. C. Richter-Landsberg and M. Heinrich, *J. Neurosci. Res.*, 1995, **42**, 657–665.
31. J. Steiner, H. G. Bernstein, B. Bogerts, T. Gos, C. Richter-Landsberg, M. T. Wunderlich and G. Keilhoff, *Neuroscience*, 2008, **154**, 496–503.
32. B. W. Schafer and C. W. Heizmann, *Trends Biochem. Sci.*, 1996, **21**, 134–140.
33. H. D. Abraha, R. J. Butterworth, P. M. Bath, W. S. Wassif, J. Garthwaite and R. A. Sherwood, *Ann. Clin. Biochem.*, 1997, **34**(Pt 4), 366–370.
34. A. Aurell, L. E. Rosengren, B. Karlsson, J. E. Olsson, V. Zbornikova and K. G. Haglid, *Stroke*, 1991, **22**, 1254–1258.
35. T. Buttner, S. Weyers, T. Postert, R. Sprengelmeyer and W. Kuhn, *Stroke*, 1997, **28**, 1961–1965.
36. C. Foerch, M. T. Wunderlich, F. Dvorak, M. Humpich, T. Kahles, M. Goertler, J. Alvarez-Sabín, C. W. Wallesch, C. A. Molina, H. Steinmetz, M. Sitzer and J. Montaner, *Stroke*, 2007, **38**, 2491–2495.
37. J. F. Arenillas, A. Rovira, C. A. Molina, E. Grive, J. Montaner and J. Alvarez-Sabin, *Stroke*, 2002, **33**, 2197–2203.
38. L. C. Lin, T. H. Lee, C. H. Chang, Y. J. Chang, C. W. Liou, K. C. Chang, J. D. Lee, T. Y. Peng, J. Chung, S. C. Chen and J. T. Yang, *Eur. Neurol.*, 2012, **67**, 186–192.
39. J. Castillo, *Cerebrovasc Dis.*, 1999, **9**(Suppl 3), 1–8.
40. J. Roquer, A. Rodriguez-Campello, M. Gomis, J. Jimenez-Conde, E. Cuadrado-Godia, R. Vivanco, E. Giralt, M. Sepulveda, C. Pont-Sunyer, G. Cucurella and A. Ois, *J. Neurol.*, 2008, **255**, 1012–1017.

41. A. Davalos, R. Leira, J. Serena, M. Castellanos, A. Aneiros and J. Castillo, *Stroke*, 2000, **32**, 370.
42. A. Davalos, J. Castillo, J. Marrugat, J. M. Fernandez-Real, A. Armengou, P. Cacabelos and R. Rama, *Neurology*, 2000, **54**, 1568–1574.
43. N. Vila, J. Castillo, A. Davalos and A. Chamorro, *Stroke*, 2000, **31**, 2325–2329.
44. W. K. Seo, H. Y. Seok, J. H. Kim, M. H. Park, S. W. Yu, K. Oh, S. B. Koh and K. W. Park, *J. Stroke Cerebrovasc. Dis.*, 2012, **21**, 181–186.
45. M. B. Pepys and G. M. Hirschfield, *J. Clin. Invest.*, 2003, **111**, 1805–1812.
46. M. Di Napoli, F. Papa and V. Bocola, *Stroke*, 2001, **32**, 917–924.
47. P. J. Hand, J. Kwan, R. I. Lindley, M. S. Dennis and J. M. Wardlaw, *Stroke*, 2006, **37**, 769–775.
48. G. A. Donnan and S. M. Davis, *Lancet Neurol.*, 2002, **1**, 417–425.
49. S. A. Dambinova, G. A. Khounteev, G. A. Izykenova, I. G. Zavolokov, A. Y. Ilyukhina and A. A. Skoromets, *Clin. Chem.*, 2003, **49**, 1752–1762.
50. S. A. Dambinova, K. Bettermann, T. Glynn, M. Tews, D. Olson, J. D. Weissman and R. L. Sowell, *PLoS One*, 2012, **7**, e42362.
51. M. A. Reynolds, H. J. Kirchick, J. R. Dahlen, J. M. Anderberg, P. H. McPherson, K. K. Nakamura, D. T. Laskowitz, G. E. Valkirs and K. F. Buechler, *Clin. Chem.*, 2003, **49**, 1733–1739.
52. T. N. Bongers, M. P. de Maat, M. L. van Goor, V. Bhagwanbali, H. H. van Vliet, E. B. Gomez Garcia, D. W. Dippel and F. W. Leebeek, *Stroke*, 2006, **37**, 2672–2677.
53. Y. Chen, J. M. Hallenbeck, C. Ruetzler, D. Bol, K. Thomas, N. E. Berman and S. N. Vogel, *J. Cereb. Blood Flow Metab.*, 2003, **23**, 748–755.
54. J. R. Lynch, R. Blessing, W. D. White, H. P. Grocott, M. F. Newman and D. T. Laskowitz, *Stroke*, 2004, **35**, 57–63.
55. D. T. Laskowitz, R. Blessing, J. Floyd, W. D. White and J. R. Lynch, *Ann. N. Y. Acad. Sci.*, 2005, **1053**, 30.
56. D. T. Laskowitz, S. E. Kasner, J. Saver, K. S. Remmel, E. C. Jauch and B. S. Group, *Stroke*, 2009, **40**, 77–85.
57. M. Rafferty, M. R. Walters and J. Dawson, *Curr. Med. Chem*, 2010, **17**, 4578–4586.
58. A. R. Folsom, J. A. Delaney, P. L. Lutsey, N. A. Zakai, N. S. Jenny, J. F. Polak and M. Cushman, *Am. J. Hematol.*, 2009, **84**, 349–353.
59. E. Haapaniemi and T. Tatlisumak, *Acta Neurol. Scand.*, 2009, **119**, 141–150.
60. M. Barber, P. Langhorne, A. Rumley, G. D. Lowe and D. J. Stott, *Stroke*, 2004, **35**, 1421–1425.
61. E. M. Golebiewska and A. W. Poole, *Blood Rev.*, 2015, **29**, 153–162.
62. R. Joseph, G. D'Andrea, S. B. Oster and K. M. Welch, *Stroke*, 1989, **20**, 38–44.
63. M. Dobaczewski, M. Nocun, I. Zavodnik, O. Ulicna, E. Lapshina, L. Zavodnik, J. Golanski, P. Kazmierczak, Z. Durackova, B. Kostka, L. Markuszewski and C. Watala, *Blood Coagulation Fibrinolysis*, 2008, **19**, 421–428.

64. H. Worthmann, R. Dengler, H. Schumacher, A. Schwartz, W. G. Eisert, R. Lichtinghagen and K. Weissenborn, *Int. J. Mol. Sci.*, 2012, **13**, 8670–8678.
65. B. S. Yildiz, E. Ozkan, F. Esin, Y. I. Alihanoglu, H. Ozkan, M. Bilgin, I. D. Kilic, A. Ergin, H. A. Kaftan and H. Evrengul, *Blood Coagulation Fibrinolysis*, 2016, **27**, 412–418.
66. B. F. Becker, *Free Radic. Biol. Med.*, 1993, **14**, 615–631.
67. X. W. Wu, D. M. Muzny, C. C. Lee and C. T. Caskey, *J. Mol. Evol.*, 1992, **34**, 78–84.
68. H. U. Hink, N. Santanam, S. Dikalov, L. McCann, A. D. Nguyen, S. Parthasarathy, D. G. Harrison and T. Fukai, *Arterioscler. Thromb. Vasc. Biol.*, 2002, **22**, 1402–1408.
69. B. N. Ames, R. Cathcart, E. Schwiers and P. Hochstein, *Proc. Natl. Acad. Sci. U S A*, 1981, **78**, 6858–6862.
70. R. C. Seet, K. Kasiman, J. Gruber, S. Y. Tang, M. C. Wong, H. M. Chang, Y. H. Chan, B. Halliwell and C. P. Chen, *Atherosclerosis*, 2010, **209**, 215–219.
71. A. Chamorro, V. Obach, A. Cervera, M. Revilla, R. Deulofeu and J. H. Aponte, *Stroke*, 2002, **33**, 1048–1052.
72. S. Amaro, X. Urra, M. Gomez-Choco, V. Obach, A. Cervera, M. Vargas, F. Torres, J. Rios, A. M. Planas and A. Chamorro, *Stroke*, 2011, **42**, S28–S32.
73. S. Amaro, D. Soy, V. Obach, A. Cervera, A. M. Planas and A. Chamorro, *Stroke*, 2007, **38**, 2173–2175.
74. A. Chamorro, S. Amaro, M. Castellanos, T. Segura, J. Arenillas, J. Marti-Fabregas, J. Gallego, J. Krupinski, M. Gomis, D. Canovas, X. Carne, R. Deulofeu, L. S. Roman, L. Oleaga, F. Torres, A. M. Planas and U.-I. Investigators, *Lancet Neurol.*, 2014, **13**, 453–460.
75. H. P. Adams Jr., B. H. Bendixen, L. J. Kappelle, J. Biller, B. B. Love, D. L. Gordon and E. E. Marsh 3rd, *Stroke*, 1993, **24**, 35–41.
76. J. P. Mohr, *N. Engl. J. Med.*, 1988, **318**, 1197–1198.
77. G. W. Petty, R. D. Brown Jr., J. P. Whisnant, J. D. Sicks, W. M. O'Fallon and D. O. Wiebers, *Stroke*, 2000, **31**, 1062–1068.
78. W. Ageno, S. Finazzi, L. Steidl, M. G. Biotti, V. Mera, G. Melzi D'Eril and A. Venco, *Arch. Intern. Med.*, 2002, **162**, 2589–2593.
79. T. Tombul, C. Atbas and O. Anlar, *J. Clin. Neurosci.*, 2005, **12**, 429–434.
80. J. Isenegger, N. Meier, B. Lammle, L. Alberio, U. Fischer, K. Nedeltchev, J. Gralla, H. P. Kohler, H. P. Mattle and M. Arnold, *Cerebrovasc. Dis.*, 2010, **29**, 82–86.
81. V. Llombart, A. Antolin-Fontes, A. Bustamante, D. Giralt, N. S. Rost, K. Furie, K. Shibazaki, M. Biteker, J. Castillo, M. Rodriguez-Yanez, A. C. Fonseca, T. Watanabe, F. Purroy, W. Zhixin, T. Etgen, N. Hosomi, S. R. Jafarian Kerman, J. C. Sharma, C. Knauer, E. Santamarina, G. Giannakoulas, T. Garcia-Berrocoso and J. Montaner, *Stroke*, 2015, **46**, 1187–1195.
82. G. C. Jickling, H. Xu, B. Stamova, B. P. Ander, X. Zhan, Y. Tian, D. Liu, R. J. Turner, M. Mesias, P. Verro, J. Khoury, E. C. Jauch, A. Pancioli, J. P. Broderick and F. R. Sharp, *Ann. Neurol.*, 2010, **68**, 681–692.

83. T. Sudoh, K. Kangawa, N. Minamino and H. Matsuo, *Nature*, 1988, **332**, 78–81.
84. N. Minamino, M. Aburaya, S. Ueda, K. Kangawa and H. Matsuo, *Biochem. Biophys. Res. Commun.*, 1988, **155**, 740–746.
85. G. Giannakoulas, A. Hatzitolios, H. Karvounis, G. Koliakos, A. Charitandi, T. Dimitroulas, C. Savopoulos, E. Tsirogianni and G. Louridas, *Angiology*, 2005, **56**, 723–730.
86. B. C. Stein and R. I. Levin, *Am. Heart J.*, 1998, **135**, 914–923.
87. P. J. Hunt, T. G. Yandle, M. G. Nicholls, A. M. Richards and E. A. Espiner, *Biochem. Biophys. Res. Commun.*, 1995, **214**, 1175–1183.
88. P. B. Gorelick, *Am. J. Cardiol.*, 2008, **101**, 34F–40F.
89. M. S. Elkind, W. Tai, K. Coates, M. C. Paik and R. L. Sacco, *Cerebrovasc. Dis.*, 2009, **27**, 42–50.
90. C. M. Ballantyne, R. C. Hoogeveen, H. Bang, J. Coresh, A. R. Folsom, L. E. Chambless, M. Myerson, K. K. Wu, A. R. Sharrett and E. Boerwinkle, *Arch. Intern. Med.*, 2005, **165**, 2479–2484.
91. A. Wang, Y. Yang, Z. Su, W. Yue, H. Hao, L. Ren, Y. Wang, Y. Cao and Y. Wang, *Stroke*, 2016, DOI: 10.1161/strokeaha.116.014816.
92. S. Bandeali and J. Farmer, *Curr. Atheroscler. Rep.*, 2012, **14**, 101–107.
93. R. S. Rosenson and G. D. Lowe, *Atherosclerosis*, 1998, **140**, 271–280.
94. P. J. Barter, R. Puranik and K. A. Rye, *Curr. Cardiol. Rep.*, 2007, **9**, 493–498.
95. P. N. Durrington, B. Mackness and M. I. Mackness, *Arterioscler. Thromb. Vasc. Biol.*, 2001, **21**, 473–480.
96. P. A. Wolf, R. D. Abbott and W. B. Kannel, *Stroke*, 1991, **22**, 983–988.
97. R. C. Seet, Y. Zhang, E. F. Wijdicks and A. A. Rabinstein, *Arch. Neurol.*, 2011, **68**, 1454–1458.
98. Z. Hijazi, J. Lindback, J. H. Alexander, M. Hanna, C. Held, E. M. Hylek, R. D. Lopes, J. Oldgren, A. Siegbahn, R. A. Stewart, H. D. White, C. B. Granger, L. Wallentin, Aristotle and S. Investigators, *Eur. Heart J.*, 2016, **37**, 1582–1590.
99. G. Y. Lip, J. V. Patel, E. Hughes and R. G. Hart, *Stroke*, 2007, **38**, 1229–1237.
100. R. C. Seet, A. M. Quek, E. C. Lim and B. Halliwell, *Free Radical Res.*, 2013, **47**, 283–290.
101. P. Pignatelli, D. Pastori, R. Carnevale, A. Farcomeni, R. Cangemi, C. Nocella, S. Bartimoccia, T. Vicario, M. Saliola, G. Y. Lip and F. Violi, *Thromb. Haemost.*, 2015, **113**, 617–624.
102. R. C. Seet, P. A. Friedman and A. A. Rabinstein, *Circulation*, 2011, **124**, 477–486.
103. A. A. Rabinstein, J. E. Fugate, J. Mandrekar, J. D. Burns, R. C. Seet, S. A. Dupont, T. J. Kauffman, S. J. Asirvatham and P. A. Friedman, *J. Stroke Cerebrovasc. Dis.*, 2013, **22**, 1405–1411.
104. M. Rodriguez-Yanez, S. Arias-Rivas, M. Santamaria-Cadavid, T. Sobrino, J. Castillo and M. Blanco, *Neurology*, 2013, **81**, 444–447.

105. J. D. Morrow, K. E. Hill, R. F. Burk, T. M. Nammour, K. F. Badr and L. J. Roberts 2nd, *Proc. Natl. Acad. Sci. U. S. A.*, 1990, **87**, 9383–9387.
106. R. C. Seet, C. Y. Lee, B. P. Chan, V. K. Sharma, H. L. Teoh, N. Venketasubramanian, E. C. Lim, W. L. Chong, W. F. Looi, S. H. Huang, B. K. Ong and B. Halliwell, *Stroke*, 2011, **42**, 2326–2329.
107. R. C. Seet, C. Y. Lee, E. C. Lim, J. J. Tan, A. M. Quek, W. L. Chong, W. F. Looi, S. H. Huang, H. Wang, Y. H. Chan and B. Halliwell, *Free Radical Biol. Med.*, 2010, **48**, 560–566.
108. R. C. Seet, C. Y. Lee, W. M. Loke, S. H. Huang, H. Huang, W. F. Looi, E. S. Chew, A. M. Quek, E. C. Lim and B. Halliwell, *Free Radical Biol. Med.*, 2011, **50**, 1787–1793.
109. Y. Chen, J. D. Morrow and L. J. Roberts 2nd, *J. Biol. Chem.*, 1999, **274**, 10863–10868.
110. S. L. Zeiger, E. S. Musiek, G. Zanoni, G. Vidari, J. D. Morrow, G. J. Milne and B. McLaughlin, *Free Radical Biol. Med.*, 2009, **47**, 1422–1431.
111. L. J. Janssen, *Am. J. Physiol.: Lung Cell. Mol. Physiol.*, 2001, **280**, L1067–L1082.
112. R. C. Seet, C. Y. Lee, E. C. Lim, A. M. Quek, H. Huang, S. H. Huang, W. F. Looi, L. H. Long and B. Halliwell, *Atherosclerosis*, 2011, **219**, 231–239.
113. C. Sanchez-Moreno, J. F. Dashe, T. Scott, D. Thaler, M. F. Folstein and A. Martin, *Stroke*, 2004, **35**, 163–168.
114. N. C. Ward, K. D. Croft, D. Blacker, G. J. Hankey, A. Barden, T. A. Mori, I. B. Puddey and C. D. Beer, *Clin. Sci.*, 2011, **121**, 501–507.
115. G. Kerr, G. Ray, O. Wu, D. J. Stott and P. Langhorne, *Cerebrovasc. Dis.*, 2009, **28**, 220–226.
116. P. James, C. J. Ellis, R. M. Whitlock, A. R. McNeil, J. Henley and N. E. Anderson, *Br. Med. J.*, 2000, **320**, 1502–1504.
117. J. K. Jensen, S. R. Kristensen, S. Bak, D. Atar, P. F. Hoilund-Carlsen and H. Mickley, *Am. J. Cardiol.*, 2007, **99**, 108–112.
118. M. Kral, D. Sanak, T. Veverka, M. Hutyra, D. Vindis, A. Kuncarova, A. Bartkova, T. Dornak, M. Svabova, V. Kubickova, J. Zapletalova, R. Herzig and D. Skoloudik, *Am. J. Cardiol.*, 2013, **112**, 117–121.
119. J. A. Alawneh, R. R. Moustafa and J. C. Baron, *Stroke*, 2009, **40**, e443–e450.
120. B. Thanvi, S. Treadwell and T. Robinson, *Postgrad. Med. J.*, 2008, **84**, 412.
121. V. Nadarajan, R. J. Perry, J. Johnson and D. J. Werring, *Pract. Neurol.*, 2014, **14**, 23.
122. E. C. Jauch, J. L. Saver, H. P. Adams, A. Bruno, J. J. Connors, B. M. Demaerschalk, P. Khatri, P. W. McMullan, A. I. Qureshi, K. Rosenfield, P. A. Scott, D. R. Summers, D. Z. Wang, M. Wintermark and H. Yonas, *Stroke*, 2013, **44**, 870.

CHAPTER 18

Development of Novel Test Platforms for the Assessment of Brain Injury

GALINA A. IZYKENOVA,*[a] RICHARD BALDWIN[b] AND STEVEN J. OLDENBURG[b]

[a] GRACE Laboratories, LLC, Atlanta, GA, USA; [b] Nanocomposix, San Diego, CA, USA
*Email: izykenova@yahoo.com

18.1 Introduction

At present, assay techniques for detecting brain biomarkers (peptide and antibodies [Abs]) are at a relatively early stage of development for stroke and brain injuries. The identification of unique reagents for promising brain biomarkers is critical for the development of successful diagnostics. Requirements for a biomarker to become a part of an immunoassay are that it must: (i) contain at least one active immune epitope; (ii) produce primary Abs (capture Abs) of high affinity for high-level sensitivity and specificity of the assay; (iii) be able to bind Abs or various surfaces through hydrogen or electrostatic bonds, van der Waals forces, and hydrophobic interactions; (iv) have a change in concentration of at least two-fold in a disease state; and (v) have limit of detection above 0.1 ng mL^{-1}.[1]

A body of scientific data related to a definite pathological process associated with a brain injury should correlate with changes in amounts of a new biomarker. In general, this biomarker will be a protein or a fragment of a

Drug Discovery Series No. 62
Acute Brain Impairment: Scientific Discoveries and Translational Research
Edited by Philip V. Peplow, Svetlana A. Dambinova, Thomas A. Gennarelli and Bridget Martinez

Published by the Royal Society of Chemistry, www.rsc.org

protein that is present either in different amounts and/or in different locations of a diseased central nervous system. Then the biomarker needs to be detected in body fluids using an assay.

Immunoassays, which rely on the exquisite binding selectivity of Abs to capture the analyte of interest, allow for the measurement of specific biomolecules in the biological fluid milieu. An appropriate immunoassay platform should be chosen and optimized to emphasize assay performance, accuracy, and productivity (cost analysis, efficiency, throughput, and turnaround time).

In this chapter, the roadmap for immunoassay development is discussed, including a unique set of reagents generated for current and upcoming technologies to detect biomarkers for brain impairments.

18.2 Diagnostic Approaches to Brain Injury

18.2.1 Currently Available Brain Injury Assays

The portable device Triage Stroke Test detects biomarkers (matrix metalloproteinase-9, brain natriuretic factor, D-dimer, and S100) values and presents a composite score that assesses acute ischemic stroke. It showed a sensitivity of 86% for detecting all strokes and a sensitivity of 94% for detecting hemorrhagic stroke at the threshold set at the first quartile. However, results were reproducible in a separate cohort tested in clinical trials on a point-of-care platform only in stroke *vs.* healthy control.[2]

The Simoa HD-1 Analyzer by Quanterix is a fully automated lab-scale instrument for running single-molecule array immunoassays (www.quanterix.com). This technology is a digital form of enzyme-linked immunosorbent assay (ELISA), capturing immunocomplexes on paramagnetic beads in femtoliter-sized wells of arrays to precisely measure a few molecules of protein biomarkers. This approach is better suited for toxins and cancer detection.

There is a need for less expensive, rapid, and user-friendly laboratory testing based on direct detection of biomarkers in biospecimens to assess brain impairments. Troponin or brain natriuretic peptide biomarker testing has been used for the diagnosis of heart failure. D-dimer assays assess pulmonary embolism, and the combination of D-dimer and glial fibrillary acidic protein tests for cerebral hemorrhages. More sophisticated multiplexed test panels expanded to specific brain biomarkers and their Abs would assist in determining the severity and location of brain damage.

18.2.2 Roadmap for Assay Development

Brain biomarkers and strategies to detect these biomarkers are guiding development of clinical immunodiagnostics that can help determine cerebrovascular risk as well as provide an assessment of acute conditions in humans. While basic concepts and requirements for immunoassay development have been documented in a number of books,[3–5] no informative "roadmap" exists that begins with biomarker discovery and ends with the launch of a diagnostic blood test. Bringing an immunoassay from the laboratory bench to the end user requires a diverse range of skills and experience, starting with basic biological research, through to assay development and transfer to manufacturing, as well as a deep understanding of the clinical rationale for the diagnostic.

Immunoassay design considerations must focus on the end user and clinical site considerations. While in some cases the clinical case for the marker might not require a rapid readout and sending samples for measurement in a carefully controlled laboratory environment where readout may take hours to days is acceptable, in other cases a rapid response is desirable. The ideal immunoassay would provide rapid data output (from seconds to a few minutes), be user friendly, and require minimal sample preparation. Portable, reliable, and low-cost assays are desirable to improve the diagnostic certainty of brain impairment. Instruction and training materials included with the assay should take into account these factors and the respective audiences.

Before a development effort begins, it is important to use clinical considerations to define the specifics of the diagnostic to be developed. These would include, among others, the clinically relevant concentration range of the biomarker that must be detected, the expected user of the diagnostic assay, and the acceptable specificity, selectivity, and accuracy of the assay to be utilized as a diagnostic test.

Once a particular biological target has been identified, the first step in an immunodiagnostic test is designing the unique affinity reagents (antigens and Abs) required to capture the analyte of interest and generate a signal. Inherent in this first step is the development of a prototype assay that utilizes these reagents and confirms their function. This step usually engages the analyte, often of recombinant or synthetic origin, spiked into a buffer. Once functional in the model system, the prototype must be optimized to detect the necessary analytes in the correct sample matrix (see Chapter 16).

The importance of access to a wide range of validated clinical samples cannot be underestimated. This process of identifying the affinity reagents is time-consuming and requires extensive research. In the case of brain injury, initial pinpointing of immunoactive peptide fragments associated with brain damage and generating respective Abs will be essential. This research and development effort demands multidisciplinary skills in biochemistry, immunology, neurology, and clinical chemistry, as well as sizeable financial capital.

Once the assay has been proven to work in real samples and meets the design specifications, the transfer to manufacturing begins. This transfer to manufacturing will involve incorporating a quality system, testing multiple lots of the various components, placing appropriate quality controls over the reagents and components employed, testing the reproducibility and accuracy of the assay, formalizing the assay method and packaging, creating the initial marketing plan based on primary marketing research, and selecting sites for clinical trials.

The final phase of development is to build and verify the assay manufacturing prototype under the appropriate controls. By the end of this development step, the immunodiagnostic product performance and format should be nearing completion and be ready for feasibility translational studies. At this point, the manufacturer should initiate discussion with the regulatory agency of the country where the product will be marketed and a clinical trial for certain test indications should be conducted. Securing requested clearances from the appropriate regulatory body after positive results of trials allows the product to be presented to the healthcare market (see Chapter 8).

18.3 Immunochemical Assays Development

18.3.1 Unique Reagents for Brain Biomarker Detection

There are some peculiarities associated with the unique reagents and biomarkers for the assessment of brain injury. Compared to antigens related to infectious disorders that rapidly activate the immune system to generate Abs, the proteins and protein fragments for nervous diseases usually have weak antigenicity and multiple epitopes. To generate the required affinity reagents, it is extremely important to select the amino acid sequences where the immune epitopes strongly correlate with pathological mechanisms and the manifestation of nervous disorder symptoms. The presence of the peptides and Abs against these peptides in the circulation should be indicative of a disease state in order for immunoassays to be engineered.

After the functional prerequisite, the major requirement of the reagents for immunoassay development is consistency. The reagents need to have a defined structure or composition and molecular weight, and high purity. Active and stable peptides with at least two epitopes should be synthetically produced with the desired level of hydrophobicity. Often, peptides associated with the blood–brain barrier are of a lipophilic origin. This hydrophobicity has significant impact on the affinity of the reagents to various substrates, with non-covalent binding to surfaces used in assay platforms (plastic, paper, micro-magnetic, and nanoparticles) being important to control.

Experienced chemists should be involved in designing peptides intended for immunoassays due to the complexities of determining the type of liquid- or solid-phase synthesis that should be employed, the temporary and permanent protection of active groups, and the possibility of obtaining a primary structure that will fold to a tertiary one while still maintaining epitopes accessible in the native (not denaturized) form (https://www.lifetein.com/peptide_synthesis_introduction.html).

Monoclonal or polyclonal Abs (mAbs or pAbs, respectively) against the synthetic peptide should be raised and utilized in immunoassays as reference standards. These Abs will be mostly immunoglobulin class G (IgG), as it is the most stable for non-infectious neurological disorders.

Immunoglobulin class M Abs are also present in benign stages of nervous system disease and generally are unstable. The choice of species to eliminate cross-reaction with human Abs (rabbit *vs.* hen) and consideration of pAb stability should be researched. For instance, a high-quality hen pAb (IgY) performs well in solution, but entirely loses activity if dried down, as is required for some immunoassays. In some cases, fusion proteins—products of several genes or humanized Abs—should be used as reference standards.[6]

18.3.2 Assay Design

Immunochemical assays involve the binding of a labeled Ab to a target. Most of these assays imply the immobilization of the probe Ab on a solid substrate (ELISA or lateral flow) or on a solid particle suspended in solution (agglutination assays). There are six different immunochemical assay configurations primarily in use (Figure 18.1): (1) a sandwich or "bridging" assay where double Abs capture an unknown antigen; (2) a sandwich assay, in which double antigens detect unknown Abs; (3) a competitive assay where an unknown antigen competes for attachment to solid surface Abs with a known amount of a labeled antigen; (4) a competitive assay where unknown Abs compete for attachment to a solid surface antigen with a known concentration of labeled Abs; (5) a serological assay in which the human Ab to be determined is bound to the solid phase's antigen and then revealed with a labeled secondary Ab; and (6) an Ab-specific assay in which an unknown specific Ab reacts with capture Abs attached to the solid phase and the reaction is revealed by a labeled antigen.

Figure 18.1 provides schematics for each of these assays. In all cases, the combination of a solid-phase affinity reagent and an affinity reagent associated with a probe gives rise to a signal at a certain location. The assay format is determined by the nature of the analyte (epitope structure, molecular weight, and origin), and immunochemical agents (mAbs or pAbs).

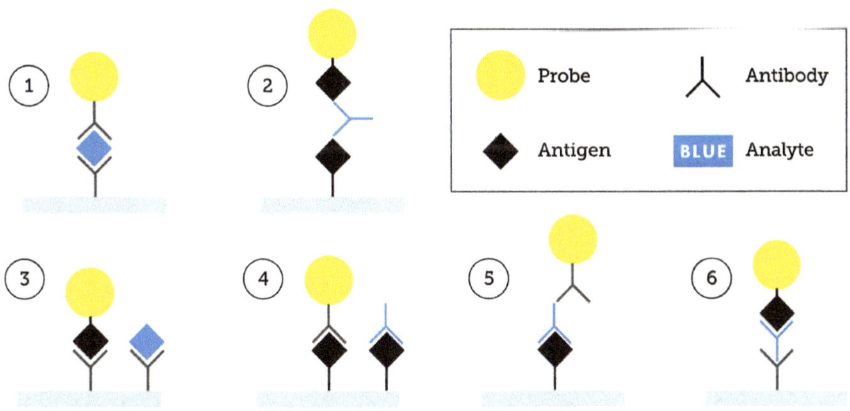

Figure 18.1 Six different modes for immunoassays.

In general, for smaller analytes the competitive-style assays are required, while for larger analytes sandwich assays usually are utilized. Each method demands the presence of a solid phase to bind a reactant capturing the analyte from unknown samples and a labeled detection agent and, in most cases, includes washing steps.

In some assays, the solid phase is already coated or bound to one reactant; in others, the coating, separation, and concentration occur in one step if the solid phase has a layer that binds the formed immunocomplex. The probe employed for the readout of the assay is also important. The probe can be an enzyme that gives rise to a reaction that generates light or a color change. It can be a fluorophore that generates a fluorescent signal or it can simply be a highly colored material that generates a color change at the location of the immobilized reagent.

A diagnostic test requires well-defined, unique reagents of high quality, purity, and stability as controls and calibrators. Control samples (reference standards) should be natural peptides or Abs that ideally would be isolated from human biological fluids. However, with concerns such as moral and ethical codes of sample drawing from patients, low concentrations in normal biological fluids, reproducibility, standardization, *etc.*, substantial amounts of reference standards should be produced synthetically or as fusion human proteins.

Multiple quality controls should be initiated in order to standardize the obtained reference standards: purity, quality, and stability, addition of anti-proteases, and bactericidal or bacteriostatic compounds for storage. The reference standards presented as controls and calibrators are assembled with standard immunochemical reagents (formulated buffer solutions, detection Abs, and substrates) into a reagent kit for the test prototype. This should be produced in several lots and consistency checked. The analytical accuracy—a number of parameters that determine the performance of an assay (detection limit, coefficient of variation, analytical range, cross-reactions with other components of blood, analyte, standard, and reagent stability, dilution recovery, and results reproducibility)—is measured.[7] This assesses the test precision and trueness (random and systematic influences).

The stability of the test also needs to be examined over time. Physical appearance (color and lucidity) and functional test performance stabilities are important for immunoassay test kits. Functional changes can involve protein and Ab degradation, whereas chemical changes may affect substrate quality due to oxidation or reduction. Therefore, stability testing of kits to determine an expiration date and conditions of transportation are performed in real or accelerated time by subjecting the product to varying temperatures and with 3–12 months of storage. The result is extrapolated to the anticipated storage temperature to determine the shelf life with ideal expiration over a 1 year period for laboratory-based testing.

The considerations of immunoassay engineering described above are of relevance for every format, from ELISAs to nanodiagnostics, and are generally considered as necessary requirements to be presented in internal documentation at *in vitro* diagnostic (IVD) companies.[7]

18.3.3 Assay Platform

An ELISA—the most affordable and widely used format in clinical laboratories—is often the first proof-of-concept assay for the validation of a biomarker and the discovery of affinity reagents. The same affinity reagents can then often be transferred (especially unique reagents) across various platforms, from automatic analyzers to point-of-care devices such as lateral flow tests.

Examples of ELISA tests that the authors have developed and are in ongoing development are peptide and Ab biomarkers assessing TIA/ischemic stroke. The ELISA format uses a microtiter plate with absorbed peptide for Ab detection and magnetic particle (MP) ELISAs containing MP-bound Abs for peptide detection (www.cisbiotech.com). In the Ab assay, the principle of capturing Abs was changed to horseradish peroxidase-labeled protein A to bind IgG in patient samples in order to maximize reproducibility.

An automated analyzer platform is utilized for ELISAs that use paramagnetic particles and chemiluminescent technology. Ready-to-use reagent packs with streptavidin supermagnetic particles (0.8 μm) bound to biotinylated carrier peptides were used for the quantitative detection of Abs in human sera. Chemiluminescent-labeled protein A ("sandwich" assay) was detected at high sensitivity. For peptide detection, the analyte in plasma competed with chemiluminescent-labeled synthetic peptides bound to MP-pAbs (competitive assay). Translation from the flat surface of the microplate to microparticles increases the reaction surface, with a simultaneous reduction in the quantity of reagents required. This resulted in a decrease in the turnaround time of laboratory-based testing from 1.5–3.0 hours to 30 minutes (Advia Centaur stroke biomarkers project).

Recognizing the medical need for bedside testing for brain damage (largely for neurocritical care), is the goal of the development program, aiming to produce a point-of-care testing (POCT) platform.[8] POCTs have the advantages of operational simplicity, speedy readout (in minutes), low price (less than a dollar a strip to manufacture), and a long lifetime without refrigeration (all reagents are dried down).

Lateral flow-based immunostrips usually employ nanoparticles as the signal-generating species with the ability to visualize particles that bind to strips. The reporter particles can be fluorescent particles, dye-loaded polymer particles, or plasmonic metal particles. The most common probe particle employed for lateral flow assays are 40 or 150 nm gold nanoshells with a strong absorption power providing a 20-fold sensitivity gain.[9]

By using gold nanoshells, a lateral flow sandwich assay was constructed to detect Abs that are generated when brain damage-specific peptides are released into the circulation. The lateral flow strip consists of a blood-filtering sample pad, a glass fiber conjugate pad on which gold nanoshells

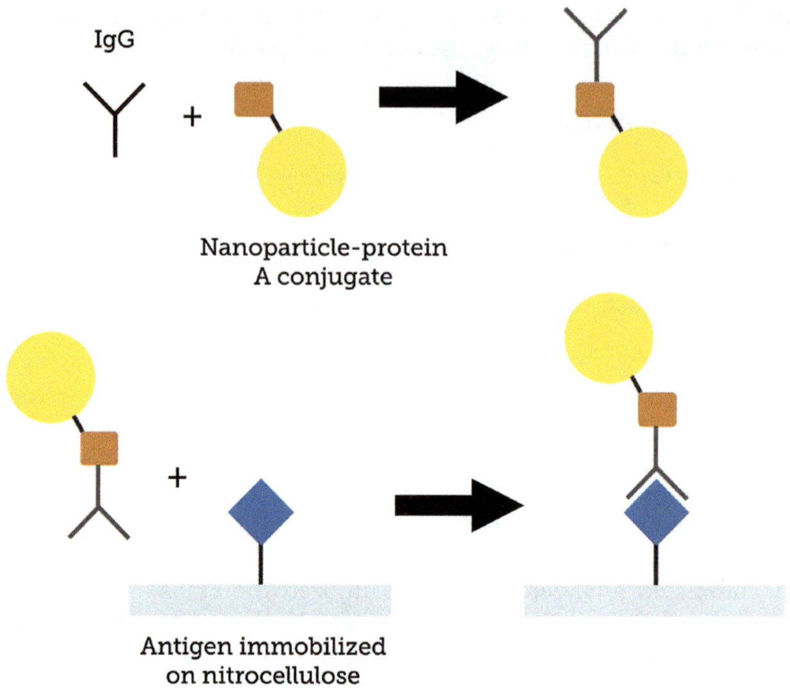

Figure 18.2 Lateral flow assay schematic for the detection of antibodies in serum samples.

conjugated to protein A are dried down, and a nitrocellulose strip on which an immobilized test line of the peptide of interest and a control line of an immobilized IgG is striped, as well as a cellulose wick. Figure 18.2 provides a schematic for the proposed assay.

When a blood sample is applied to the blood filter, the erythrocytes are removed and the remainder is passed to the conjugate pad. At the conjugate pad, the sample fluid re-suspends the dried down conjugate and the protein A on the particles captures any IgG molecules that are present in the sample. The sample fluid, now with suspended gold nanoshells that have captured IgG, runs through the nitrocellulose. When the gold nanoshells reach the test line, any that are bound to IgGs, which recognize the peptide at the test line, are captured, leading to a colored signal at the test line. The control line then captures a portion of the remaining nanoshells before the remaining fluid is drawn into the wick. The colored nanoparticles give rise to a visible signal with intensity proportional to the concentration of the Abs present in the sample. The control line will generate a signal regardless of the presence or absence of the peptide. Figure 18.3 provides a photograph of a positive and negative lateral flow assay in which the presence of Abs against the peptide of interest clearly give rise to a signal. The assay runs in minutes with a single drop of blood and requires no more expertise to run than a consumer pregnancy test.

Figure 18.3 Lateral flow assay to detect antibodies. No binding is present at the test line in the absence of antibodies (left), and strong binding is present at the test line in the presence of antibodies in the sample (right).

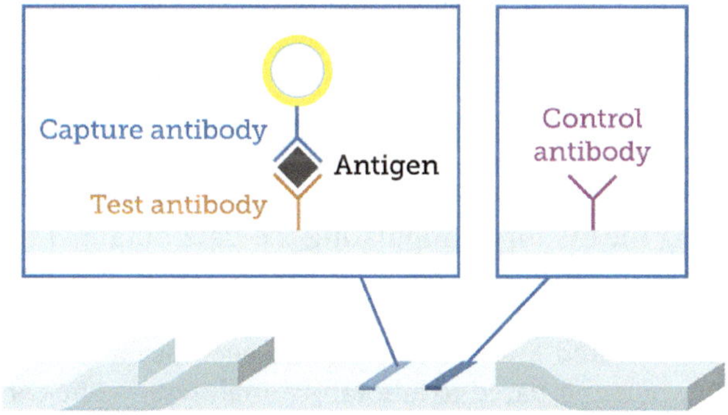

Figure 18.4 Format of peptide detection by the lateral flow assay.

In parallel with the Ab assay, an assay that works on a similar principle in which the peptide itself is the analyte is also being developed. This assay captures the peptide of interest between two different Abs, one immobilized on the nitrocellulose and one on the gold nanoshell (Figure 18.4). This sandwiching of the analyte leads to the particles again being immobilized at the test line, giving rise to a signal.

18.4 POCT for Brain Injury Assessments

18.4.1 Laboratory-based Testing and POCT

Testing in a laboratory is the general worldwide model for diagnostics, employing automated analytical processes of significant test volume at relatively acceptable costs and turnaround times that are performed in centralized laboratories. However, distance from patient, time spent for blood draw and processing, a shortage of qualified laboratory professionals at clinical labs, and the need for monitoring on an hourly basis (post-surgical complications) demand bedside assessments. Besides, ambulances, emergency rooms, and intensive care units request rapid diagnosis of life-threatening conditions (heart attack, stroke, and TBI) in order to stabilize the patient and instigate timely treatment. It would be desirable to be able to triage patients in urgent care situations using POCT rather than utilizing a central laboratory. Being able to distinguish between stroke and stroke mimics while still in the ambulance would be of tremendous utility and would allow for appropriate interventions at the earliest stage possible.

Other areas where the slow turnaround time of laboratory testing precludes its use include sideline assessments for sport-related concussions. Supplying trainers with reliable, simple and rapid tests would make a difference in return-to-play decisions. There is also a high demand for portable POCT for mild traumatic brain injury diagnosis in military combat training and warfare.

POCT eliminates the need for specimen collection and preservation and transportation to the lab and may greatly reduce the effort required in order to generate a diagnosis. Yes/no lateral flow tests have no need for an instrumental readout, being simply read by eye. Alternatively, where probes are fluorescent or the assay requires quantification, a device to read the strip is required. Reader devices based on a simple smartphone platform have the potential to bring near laboratory-level reproducibility to the point of care, as well as providing means for manipulating the data obtained in a handheld platform that can be deployed anywhere. This speed and efficiency often greatly improve patient outcomes and satisfaction. Nevertheless, to date, POCT has demonstrated lower precision and accuracy and lower sensitivity and specificity when compared to laboratory analyzers.[10] The correct balance between POCT's rapid turnaround and lab-based testing should be considered according to their importance to patient care. POCT might initially be used as a screening tool in emergent situations to triage patients for further laboratory and radiological testing.

18.4.2 Emerging Assay Technologies

The further miniaturization of technologies in the POCT of biomarker detection is progressing in two major directions—microfluidics and microchips[11]—with possibilities for lab-on-a-chip development that might replicate laboratory-based testing across several diseases.[12]

IVD developers adapted the microelectronic industry's fabrication techniques in microfluidic technologies that allow precise control and manipulation of fluids within a constrained space. For biomarker detection, this approach is used to process low volumes of human samples and reagents, achieving multiplexing, automation, and high-throughput screening. Paper-based microfluidics incorporating cellulosic and nitrocellulose fibers seems to be the best platform for peptide- and protein-based brain biomarkers.[11,13] Combined with 3D printing, this platform might enable low-cost POCT.[14]

Nanoparticles combined with microfluidics form mini-sensors, allowing for detection using high-performance, low-cost imaging systems incorporated into portable devices. These are now in development and testing, and some are undergoing clinical validation.[15]

18.5 Conclusion

The development and commercialization of a new standalone medical diagnostic device in the USA and Europe may cost between $25 million and $100 million and take up to 10–15 years between discovery of biomarkers, laboratory testing of the prototype, feasibility studies, translational research to clinical trials, and regulatory approval.[16]

Research and development is a function of time, available capital, and allowable risks. Careful initial planning and implementation can significantly reduce the time and cost of developing a brain biomarker immunoassay; for example, if prototype trials are positive, costs can be decreased.

Once a manufacturing prototype has been built and verified, actual testing of the assay can begin. Translational research and clinical trials represent the most costly and time-consuming processes in new blood test development. Since the goal is to obtain FDA regulatory approval, results using the assay must meet several criteria, including high sensitivity, high specificity, and clinical feasibility relative to the gold standard, which, in the case of TIA/stroke and TBI, is neuroimaging.

Advances in materials science and miniaturization techniques will expand biomarker analysis to lab-on-a-chip technology. Lab-on-a-chip technology has already demonstrated advantages in the use of extremely small sample volumes and multi-process integration with satisfactory performance characteristics. The compact package of the single chip platform and its portability will assist in making treatment for brain damage much more accessible at the bedside, on the sideline of a sports field, in the military theater, and in other point-of-care settings.

References

1. S. B. Nimse, M. D. Sonawane, K. S. Song and T. Kim, *Analyst*, 2016, **141**, 740.

2. Clinical Trials Registry, Biomarker Rapid Assessment of Ischemic Injury (BRAIN) [http://www.strokecenter.org/trials/clinical studies/brain-biomarker-rapid-assessment-of-ischemic-injury/results], (accessed May 2, 2017).
3. *The Immunoassay Handbook. Theory and Applications of Ligand Binding, ELISA and Related Techniques*, ed. D. Wild, Elsevier Science, Amsterdam, 4th edn, 2013.
4. Assay Guidance Manual, Eli Lilly Online, https://www.ncbi.nlm.nih.gov/books/NBK53196/pdf/Bookshelf_NBK53196.pdf (accessed May 1, 2017).
5. G. Wu, *Assay Development: Fundamentals and Practices*, John Willey & Sons, Hoboken, 2010.
6. A. Malik, *3 Biotech*, 2016, **6**, 44.
7. A. St John and C. P. Price, *Clin. Biochem. Rev.*, 2014, **35**, 155.
8. K. King, L. P. Grazette, D. N. Paltoo, J. T. McDevitt, S. K. Sia, P. M. Barrett, F. S. Apple, P. A. Gurbel, R. Weissleder, H. Leeds, E. J. Iturriaga, A. Rao, B. Adhikari, P. Desvigne-Nickens, Z. S. Galis and P. Libby, *JACC Basic Transl. Sci.*, 2016, **1**, 73.
9. S. J. Oldenburg, R. D. Averitt, S. L. Westcott and N. J. Halas, *Chem. Phys. Lett.*, 1998, **288**, 243.
10. E. Ter Avest, A. Visser, B. Reitsma, R. Breedveld and A. Wolthuis, *Eur. J. Emerging Med.*, 2016, **23**, 95.
11. J. V. Pagaduan, V. Sahore and A. T. Woolley, *Anal. Bioanal. Chem.*, 2015, **407**, 6911.
12. M. P. McRae, G. Simmons, J. Wong and J. T. McDevitt, *Acc. Chem. Res.*, 2016, **49**, 1359.
13. S. T. Sanjay, G. Fu, M. Dou, F. Xu, R. Liu, H. Qi and X. Li, *Analyst*, 2015, **140**, 7062.
14. N. Bhattacharjee, A. Urrios, S. Kang and A. Folch, *Lab Chip*, 2016, **16**, 1720.
15. P. C. Guest, *Methods Mol. Biol.*, 2017, **1546**, 311.
16. C. D. Shimasaki, *The Business of Bioscience: What Goes into Making a Biotechnology Product*, Springer, New York, 2009.

CHAPTER 19

Advancements and Challenges in Hyperacute Stroke Translational Research

SVETLANA A. DAMBINOVA,*[a] PHILIP V. PEPLOW[b] AND THOMAS A. GENNARELLI[c,d]

[a] DeKalb Medical, Decatur, GA, USA; [b] Department of Anatomy, University of Otago, Dunedin, New Zealand; [c] Medical College of Wisconsin, West Chester, PA, USA; [d] George Washington University, Washington, DC, USA
*Email: dambinova@aol.com

19.1 Introduction

As the population ages, stroke has the potential to become an even heavier medical and economic burden on the healthcare system than it is currently.[1] The public misperception that nothing can be done about stroke has prevailed for a long time. Recent advancements in neurological research have completely changed awareness in hyperacute stroke management. The goal of hyperacute stroke (moderate to severe) care is early onset recognition and the mobilization of multidiscipline efforts for prompt diagnosis and emergent intervention within 6 hours of onset.

Hyperacute stroke, known as a cerebrovascular accident, is now characterized as a "brain attack", which is both the most descriptive and realistic image of multi-origin stroke. As with a heart attack, the appropriate response to a brain attack is emergency action, both by the person it strikes and the

Drug Discovery Series No. 62
Acute Brain Impairment: Scientific Discoveries and Translational Research
Edited by Philip V. Peplow, Svetlana A. Dambinova, Thomas A. Gennarelli and Bridget Martinez

Published by the Royal Society of Chemistry, www.rsc.org

medical community. Urgent care should involve rapid assessment of biological fluids and differential diagnosis by use of instrumental methods such as mass spectrometry (MS)-based metabolomics, proteomics, peptidomics, or neurodegradomics.[2–4] These biological techniques might capture prompt changes of diverse biomarkers associated with cerebral ischemic and hemorrhagic injuries.

Accurate diagnosis enables clinicians to select the most effective emergency intervention, including thrombolysis or neuroprotective or neuroregenerative therapies. The importance of prompt diagnosis and treatment after a patient presents with cerebrovascular symptoms cannot be overstated. Delays in diagnosis and medical intervention beyond 6 hours of stroke onset may contribute to clinical deterioration and disability. Timely recanalization by thrombolytic therapy is the strongest predictor of a positive outcome;[5] however, it has unresolved issues, particularly for impaired small-, medium-, and large-sized cerebral vessels.

A combined use of intra-arterial thrombolysis and a self-expanding stent allows an increased favorable outcome after hyperacute stroke in up to 60% of cases.[6] It seems plausible to consider a combination therapy consisting of neuroprotective and neuroregenerative strategies together with thrombolytic treatment.

A more contemporary algorithm of hyperacute stroke treatment should include subgrouping of patients triaged on the basis of brain biomarker assay(s) thresholds and advanced neuroimaging. Biomarkers detected in biological fluid should be coupled with neurotoxicity cascades, oxidative stress reactions, and immune responses as consequences of impaired vascular viability. An integrative approach with functional assessments, advanced neuroimaging, and brain biomarker laboratory detection is needed in order to optimize hyperacute stroke management.

This chapter describes the advances in metabolic biomarkers detected by instrumental techniques for hyperacute strokes. The diversity of challenges in thrombolytic, neuroprotective, and neuroregenerative therapy in conjunction with neuroimaging are reviewed.

19.2 Metabolic Impairments and Diagnostic Imaging

Hyperacute ischemic stroke (IS) is characterized as a sudden impediment in glucose and oxygen supply to medium- and large arteries within minutes of their occlusion. Hyperacute cerebral hemorrhage occurs as a bleed within the brain tissue and is commonly found near gray–white matter junctions in cerebral lobes and subcortical structures (basal ganglia, brainstem, and deep cerebellar nuclei).[7] Subarachnoid hemorrhage (SAH) is bleeding into the area between the arachnoid membrane and the pia mater surrounding the brain and occurs usually from a ruptured cerebral vascular anomaly (aneurysm or vascular malformation), or may result from a traumatic brain injury. Both hyperacute IS and hemorrhagic strokes are serious medical conditions that require emergency care.

19.2.1 Selected Metabolic Biomarkers and Emerging Technologies

Metabolomics offers the assessment of metabolic profiles derived from certain cerebral areas and vascular territories associated with individual phenotypes of ischemic penumbra and lesions. The metabolome (small molecules of 50–1500 Da resulting from metabolic reactions) is highly liable, depending on time and surrounding biological fluids. Polarity, charge, solubility, and solution acidity influence the stability and reactivity of metabolites. There are more than 8000 metabolites verified by MS and nuclear magnetic resonance (NMR) spectroscopy registered at the Human Metabolome Database.[8]

Cerebral microdialysis (CMD) can be used for monitoring nervous tissue biochemistry of brain-impaired patients.[9] Based on the results, clinicians could adjust the treatment strategy to optimize the outcome in different clinical scenarios. *In vivo* CMD is a principal method of metabolome detection in brain structures achieved by withdrawing intracellular fluids, arterial and venous blood, and cerebral spinal fluid (CSF) for MS/NMR and liquid chromatography.

In microdialysis studies of ischemia induced by bilateral carotid artery occlusions in rats, release of glutamate and aspartate in both global and focal ischemia demonstrated up to 20-fold higher neurotransmitter concentrations when compared with controls.[10] Within the first 6 hours of onset, a very large accumulation of aspartate and glutamate was found in the CSF of patients. Glutamate concentrations of more than 200 pmol L^{-1} in plasma and of more than 8.2 pmol L^{-1} in CSF correctly classified a progressing course of stroke with probabilities of 92% and 93%, respectively. Glutamate concentrations correlated significantly with infarct volume and stroke severity in cortical infarcts, but not in deep infarcts, most likely indicating a higher density of glutamatergic neurons in some areas, such as the cortex.[11] Hyperacute IS requires monitoring due to complications that may cause neurological worsening. These biomarkers are up-regulated by ischemic events while being down-regulated in the first minutes of hemorrhage.[12] Indeed, there is a reduction of glutamate, glycerol, and the lactate–pyruvate ratio (LPR) after treatment with hemicraniectomy and hypothermia in hyperacute stroke therapy.

Hyperacute SAH is often complicated by vasospasm, resulting in poor outcome. Human microdialysis studies of SAH in the intraoperative period showed delayed ischemic neurological deficit with significant changes in lactate and glutamate concentrations.[13] In addition, the LPR was found to be the best prognostic marker of the 12 month outcome.[14] A low level of glucose <0.7 mmol L^{-1} and a high LPR >45 indicated a metabolic crisis and was associated with poor outcome.[15]

Several metabolic biomarkers for hyperacute IS have been identified by MS techniques. Elevated levels of lactate, carbonate, and glutamate were found in the hyperacute IS serum.[3] Tyrosine, lactate, and tryptophan were screened as a panel of potential biomarkers of hyperacute IS, offering high precision (91.7%).[3]

Oxidative/nitrosative stress is involved in glutamate receptor-mediated excitotoxic brain damage produced by quinolinic acid (QUIN), a tryptophan catabolite. QUIN, as a competitive agonist of the NR2A and NR2B subunits of *N*-methyl-D-aspartate (NMDA) receptors, activates reactive oxygen species, with intracellular Ca^{2+} influx enhancing peroxynitrite levels and damaging the cells.[16] QUIN exerts the greatest damage on cortical microvessel surfaces where NR2A/2B subunits are located.[17] In addition to modulating neuronal activity, QUIN can also act as an initiator and promoter of local inflammation within the central nervous system (CNS), increasing the expression and secretion of chemokines such as monocyte chemoattractant protein-1 and regulating the expression of normal T cells.[18]

In spontaneously hypertensive rats, the levels of homocysteine were raised and showed depressed cardiac contractility as compared to control animals.[19] The results were consistent with a view that homocysteine can produce a detrimental effect in the brain following the occlusion of small cerebral vessels. As a proposed risk factor for stroke, homocysteine is hypothesized to induce oxidative injury to vascular endothelial cells, to reduce the production of nitric oxide, to enhance platelet adhesion to endothelial cells, and to promote the growth of vascular smooth muscle cells.[20]

NMR analysis of plasma and urine in human small vessel occlusion revealed significant differences in sulfuric derivatives of aspartate compared to controls.[20] Increased serum concentrations of cysteine, *S*-adenosyl homocysteine, oxidized glutathione, hydroxyeicosatetraenoic acid, and hydroxyoctadecadienoic acid were found in an MS analysis of hyperacute stroke and controls matched for vascular risk factors.[21] Homocysteine levels greater than 20 μM have toxic effects on the arterial endothelium and are associated with small- to medium-sized cerebrovascular vessel occlusion.[22]

As for large cerebrovascular vessels occlusion, kynurenic acid, as an endogenous antagonist of α-amino-3-hydroxy-5-methyl-4-isoxazolepropionic acid (AMPA), NMDA, and kainate receptors,[16] plays a significant role. Epidemiological research indicates that kynurenine is associated with severe hyperacute stroke and chronic induced inflammatory response.[23]

Occlusion of a vertebral artery may be asymptomatic, but in some cases can result in large lesions with severe symptoms caused by bilateral damage to the brainstem and cerebellum. A potential biomarker of vertebral artery occlusion could be picolinic acid (PIC), a catabolite of the amino acid tryptophan through the kynurenine pathway. It has the capacity to chelate a wide range of bivalent metals, such as Cu^{2+}, Fe^{2+}, Ni^{2+}, Zn^{2+}, Cd^{2+}, and Pb^{2+}.[16] PIC may cause neurotoxicity to hippocampal, substantia nigra, and striatal cells and may stimulate the immune system through macrophage activation. It can also induce the expression of nitric oxide synthase, which possesses an anti-oxidative effect in the nucleus basalis.

Several advanced instrumental techniques have been applied in the search for metabolic biomarkers of stroke progression, uniting specific pathological processes like neurotoxicity cascades, oxidative stresses, and neuroimmune reactions. However, there are a few markers from various metabolites that change under ischemic or hypoxic events that could be selected for further research. Potential metabolic biomarkers of arterial occlusion might be presented in assays for easily accessible detection and rapid data interpretation compared to MS. These assays should then be verified by translational studies involving advanced neuroimaging in order to correlate real-time metabolite values with structural and functional alterations.

19.2.2 Diagnostic Biomarkers of Imaging

The following magnetic resonance technologies are used to distinguish ischemic from hemorrhagic stroke: computed tomography (CT) and, to a lesser extent, magnetic resonance imaging (MRI).

19.2.2.1 Computed Tomography

An admission CT scan usually evaluates the presence of ischemic signs. Follow-up CT scans performed within 22–36 hours of stroke additionally evaluate for infarct swelling and intracerebral hemorrhage. Dense arteries have been described as a manifestation of a cerebral artery thrombosis on plain CT.[24] The sensitivity of this sign for middle cerebral artery (MCA) occlusion is 27–54%, but the specificity approaches 100%.[25] CT angiography is used to visualize cerebral arterial stenosis in more than 70% of cases, whereas CT perfusion provides high specificity in the detection of ischemia and infarct tissue of the brain.[26]

Infarct-related hemorrhages are subdivided into hemorrhage (small bleeding in and out of infarction area) and parenchymal hematomas with a dense blood clot not exceeding 30% of the infarcted area or with a significant space-occupying effect exceeding that of the infarct volume.

19.2.2.2 Magnetic Resonance Imaging

Many hospitals now have MRI capability. However, MRI scans require longer image acquisition times and currently cannot readily distinguish ischemic from hemorrhagic stroke within 6 hours of onset. Therefore, MRI is not considered a primary diagnostic modality. A study has suggested that MRI is better than CT (sensitivities of 83% *vs.* 26%, respectively) for detecting hyperacute ischemia, as well as acute and chronic hemorrhage, and should therefore be the preferred test for an accurate diagnosis of suspected acute stroke.[27]

Currently, diffusion-weighted imaging (DWI) is the method of assessment of edema formation that permits *in vivo* measurements of translational water mobility in brain tissue. The degree of this mobility can be quantified by a physical parameter: the apparent diffusion coefficient (ADC). Areas of ischemic tissue that have reduced ADCs appear brighter on DWI, which is generally attributed to cytotoxic edema or depolarization, allowing differentiation between restricted diffusion (cytotoxic edema) and increased diffusion (vasogenic edema). DWI can detect regions of high-energy metabolism failure and loss of ion homeostasis within minutes of stroke onset.[28] 3 T DWI is still in research mode for hyperacute IS, in spite of its high specificity for revealing edema, due to it being a time-consuming procedure with additional costs.

Perfusion-weighted imaging (PWI) measures the hemodynamics of the brain through dynamic susceptibility of contrast-enhanced MRI. Perfusion MRI evaluates blood flow in the brain's microcirculation and can readily identify hypoperfused areas.[29] The combination of DWI and PWI provides prognostic value for acute subcortical infarction and ischemic penumbra.[30]

In studies of potential therapies for stroke, it was shown that ischemic lesion volume is reduced after intravenous tissue plasminogen activator (tPA) treatment from the detected MRI baseline by 15–30 days.[31] The thrombolysis performed within a "therapeutic window" of 4.5 hours is directed towards ischemic lesions on DWI that might be partially reversible. Otherwise, without treatment, lesion progression to complete infarction on day 3–7 occurs. A significant correlation exists between increasing lesion volume from early to delayed MRI studies and clinical outcome.[31]

In addition to MRI (DWI and PWI) being a powerful technique for the visualization of *in vivo* pathological changes due to stroke, the *in vivo* biochemical changes that can be observed by magnetic resonance spectroscopy (MRS) and are extremely valuable in making therapeutic decisions.

N-acetylaspartate (NAA) is a derivative of L-aspartate and is another abundant metabolite after glutamate. Drastic NAA decreases are characterized by acute energy failure after IS and hypoxia in both animals and humans.[32] An increase of lactate[20] and attenuation in NAA, creatine (Cr), and choline (Cho) compound concentrations have been shown by MRS studies in acute IS.[33] Lactate surge occurs immediately after cerebral blood flow is compromised due to anaerobic glycolysis that compensates for mitochondrial dysfunction.[34] Reduced NAA levels become evident within 1 hour of hypoperfusion and remain low up to 24 hours. There is a salvaged area of penumbra where NAA does not change and lactate is present in tissue.[35] Necrotic transformation in nervous tissue begins when the ratio of Cho (cell membrane marker) to Cr (energy marker) becomes greater than 1.5 compared to a normal ratio of 1.2.[36]

Decreased NAA/Cr ratios in the hippocampus were confirmed by MRS in severe asymptomatic carotid stenosis, a condition with a high risk of stroke.[3] These results have been confirmed by similar findings in stroke penumbra within 4.5 hours of symptom onset.[37] White matter atrophy 1 week after IS correlated with decreased levels of NAA and total Cr.[38]

A recent study aimed at optimizing NAA detection by *in vivo* diffusion-weighted MRS in subcortical structures was conducted.[39] The corpus callosum area was chosen for detailed assessment by use of 3 T MRI and 7 T MRI in healthy volunteers, as it is associated with the most devastating stroke consequences. The reproducibility of total NAA measurements resulted in similar values at both field strengths, provided sufficient power with fewer subjects, and demonstrated the opportunity to individualize diagnosis with further targeted therapy.

19.3 Revascularization in Hyperacute Stroke

Hyperacute cerebral artery occlusion leads to brain infarction and irreversible stroke. The revascularization and reperfusion within a time window of 6 hours is a substantiated achievement of stroke management. Prompt restoration of blood supply may minimize the extent of infarction due to reperfusion of the potentially viable penumbra and so improve neurological outcome.

Recanalization of the previously occluded medium- to large vessels and subsequent reperfusion have been proven to restore brain functions.[5] Pharmacologically induced recanalization by recombinant tPA (rtPA) is a gradual process whereby the thrombus initially softens and partially dissolves, allowing some degree of blood flow restoration. The restored bloodstream delivers more rtPA to bind with fibrinogen within the clot and leads to complete or partial lysis. The most consistent predictors of completeness of recanalization in stroke are vessel size and clot burden.[40,41]

Streptokinase is a substance released from group C β-hemolytic streptococci and was noted to dissolve blood clots. Its mechanism of action is in combination with plasminogen followed by its activation to plasmin; the drug, however, is not fibrin specific, and its use may lead to a systemic effect due to depletion of coagulation factors.[4] Use of streptokinase in stroke was associated with high rates of hemorrhagic complications. Nowadays, indications for use of streptokinase include myocardial infarction, pulmonary embolism, and deep vein thrombosis, but not acute IS.

Urokinase is also a non-fibrin-specific serine protease that directly transforms plasminogen to plasmin, which is capable of activating fibrin-bound and circulating plasminogen. Pro-urokinase, a precursor of urokinase characterized by significant fibrin specificity and a half-life of 14–20 minutes, is used as an effective modality in acute IS.[42]

Functional improvement is the earliest phenomenon associated with the salvage of the critically hypoperfused tissue, the penumbra. Both animal studies[43] and human transcranial Doppler monitoring in acute stroke patients treated with intravenous thrombolytics have demonstrated a strong association between recanalization of the previously occluded vessels and reduction of neurological deficit as defined by NIH Stroke Scores of 10 or more points.[44]

Often, the phenomenon of delayed neurological improvement associated with potential neuronal damage is caused by edema formation. Another possible independent mechanism of delayed blood supply due to hyperacute stroke is collateral vessel engagement.[45] Strategies aimed at increasing collateral flow, such as pharmacologically induced hypertension, could be beneficial in selected cases and need further evaluation.

The outcome in terms of successful clinical translation of promising preclinical research findings in animal models of IS is poor. Only three interventions for acute IS had a convincing effect out of about 500 interventions that showed a positive effect in animal models. They were thrombolysis with tPA within 4.5 hours of stroke onset, aspirin given within 48 hours, and management in a dedicated stroke unit.[46] This discrepancy is despite efforts to define the most important criteria that have to be considered in order to substantially increase the likelihood that a given preclinical therapy will successfully translate into a clinically approved treatment.

19.4 Pitfalls in Neuroprotective Translational Research

Translational research aims to improve the probability of success of neuroprotective drug candidates. To achieve this, it is important to identify the factors contributing to the high failure rate of neuroprotective drug candidates. Such factors include: (1) failure to cross the blood–brain barrier and enter the brain interstitial fluid in order to interact with its target molecule; (2) heterogeneity in the patient population; and (3) the dose or dosing schedule being suboptimal. In addition to solubility, permeability, metabolic stability, and protein binding properties, a neuroactive compound needs to have the correct physicochemical properties to permit movement through a structural and dynamic barrier that separates the blood from the CNS.

The selection of compounds with properties that are favorable for movement from blood into the CNS has the potential of increasing the chances of candidates being successful as neuroprotective agents. Methods to assess the probability of compounds crossing the blood–CNS barrier in humans may involve *in vitro* studies using human brain endothelial cells co-cultured with astrocytes. Immortalized brain endothelial cell lines have also been used.[47] Several computer models have been established to predict the ability of neuroactive compounds to permeate the blood–CNS barrier on the basis of molecular properties.[48,49] The key properties include the octanol–water partition coefficient, molecular weight, polar surface area, and hydrogen bonding properties. Such models consider passive permeation only and do not take account of transport mechanisms.[49]

A large number of clinical stroke trials have examined a non-homogeneous patient population in which some patients have a large cortical infarction while others have a lacunar infarction. Elimination of patients with lacunar stroke therefore increases the power of the clinical trials. It is likely that some or all of the neuroprotective drug candidates showing efficacy in animal models of stroke but not in clinical trials possess poor therapeutic ratios due to reducing the dose to avoid unwanted side effects. This has led to the administration of subtherapeutic doses.[50] A low therapeutic efficacy can also be due to late time of administration or to treatment duration being insufficient. In practice, the delay to therapy has been variable (4–48 hours).[51]

Following a stroke, it appears that ischemia-induced pathology continues for 6 days or more,[52] suggesting that treatment over a prolonged period (>6 days) is necessary to achieve maximum therapeutic benefit. Clearly, studies should be undertaken to assess the ability of neuroprotective candidates to permeate the blood–CNS barrier. Also, a therapeutic dose needs to be defined in suitable animal models, possibly also using non-human primates, before proceeding to clinical trials.

Differences between human stroke and animal stroke models include the following: (1) most animal studies report successful neuroprotection as soon as a decrease in infarct volume (determined by histology or MRI) is found; (2) there is a large discrepancy in terms of the average animal age and the average age of patients; (3) the ischemic territory is rather uniform in the majority of animal studies (involving occlusion of the MCA); (4) there is variation in the time of applying neuroprotective therapies, which, in stroke animal models, is done very early (*i.e.* <1 hour) after induction of the ischemia, and in humans is initiated within a time window of 24 hours after ischemia;[46] (5) in animal stroke studies, assessments of neurological and behavioral function, which frequently contribute substantially to disability, are lacking; (6) it is rare in animal studies to examine pharmacokinetic or pharmacodynamic profiles of the drugs being examined, or levels of drugs in the brain in relation to therapeutic actions; (7) comorbidities are not properly taken into consideration in animal models, as hypertension and diabetes could influence the outcome of neuroprotective treatment, but are not included in respective animal models; and (8) animal stroke models employ uniform lesion mechanisms that do not adequately reflect the etiology of stroke in humans.

Primary large and small vessel occlusions as well as arterial embolisms cause IS. However, only vessel occlusion models are used as reliable stroke models. Male gender preferences are observed in stroke models, and large animal models (preferably non-human primates) are desirable, but to date very few published studies exist in this context. Animal studies often seem to lack certain aspects of study design that are fully established in clinical trials (*e.g.* randomization of test subjects to treatment or control groups, blinded performance of treatment, and blinded assessment of outcome).

Such study design aspects, if neglected, seem to lead to overestimated drug efficacy in animal studies.[46,53] The various sources of bias observed in animal studies impact on internal validity. Adequate internal validity of an animal study implies that the differences observed between groups of animals allocated to different treatments may, apart from random error, be attributed to the treatment under investigation.[54]

A number of animal studies report small group sizes and are often underpowered, and so it is not possible to reliably detect group differences with high enough probability. In an attempt to outweigh power problems due to small group sizes, test parameters, including test animals, are standardized as much as possible to reduce data variance (*e.g.* test animals are often inbred).

Taken together, issues inherent to stroke models support the notion that animals fail to reflect human stroke. The animal models can at best provide proof of principle that a neuroprotective or neuroregenerative therapy may be effective.[55] Better design of animal models together with higher general study quality, such as blinding, randomization, and avoiding underpowered studies, would meet the highest ethical standards, even if individual experimental group sizes might be larger. Publication bias, in which publication of positive or confirming data over negative or inconclusive data is favored by journal editors and peer reviewers, leads to a subsequent overestimation of effect sizes in meta-analyses and systematic reviews.[56]

Failure of translation to the clinic may also be explained in part by shortcomings of the studies. For instance, they may have had insufficient statistical power to detect a true benefit of the treatment under study. The designs of some clinical trials have also failed to acknowledge the limitations of efficacy observed in animal studies, such as by allowing therapy at later time points when the window of opportunity has passed. Moreover, failure of apparently promising interventions may be caused by inadequate animal data and overoptimistic conclusions about efficacy from methodologically flawed animal studies. Another possible explanation is the lack of external validity of some animal models for which the results do not provide a correct basis for generalization to other circumstances (*i.e.* do not sufficiently reflect disease in humans).[54,57]

19.5 Challenges in Neuroregenerative Translational Research

In cell culture models of oxygen–glucose deprivation and animal models of focal cerebral ischemia, blockade of various excitotoxic, oxidative stress, and cell death mechanisms appears to significantly reduce neuronal injury. However, none of these neuroprotective strategies have proven to be effective in clinical trials.[58] This does not mean that the basic molecular and cellular mechanisms are incorrect or irrelevant.[59,60] Adaptive and maladaptive signaling occurs between multiple cell types in the brain from neuronal, glial, and vascular compartments. The blood–brain barrier function requires crosstalk between the cerebral endothelium and pericytes and astrocytic end-feet. Renewal and homeostasis in white matter require crosstalk between endothelial cells and oligodendrocyte precursors. There is crosstalk between central and peripheral responses from immune, hormonal, and cardiac systems. There are also overlapping pathways involving crosstalk between all cell types in the brain during initial injury and subsequent repair. These numerous feed-forward and feed-back interactions have made it difficult to translate experimental discoveries into clinical applications. An emerging hypothesis in biomedical research now suggests that translation may not be efficiently obtained without a rigorous understanding of mechanisms. Hence, to optimize diagnostics and therapeutics for stroke patients, it is necessary to identify and define causal mechanisms.[61]

Communication between the CNS and peripheral responses may provide the potential signals and substrates that permit the influence of a number of cofactors and comorbidities (*e.g.* hypertension and diabetes). These various modifying factors alter stroke risk and contribute significantly to the acute process of brain injury, as well as the delayed process of potential brain repair. From the experimental side, animal model systems will have to be adapted to incorporate these important modifying factors. From the clinical side, stroke trials may have to more carefully identify and separate patient cohorts depending on how the mechanisms being targeted are affected by various risk factors.[61] The use of novel neuroimaging technologies to visualize the penumbra and the location and extent of vascular obstruction will identify those patients with acute IS who are most likely to respond to thrombolytic and neuroprotective therapy.[62]

19.6 Conclusion

The possibility of establishing a differential diagnosis of hyperacute stroke will aid in triaging for urgent thrombolysis and endovascular therapy depending on the vascular territory occluded. The coordinated efforts of an interdisciplinary team in the emergency department, including a portable neuroimaging system, stroke clinicians, and rapid assessment of metabolic biomarkers, might minimize the mortality and long-term disability caused by hyperacute stroke.

Mobile laboratory techniques for detecting metabolic and other biomarker profiles of hyperacute IS in conjunction with neurovascular imaging could ensure the best choice of therapy for medium- and large-sized vessel revascularization. For cerebral hemorrhage, prompt endovascular intervention is recommended (NIH Stroke Consensus Committee, https://www.ncbi.nlm.nih.gov/pmc/articles/PMC4871176/).

Stroke translation research has demonstrated biologically plausible derangement of metabolites. The range of challenges and pitfalls related to study design and protocols has been highlighted to enhance the probability of observing diagnostic and therapeutic efficacy. Both metabolic biomarker candidates as well as potential neuroprotective and neuroregenerative pharmacological agents might facilitate a favorable outcome.

Additionally, educating non-stroke specialists globally on advanced approaches in rapid blood testing for stroke diagnosis will help save lives and decrease disabilities from recurrent strokes. The active role of clinical laboratories in the rapid and accurate assessment of hyperacute stroke cannot be underestimated in emergency rooms, the hospital, and primary care settings. A targeted hyperacute stroke group should be immediately directed towards advanced neuroimaging and the administration of novel multi-therapies (thrombolytics and neuroprotective and neuroregenerative medications).

References

1. E. S. Roach, K. Bettermann and J. Biller, *Toole's Cerebrovascular Disorders*, Cambridge University Press, New York, 6th edn, 2010.
2. R. Cunningham, D. Ma and L. Li, *Front Biol.*, 2012, **7**, 313.
3. D. Wang, J. Kong, J. Wu, X. Wang and M. Lai, *Neurosci. Lett.*, 2017, **642**, 7.
4. S. A. Dambinova, in *Biomarkers for TBI*, ed. S. A. Dambinova, R. L. Hayes and K. K. W. Wang, Royal Society of Chemistry, London, 2012, ch. 4, p. 66.
5. S. Rangaraju, M. Frankel and T. G. Jovin, *Interv. Neurol.*, 2016, **4**, 120.
6. D. Šaňák, M. Köcher, T. Veverka, M. Černá, M. Král, S. Buřval, D. Školoudík, V. Prášil, J. Zapletalová, R. Herzig and P. Kanovsky, *J. Vasc. Interv. Radiol.*, 2013, **24**, 1273.
7. R. Babu, J. H. Bagley and C. Di, *Neurosurg. Focus*, 2012, **32**, E8.
8. I. Aretz and D. Meierhofer, *Int. J. Mol. Sci.*, 2016, **17**, E632, DOI: 10.3390/ijms17050632.
9. C. Mahajan and G. P. Rath, *J. Neuroanaesthesiol. Crit Care*, 2015, **2**, 232.
10. L. L. Sternau, M. Y.-T. Globus, W. D. Dietrich, E. Martinez, R. Busto and M. D. Ginsberg, in *The Role of Neurotransmitters in Brain Injury*, ed. M. Y.-T. Globus and W. D. Dietrich, Plenum Press, New York, 1992, pp. 33–38.
11. M. Fisher and J. Bogousslavsky, *Current Review of Cerebrovascular Disease*, Springer, Philadelphia, 4th edn, 2001.
12. T. W. Lai, S. Zhang and Y. T. Wang, *Prog. Neurobiol.*, 2014, **115**, 157.
13. C. T. Ulrich, C. Fung, H. Vatter, M. Setzer, E. Gueresir, V. Seifert, J. Beck and A. Raabe, *PLoS One*, 2013, **8**, e62754.
14. A. Sarrafzadeh, D. Haux, I. Küchler, W. R. Lanksch and A. W. Unterberg, *J. Neurosurg.*, 2004, **100**, 400.
15. J. M. Schmidt, S. B. Ko, R. Helbok, P. Kurtz, R. M. Stuart, M. Presciutti, L. Fernandez, K. Lee, N. Badjatia, E. S. Connolly, J. Claassen and S. A. Mayer, *Stroke*, 2011, **42**, 1351.
16. M. I. Cuartero, J. de la Parra, A. García-Culebras, I. Ballesteros, I. Lizasoain and M. Á. Moro, *Curr. Pharm. Des.*, 2016, **22**, 1060.
17. C. D. Sharp, M. Fowler, T. H. Jackson 4th, J. Houghton, A. Warren, A. Nanda, I. Chandler, B. Cappell, A. Long, A. Minagar and J. S. Alexander, *BMC Neurosci.*, 2003, **4**, 28.
18. G. J. Guillemin, J. Croitoru-Lamoury, D. Dormont, P. J. Armati and B. J. Brew, *Glia*, 2003, **41**, 371.
19. P. Ganguly and S. F. Alam, *Nutr. J.*, 2015, **14**, 6.
20. J. Y. Jung, H. S. Lee, D. G. Kang, N. S. Kim, M. H. Cha, O. S. Bang, D. H. Ryu and G. S. Hwang, *Stroke*, 2011, **42**, 1282.
21. M. I. Qureshi, P. A. Vorkas, A. P. Coupland, I. H. Jenkins, E. Holmes and A. H. Davies, *Neuroscientist*, 2016, DOI: 10.1177/1073858416673327.
22. H. Škovierová, E. Vidomanová, S. Mahmood, J. Sopková, A. Drgová, T. Červeňová, E. Halašová and J. Lehotský, *Int. J. Mol. Sci.*, 2016, **17**, E1733.

23. G. Liu, S. Chen, J. Zhong, K. Teng and Y. Yin, *Oxid. Med. Cell. Longevity*, 2017, **2017**, 1602074.
24. D. Birenbaum, L. W. Bancroft and G. J. Felsberg, *West J. Emerging Med.*, 2011, **12**, 67.
25. T. Kharitonova, N. Ahmed, M. Thorén, J. M. Wardlaw, R. von Kummer, J. Glahn and N. Wahlgren, *Cerebrovasc. Dis.*, 2009, **27**, 51.
26. A. Sabarudin, C. Subramaniam and Z. Sun, *Quant. Imaging Med. Surg.*, 2014, **4**, 282.
27. J. A. Chalela, C. S. Kidwell, L. M. Nentwich, M. Luby, J. A. Butman, A. M. Demchuk, M. D. Hill, N. Patronas, L. Latour and S. Warach, *Lancet*, 2007, **369**, 293.
28. M. M. Kunst and P. W. Schaefer, *Radiol. Clin. North Am.*, 2011, **49**, 1.
29. B. J. Kim, H. G. Kang, H.-J. Kim, S.-H. Ahn, N. Y. Kim, S. Warach and D.-W. Kang, *J. Stroke*, 2014, **16**, 131.
30. F. Chen and Y.-C. Ni, *World J. Radiol.*, 2012, **4**, 63.
31. A. Purushotham, B. C. Campbell, M. Straka, M. Mlynash, J. M. Olivot, R. Bammer, S. M. Kemp, G. W. Albers and M. G. Lansberg, *Int. J. Stroke*, 2015, **10**, 348.
32. H. Igarashi, Y. Suzuki, N. J. Huber, M. Ida and T. Nakada, *Magn. Reson. Med. Sci.*, 2015, **14**, 13.
33. K. A. Dani and S. Warach, *Am. J. Neuroradiol.*, 2014, **35**, S37.
34. A. Bivard, V. Krishnamurthy, P. Stanwell, N. Yassi, N. J. Spratt, M. Nilsson, C. R. Levi, S. Davis and M. W. Parsons, *J. Cereb. Blood Flow Metab.*, 2014, **34**, 1944.
35. K. A. Dani, L. An, C. Henning, J. Shen and S. Warach, *Stroke*, 2012, **43**, 2962.
36. Neuroradiology Teaching File Database online, J. R. Hesselink, http://spinwarp.ucsd.edu/neuroweb/Text/mrs-TXT.htm, UCSD (accessed May 7, 2017).
37. A. Bivard, N. Yassi, V. Krishnamurthy, L. Lin, C. Levi, N. J. Spratt, F. Mittef, S. Davis and M. Parsons, *Neuroradiology*, 2016, **58**, 409.
38. N. Yassi, B. C. Campbell, B. A. Moffat, C. Steward, L. Churilov, M. W. Parsons, G. A. Donnan, P. M. Desmond, S. M. Davis and A. Bivard, *Neuroradiology*, 2016, **58**, 3.
39. E. T. Wood, A. E. Ercan, F. Branzoli, A. Webb, P. Sati, D. S. Reich and I. Ronen, *NMR Biomed.*, 2015, **28**, 976.
40. J. Schröder and G. Thomalla, *Front. Neurol.*, 2016, **7**, 245.
41. K. Butcher, A. Shuaib, J. Saver, G. Donnan, S. M. Davis, B. Norrving, K. S. Wong, F. Abd-Allah, R. Bhatia and A. Khan, *Int. J. Stroke*, 2013, **8**, 560.
42. R. S. Lee, Y. C. Ok, J. S. Lim, B. C. Lim, K. Y. Cho and M. C. Lee, *Chonnam Med. J.*, 2012, **48**, 52.
43. A. R. Young, O. Touzani, J. M. Derlon, G. Sette, E. T. MacKenzie and J. C. Baron, *Stroke*, 1997, **28**, 632.
44. L. L. Yeo, P. Paliwal, H. L. Teoh, R. C. Seet, B. P. Chan, B. Wakerley, S. Liang, R. Rathakrishnan, V. F. Chong, E. Y. Ting and V. K. Sharma, *J. Stroke Cerebrovasc. Dis.*, 2013, **22**, e590.

45. E. Cuccione, G. Padovano, A. Versace, C. Ferrarese and S. Beretta, *Exp. Transl. Stroke Med.*, 2016, **8**, 2.
46. E. Sena, H. B. van der Worp, D. Howells and M. Macleod, *Trends Neurosci.*, 2007, **30**, 433.
47. Y. He, Y. Yao, S. E. Tsirka and Y. Cao, *Stroke*, 2014, **45**, 2514.
48. W. A. Banks, *BMC Neurol.*, 2009, **9**(Suppl. 1), S3.
49. A. M. Palmer and M. S. Alavijeh, *Drug Discovery Today*, 2012, **17**, 1068.
50. D. A. Dawson, G. Wadsworth and A. M. Palmer, *Brain Res.*, 2001, **892**, 344.
51. J. De Keyser, G. Sulter and P. G. Luiten, *Trends Neurosci.*, 1999, **22**, 535.
52. D. E. Saunders, F. A. Howe, A. van den Boogaart, M. A. McLean, J. R. Griffiths and M. M. Brown, *Stroke*, 1995, **26**, 1007.
53. N. A. Crossley, E. Sena, J. Goehler, J. Horn, B. van der Worp, P. M. Bath, M. Macleod and U. Dirnagl, *Stroke*, 2008, **39**, 929.
54. P. Jun, D. G. Altman and M. Egger, *Br. Med. J.*, 2001, **323**, 42.
55. S. Franz, A. Hug and N. Weidner, in *Oxford Textbook of Neurorehabilitation*, ed. V. Dietz and N. Ward, Oxford University Press, Oxford, 2015, p. 112.
56. M. I. Martic-Kehl, R. Schibli and P. A. Schubiger, *Eur. J. Nucl. Med. Mol. Imaging*, 2012, **39**, 1492.
57. H. B. van der Worp, D. W. Howells, E. S. Sena, M. J. Porritt, S. Rewell and V. O'Collins, *PLoS Med.*, 2010, **7**, e1000245.
58. V. E. O'Collins, M. R. Macleod, G. A. Donnan, L. L. Horky, B. H. van der Worp and D. W. Howells, *Ann. Neurol.*, 2006, **59**, 467.
59. E. H. Lo, *Stroke*, 2014, **45**, 305.
60. M. Fisher, *Neurol. Res.*, 2013, **35**, 443.
61. E. H. Lo and M. M. Ning, *J. Investig. Med.*, 2016, **64**, 827.
62. C. Leiva-Salinas, M. Wintermark and C. S. Kidwell, *Neurotherapeutics*, 2011, **8**, 19.

Subject Index